Current Topics in
Microbiology
269 and Immunology

Editors

R.W. Compans, Atlanta/Georgia
M.D. Cooper, Birmingham/Alabama · Y. Ito, Kyoto
H. Koprowski, Philadelphia/Pennsylvania · F. Melchers, Basel
M.B.A. Oldstone, La Jolla/California · S. Olsnes, Oslo
M. Potter, Bethesda/Maryland
P.K. Vogt, La Jolla/California · H. Wagner, Munich

Springer
Berlin
Heidelberg
New York
Barcelona
Hong Kong
London
Milan
Paris
Tokyo

Viral Proteins Counteracting Host Defenses

Edited by U.H. Koszinowski
and H. Hengel

With 32 Figures and 8 Tables

Springer

Professor Dr. Ulrich H. Koszinowski
Max von Pettenkofer Institute
University of Munich
Pettenkoferstr. 9a
80336 Munich
Germany
E-mail: koszinowski@m3401.mpk.med.uni-muenchen.de

Privatdozent Dr. Hartmut Hengel
Robert Koch-Institut
Fachgebiet 1.2 Virale Infektionen
Nordufer 20
13353 Berlin
Germany
E-mail: hengelh@rki.de

Cover Illustration: Model of TAP
TAP forms a transmembrane pore in the ER membrane. The pore is followed by a peptide binding domain located to one side of the pore at the cytoplasmic side. The structure is concluded by the two ATP-binding domains. In this model, both peptides and ICP47 approach the binding site of TAP from the cytosolic side, while US6 interacts with TAP from the ER luminal side.
By J. Neefjes (this volume)

ISSN 0070-217X
ISBN 3-540-43261-2 Springer-Verlag Berlin Heidelberg New York

This work is subject to copyright. All rights are reserved, whether the whole or part of the material is concerned, specifically the rights of translation, reprinting, reuse of illustrations, recitation, broadcasting, reproduction on microfilm or in any other way, and storage in data banks. Duplication of this publication or parts thereof is permitted only under the provisions of the German Copyright Law of September 9, 1965, in its current version, and permission for use must always be obtained from Springer-Verlag. Violations are liable for prosecution under the German Copyright Law.

Springer-Verlag Berlin Heidelberg New York
a member of BertelsmannSpringer Science + Business Media GmbH

http://www.springer.de

© Springer-Verlag Berlin Heidelberg 2002
Library of Congress Catalog Card Number 15-12910
Printed in Germany

The use of general descriptive names, registered names, trademarks, etc. in this publication does not imply, even in the absence of a specific statement, that such names are exempt from the relevant protective laws and regulations and therefore free for general use.

Product liability: The publishers cannot guarantee the accuracy of any information about dosage and application contained in this book. In every individual case the user must check such information by consulting other relevant literature.

Cover Design: *design & production GmbH*, Heidelberg
Typesetting: Scientific Publishing Services (P) Ltd, Madras
Production Editor: Christiane Messerschmidt, Rheinau
Printed on acid-free paper   SPIN: 10765000   27/3020 5 4 3 2 1 0

Preface

The first report on MHC class I modulation by a virus dates back to an observation with adenovirus in the middle of the 1980s. Only a few years later, a similar observation was made for mouse cytomegalovirus, a herpesvirus. The *Herpesviridae* comprise an extensive family of large DNA viruses which infect a vast range of species from invertebrates to mammals and humans. *Herpesviridae* are divided into three subfamilies, α-, β-, and γ-*Herpesviridae*. Because of their distinct biological and genetic properties, herpesvirus genomes come in different sizes ranging from 120 to 240 kbp. A common ancestor of herpesviruses must be dated about 200 million years before our time, and it is quite likely that the herpes virus subfamilies occurred in association with the host speciation processes of the mammalian radiation about 60–80 million years ago. Herpesviruses are species-specific viruses. To be maintained in nature, they must not only co-speciate but also need to co-evolve with the immune system of the host. Since the evolution of immunoglobulin light chains and the evolution of the T cell receptor signal transduction units occurred more than 100 million years ago, the evolution of mammalian herpesviruses must have occurred in the presence of an active and already increasingly complex immune system. Therefore, it does not come as a surprise that α-, β-, and γ-herpesviruses address the same immune control mechanisms.

Since the first observations of viral interference with antigen presentation in the MHC class I pathway, the field has advanced to detailed analysis. We know numerous genes, and for some of them we have profound information on mechanistic function. The antigen presentation pathway is affected at all stages starting with proteasomal degradation of an antigenic viral protein, as shown for EBV, transfer of the proteasomal cleavage products as peptides into the ER by specific transporters, the loading of the nascent MHC class I molecule, and finally the transport of the complex to the surface and presentation in a normal or deranged form. All these different steps of the MHC class I antigen presentation pathway are targets for viral proteins. Although different viruses have proteins with similar molecular functions, a direct relationship between the viral proteins is lacking.

Not only MHC class I but also MHC class II proteins are a target of viral influence, either by direct downregulation and degradation of proteins or by interference of signal transduction pathways which affect the real abundance of these proteins in cells.

NK cells are important constituents of the primary natural immune system. NK cell function is modulated by the surface expression of MHC molecules. Unlike T cells, NK cells form a first line of defense and kill target cells without prior sensitization. In addition, stimulatory and inhibitory receptors signal and control NK cell function. Therefore, it is plausible that herpesviruses also address this aspect of natural immunity. The status of this emerging field of research is presented in two reviews. An even more recent addition to the field is the recognition of the importance of chemokines, cytokines and their receptors. As expected from a virus which has co-speciated with the host, herpesviruses use this information and divert it to their advantage. For a virus it makes no difference whether the cell itself responds to virus infection, e.g. by apoptosis or any other type of internal cellular antiviral regulation, or whether the reaction is systemic and involves several specialized cells. It is therefore not surprising that viruses have also found principles to avoid induced cell death.

This book shows the current knowledge presented by specialists in the field. The genes we know today were found either by chance or by specific gene-hunting enterprises. One chapter specifically addresses the genetic methods for identification of such genes. Most of these studies deal with isolated genes expressed in cells overexpressing the isolated protein. As with many other situations, often in science we find ourselves in the situation that by answering a number of questions, many more questions are generated. Important questions have not been addressed for many of these genes, for example, what is their function in the genomic context? How do the different gene functions interact? Where, during the complex infection and transmission cycle, do these viral genes have their major function? What is the origin of these genes and what is their degree of relatedness? Which are the cellular counterparts of the viral proteins for which no cellular homologue is known? All these areas are actively being pursued, and new ideas and concepts are emerging. We thank the contributors for sharing their present views with the community. As this is a very active area of science, the years to come will show how fast these functions, which up to now represent stones of a mosaic, can be integrated into a coherent picture.

Munich,　　　　　　　　　　　　　　　　　　Hartmut Hengel
March 2002　　　　　　　　　　　　　　and Ulrich H. Koszinowski

List of Contents

A. GUTERMANN, A. BUBECK, M. WAGNER, U. REUSCH,
C. MÉNARD, and U.H. KOSZINOWSKI
Strategies for the Identification and Analysis
of Viral Immune-Evasive Genes – Cytomegalovirus
as an Example 1

N.P. DANTUMA, A. SHARIPO, and M.G. MASUCCI
Avoiding Proteasomal Processing:
The Case of EBNA1. 23

F.J. VAN DER WAL, M. KIKKERT, and E. WIERTZ
The HCMV Gene Products US2
and US11 Target MHC Class I Molecules
for Degradation in the Cytosol 37

F. MOMBURG and H. HENGEL
Corking the Bottleneck: The Transporter Associated
with Antigen Processing as a Target
for Immune Subversion by Viruses. 57

E. REITS, A. GRIEKSPOOR, and J. NEEFJES
Herpes Viral Proteins Manipulating
the Peptide Transporter TAP. 75

D. BAUER and R. TAMPÉ
Herpes Viral Proteins Blocking
the Transporter Associated with Antigen Processing
TAP – From Genes to Function and Structure 85

D.C. JOHNSON and N.R. HEGDE
Inhibition of the MHC Class II Antigen Presentation
Pathway by Human Cytomegalovirus. 101

V.M. BRAUD, P. TOMASEC, and G.W.G. WILKINSON
Viral Evasion of Natural Killer Cells
During Human Cytomegalovirus Infection 117

H.E. Farrell, N.J. Davis-Poynter, D.M. Andrews,
and M.A. Degli-Esposti
Function of CMV-Encoded MHC Class I
Homologues 131

D.M. Miller, C.M. Cebulla, and D.D. Sedmak
Human Cytomegalovirus Inhibition
of Major Histocompatibility Complex Transcription
and Interferon Signal Transduction 153

D.A. Leib
Counteraction of Interferon-Induced Antiviral Responses
by Herpes Simplex Viruses 171

R.E. Means, J.K. Choi, H. Nakamura, Y.H. Chung,
S. Ishido, and J.U. Jung
Immune Evasion Strategies
of Kaposi's Sarcoma-Associated Herpesvirus 187

P.S. Beisser, C.-S. Goh, F.E. Cohen, and S. Michelson
Viral Chemokine Receptors and Chemokines
in Human Cytomegalovirus Trafficking and Interaction
with the Immune System 203

N. Saederup and E.S. Mocarski Jr
Fatal Attraction:
Cytomegalovirus-Encoded Chemokine Homologs 235

T. Derfuss and E. Meinl
Herpesviral Proteins Regulating Apoptosis 257

H.-G. Burgert, Z. Ruzsics, S. Obermeier,
A. Hilgendorf, M. Windheim, and A. Elsing
Subversion of Host Defense Mechanisms
by Adenoviruses............................... 273

Subject Index................................. 319

List of Contributors

(Their addresses can be found at the beginning of their respective chapters.)

ANDREWS, D.M.   131
BAUER, D.   85
BEISSER, P.S.   203
BRAUD, V.M.   117
BUBECK, A.   1
BURGERT, H.-G.   273
CEBULLA, C.M.   153
CHOI, J.K.   187
CHUNG, Y.H.   187
COHEN, F.E.   203
DANTUMA, N.P.   23
DAVIS-POYNTER, N.J.   131
DEGLI-ESPOSTI, M.A.   131
DERFUSS, T.   257
ELSING, A.   273
FARRELL, H.E.   131
GOH, C.-S.   203
GRIEKSPOOR, A.   75
GUTERMANN, A.   1
HEGDE, N.R.   101
HENGEL, H.   57
HILGENDORF, A.   273
ISHIDO, S.   187
JOHNSON, D.C.   101
JUNG, J.U.   187
KIKKERT, M.   37

KOSZINOWSKI, U.H.   1
LEIB, D.A.   171
MASUCCI, M.G.   23
MEANS, R.E.   187
MEINL, E.   257
MÉNARD, C.   1
MICHELSON, S.   203
MILLER, D.M.   153
MOCARSKI, E.S. Jr.   235
MOMBURG, F.   57
NAKAMURA, H.   187
NEEFJES, J.   75
OBERMEIER, S.   273
REITS, E.   75
REUSCH, U.   1
RUZSICS, Z.   273
SAEDERUP, N.   235
SEDMAK, D.D.   153
SHARIPO, A.   23
TAMPÉ, R.   85
TOMASEC, P.   117
WAGNER, M.   1
VAN DER WAL, F.J.   37
WIERTZ, E.   37
WILKINSON, G.W.G.   117
WINDHEIM, M.   273

Strategies for the Identification and Analysis of Viral Immune-Evasive Genes – Cytomegalovirus as an Example

A. Gutermann, A. Bubeck, M. Wagner, U. Reusch, C. Ménard, and U.H. Koszinowski

Co-evolution of herpesviruses with their hosts has resulted in multiple interactions between viral genes and cellular functions. Some interactions control genomic maintenance and replication in specific tissues, others affect the immune control at various stages. Few immunomodulatory functions of genes can be predicted by sequence homology. The majority of genes with immunomodulatory properties only become apparent in functional assays. This chapter reviews procedures which have been used for successful identification of immunomodulatory genes in the past and deals with recent methods which may be applicable for the identification of additional immunomodulatory functions unknown so far.

1	Introduction	1
2	Usage of Data Bases for Gene Identification	2
3	Infection Phenotype as Basis for Gene Identification	4
3.1	Positive Selection Procedures	6
3.1.1	From Protein Complexes to Genes	7
3.1.2	Expression of a Genome Subset Library	8
3.1.3	Stable Expression of Single Candidate Genes	9
3.2	Negative Selection Procedures	9
3.2.1	Deletion Mutants Generated by Classic Site-Directed Mutagenesis	10
3.2.2	Reverse Genetics with BAC Technology	11
3.2.3	Forward Genetics with BAC Technology and Invasive Bacteria	12
4	Functional Analysis of Immune-Evasive Genes	12
5	The Crucial Confirmation – In Vivo Studies	17
6	Future Aspects	18
	References	19

1 Introduction

Cytomegaloviruses (CMV) define the β-subgroup of herpesviruses. CMV have been identified in many mammalian species, some of which are used as animal models for the analysis of the human CMV disease. As typical for many herpesvirus

---

Max-von-Pettenkofer Institut, Ludwig-Maximilians-Universität München, 80336 München, Germany

infections, the extent of primary CMV infection is usually efficiently controlled in the immunocompetent host. However, despite the immune response of the host, the virus can persist and has the potential to reactivate. The long co-evolution of the highly species-specific cytomegaloviruses and their hosts has resulted in a complex balance between the virus and the host immune system which is controlled by intricate interactions between viral and cellular genes. In the past, the identification of viral immune-evasive genes has revealed first details of the virus–host interaction. Viral genes have been identified to interfere with the host T-cell response and the NK-cell response. The pathways associated with cytokine and chemokine functions are addressed by viral genes as well as the interferon signal transduction pathway and the regulation of apoptosis (ALCAMI and KOSZINOWSKI 2000). The detailed understanding of the function of these and other not yet detected genes will help to understand the cell biology and immune biology of CMV. As these genes probably also define virus fitness, the elimination of some of these genes may be of advantage in the development of an attenuated CMV vaccine.

The genome of cytomegalovirus comprises about 200 open reading frames (ORFs), and the function of the majority of these genes is still unknown. Because the sequence of human, mouse and rat cytomegalovirus is known, many genes of interest have been selected for further analysis on the basis of sequence comparisons. In particular, those genes that show homology with mammalian genes became immediate subjects of study. Other genes that affect the function of host cell gene products were only found after description of a specific phenotype seen in infected cells. Subsequently, the gene causing the infected cell phenotype was identified. Here, we focus on the methods that have been used for the identification of immune-evasive genes of human and murine cytomegalovirus. An overview of so-far identified immune-evasive CMV genes with a characterised function is given in Tables 1 and 2. Some of these research tools have been recently prepared and have not yet been used on a wide scale for further studies on new gene functions. The use of virus mutant libraries with transposon insertions, for example, may help to screen for new genes involved in immune evasion and other types of virus-cell interaction in the future.

2 Usage of Data Bases for Gene Identification

Because the complete genome sequences of HCMV, MCMV and recently also RCMV are available (CHEE et al. 1990a; RAWLINSON et al. 1996; VINK et al. 2000), the first step in identifying relevant CMV genes that might counteract host immune function is to compare the viral sequence with published viral and cellular sequences and to search for homologies. The success of homology search depends on the search program that is used. Nowadays, several search methods are possible. Searching with FASTA and BLAST represents an easy first access. Thereby, it is possible to search for overall homology or for homologies between conserved

Table 1. Identified immune-evasive genes of MCMV with characterised function

Gene	Mode of identification	Characterised function	Literature
m04	Deletion mutant	Binds MHC class I	KLEIJNEN et al. 1997
m06	Deletion mutant, co-immuno-precipitation	Binds MHC class I and targets the complex to the lysosome for degradation	REUSCH et al. 1999
M45	Screening of Tn library	Essential for virus replication in endothelial cells/role in apoptosis	BRUNE et al. 2001
m131/129	Homology of a motif	Chemokine homolog	FLEMING et al. 1999
m138	Random genome fragments	Fc receptor homolog	THÄLE et al. 1994
m144	Sequence homology	MHC class I homolog	RAWLINSON et al. 1996; FARRELL et al. 1997
m152	Random genome fragments	MHC class I downregulation	THÄLE et al. 1995; ZIEGLER et al. 1997

Table 2. Identified immune-evasive genes of HCMV with characterised function

Gene	Mode of identification	Characterised function	Literature
IE-1, IE-2	Expression kinetics, infection phenotype	Inhibition of TNF-induced apoptosis	ZHU et al. 1995
UL18	Sequence homology	MHC class I homolog	CHEE et al. 1990a; COSMAN et al. 1999
UL37	Cloning of small random genome fragments	Inhibition of apoptosis	GOLDMACHER et al. 1999
UL40	Database search for signal peptide	HLA-E binding, inhibition of NK cell lysis	TOMASEC et al. 2000, ULBRECHT et al. 2000
UL111A	Sequence homology	Binds human IL-10 receptor, competes with human IL-10	KOTENKO et al. 2000; LOCKRIDGE et al. 2000
UL146	Sequence homology	α-Chemokine	PENFOLD et al. 1999
US2	Deletion mutant	MHC class I downregulation, degradation of HLA-DR-α and HLA-DM-α	WIERTZ et al. 1996b; JONES and SUN 1997; TOMAZIN et al. 1999
US3	Single candidate gene expression, homology to US2	Interaction with MHC class I molecules, influence on MHC class II pathway	AHN et al. 1996; JONES et al. 1996; Hegde (this volume)
US6	Single candidate gene expression, deletion mutant	Inhibition of TAP	AHN et al. 1997; HENGEL et al. 1997; LEHNER et al. 1997
US11	Deletion mutant	MHC class I downregulation	JONES et al. 1995; WIERTZ et al. 1996a
US28	Sequence homology	Functional receptor for β-chemokines	CHEE et al. 1990a; BODAGHI et al. 1999

motifs characteristic for the gene product. For more information concerning different search tools, see http://www.ncbi.nlm.nih.gov.

The comparison of potential ORFs of the HCMV sequence with cellular genes led to the early discovery of potential homologs of G protein-coupled receptors such as *US27*, *US28* and *UL33* or the MHC class I homolog *UL18* (CHEE et al. 1990a,b). Later, the G protein-coupled receptor *UL78* was identified by comparison to human herpesvirus 6 (GOMPELS and MACAULAY 1995). Since then, various publications have dealt with the analysis of these ORFs (for review of the G protein-coupled receptors see the chapter of Beisser and colleagues, this volume).

Besides the sequence comparison of whole ORFs, the search for defined short sequence motifs will probably be the method of choice in the future. Accordingly, in HCMV the search for a signal peptide of the non-classic HLA-E molecule led to the identification of *UL40* that influences NK cell activity in CMV-infected cells (TOMASEC et al. 2000; ULBRECHT et al. 2000). The chemokine homolog in MCMV, which comprises a spliced product of the ORFs *m130* and *m129*, was identified by searching for chemokine motifs (MACDONALD et al. 1997). Further analysis demonstrated that this chemokine homolog consists of two predicted ORFs (FLEMING et al. 1999) and hence could not be found by comparison of a whole ORF.

Clearly, once a viral genome region or ORF with significant homology features has been defined, the functionality is by no means proven, because sequence homology does not necessarily predict a homologous function of the gene product.

3 Infection Phenotype as Basis for Gene Identification

Despite the fact that over 50% of the so-far identified genes of CMV and their products counteracting the immune system were discovered by sequence homology, these studies are dependent on the availability of the sequence of the virus. Sequence comparison of these viral genomes reveals that homologous viruses from different species also harbour different and individual genes. Furthermore, functional properties of a virus cannot be easily associated with sequences. In the absence of sequences (e.g. for clinical isolates) or to identify novel functions, different approaches can be applied. A prerequisite is the description of a defined phenotype or function seen during viral infection. Because of the long co-existence of viruses and their hosts, viruses interfere with the host immune system in various ways and at various stages. Therefore, the viral infection might influence nearly every pathway of the host immune system. To investigate viral genes with immune-evasive functions, it is important to investigate whether any of the different pathways of the immune system is compromised.

So far, the described phenotypes that led to the identification of viral genes interfering with the immune response are frequently the modulation of MHC class I functions that affect the CTL and NK response of the host. The description of the loss of MHC class I from the cell surface led to the identification of immune-evasive functions in MCMV and HCMV (for examples, see below). Another common strategy used by different viruses is the inhibition of apoptosis by targeting cellular pathways that trigger apoptosis. Apoptotic phenotypes are easy to define because a variety of test systems for apoptosis are commercially available. Viruses can also interfere with interferons which normally protect the cells from viral infection. The IFN-induced transcriptional responses can be blocked at various stages and, for example, can be detected by the activity of Janus kinase (JAK)/signal transducers and activators of transcription (STAT) signal transduction pathways (KALVAKOLANU 1999; GOODBOURN et al. 2000). Finally, the inhibition and modulation of cytokines and chemokines is a common observation during viral infection (ALCAMI and KOSZINOWSKI 2000; TORTORELLA et al. 2000).

Once an infection phenotype is clearly defined, two different approaches help to identify the gene or genomic region causing the phenotype. These approaches represent *positive* and *negative selection* procedures (Fig. 1, I and II). In the case of *positive selection*, genomic fragments, single candidate genes or cDNAs are expressed in suitable cell systems and tested for the infection phenotype. The *negative selection* is based on the construction of deletion mutants of the virus genome. Mutations can be inserted randomly or directly in a specific gene. The deletion can comprise a large region or a single gene. The deletion mutants are then screened for the loss of the infection phenotype. If a viral and cellular protein interact directly, the complex can be isolated and further analysed. Thus information can be gained by using *positive* or *negative selection* procedures in isolation or in combination.

So far, immune-evasive genes of CMV have been studied either individually by analysing biochemical activities and protein–protein interactions or by creating mutant viruses and analysing their phenotype. A new method which allows a rapid and parallel analysis of gene expression at the whole viral genome level is the DNA array technique (Fig. 1, III). For CMV, two approaches were followed until now. To monitor the up- or downregulation of cellular genes after HCMV infection, labelled cDNA from infected and non-infected cells was hybridised with a commercially available human gene chip (ZHU et al. 1998). In a different approach the CMV sequence was used to define and synthesise oligos of all putative ORFs. The oligos were spotted onto an array and hybridised with fluorescence-labelled cDNAs from HCMV- or non-infected cells. In this case, analysis of the array resulted in identification of up- or downregulated mRNA of viral genes (CHAMBERS et al. 1999). The microarray approach can be coupled with biochemical and genetic strategies and can enhance the functional analysis of large viral genomes like CMV (for review of DNA arrays see LOCKHART and WINZELER 2000).

In the field of proteomics, arrays can also be used to identify possible interaction partners. Here, the array is spotted with recombinant proteins or antibodies and then hybridised with labelled cell lysate or an expressed cDNA library. Additionally, techniques like two-dimensional gel electrophoresis, the identification of isolated proteins by mass spectrometry or the two-hybrid analysis are valuable tools to identify new proteins. These methods allow a large-scale study of viral and cellular proteins without knowledge of the DNA sequence (Fig. 1, IV. For review, see PANDEY and MANN 2000).

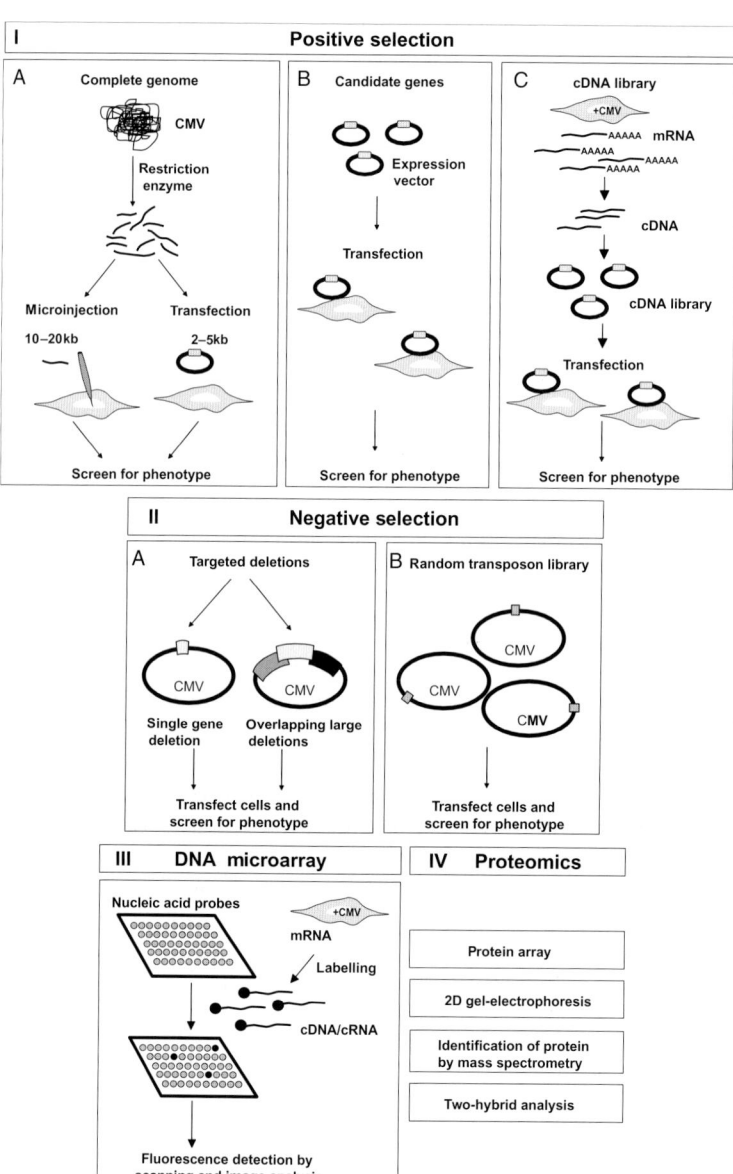

3.1 Positive Selection Procedures

The observation that the virus interferes with a certain mechanism of the host immune system leads to the search for the viral genes responsible. The description of a phenotype, as discussed in the previous section, can be strengthened by additional information on the viral protein. Sometimes, the viral protein interacts

Fig. 1. Strategies for the identification of herpesviral genes affecting cellular functions. Genes without homology to a sequence in a database can be identified by positive selection, negative selection, DNA microarray or proteomics. **I** Positive selection procedures: **A** The complete virus genome is digested with an enzyme and depending on the size, the fragments are either microinjected or transfected into cells. Screening for the infection phenotype will allow to identify the corresponding fragment, further subfragments and finally the gene causing the phenotype. **B** This method is based on previous knowledge. Candidate genes are cloned into a suitable expression vector and transfected into cells to screen for a phenotype. **C** To create a cDNA library permissive cells are infected with CMV and the mRNA is isolated to be transcribed into cDNA. The cDNA fragments are cloned into an expression vector and the cDNA library is transfected into cells to be screened for a phenotype. **II** Negative selection procedures: **A** In a reverse genetics approach, the CMV genome is modified by deletion of a single gene or large overlapping regions. The large deletions can be based on previous knowledge or be randomly distributed. After infection of cells with the mutants, a screening for alteration of phenotype will elucidate candidate genes or genome regions. **B** A forward genetics approach by using a random transposon library can be used to screen for alteration of a phenotype without previous knowledge of candidate genes or genome regions. Cells are infected with Tn-mutants of the CMV genome and screened for alteration of the phenotype. Using large and subsequently smaller pools of CMV mutants will lead to identification of the Tn-mutant responsible for the altered phenotype. **III** Microarray: A DNA microarray can be coated with PCR products, cDNA or oligonucleotides of host genes or viral genes. The array is then hybridised with fluorescence-labelled cDNA or cRNA fragments of CMV-infected cells. Up- or downregulated genes of the host or the virus can be detected. **IV** Proteomics: Further methods to identify new genes on a protein level include protein array techniques, 2D gel electrophoresis, the sequencing of a protein after isolation from a gel by mass spectrometry and the two-hybrid analysis

directly with a cellular protein, and this property allows subsequent analysis. Clues for gene identification can be provided by information like protein size and number of glycosylation sites. The complex of viral and cellular interaction partners can also be used to produce monoclonal antibodies against the unknown viral protein. If there is no direct interaction between viral and cellular protein, the expression of random genome fragments and the subsequent analysis of proteins can lead to the identification of the viral gene. These random genome fragments are expressed in cells and can be used to screen for a specific phenotype. The viral gene of interest may be narrowed down by testing overlapping fragments. This can be very cumbersome, and additional knowledge of the protein will speed up the search for the viral gene.

3.1.1 From Protein Complexes to Genes

The description of a phenotype can lead to the prediction that a viral protein directly complexes with a specific cellular protein. An antibody against the cellular target protein can then be used to co-precipitate viral proteins with the cellular protein from virus-infected cells.

On the basis of a direct interaction of a viral and a cellular protein, the Fc receptor of MCMV was identified (THÄLE et al. 1994). Observation of an interaction of immunoglobulin G (IgG) with a viral protein of HCMV suggested a similar interaction of IgG and a viral protein in MCMV. Cells infected with MCMV were immunoprecipitated with normal mouse IgG. A 65-kDa protein could be identified and in combination with data from deletion mutants revealed the MCMV Fc receptor as the protein interacting with mouse IgG. At that time, this approach could not benefit from the knowledge of the MCMV genome sequence. After publication of the MCMV sequence in 1996, the information of putative ORFs was used to identify the gene *m04* (KLEIJNEN et al. 1997). The interference with the MHC class I antigen presentation pathway suggested viral interaction partners for MHC class I. Antibodies directed against MHC class I precipitated MHC class I in complex with a viral protein. This viral protein (later identified as gp34) was biochemically characterised. The combination of the biochemical data with possible ORFs deduced from studies of deletion mutants led to identification of the MCMV gene *m04*.

A related approach was taken to identify the *m06* gene product gp48. It was assumed that unknown viral proteins complex with MHC class I molecules. Therefore, conventional antibodies directed against MHC class I were used to precipitate a complex of MHC class I and the putative MCMV proteins from infected cells. Precipitates were used to immunise mice for monoclonal antibody generation. The monoclonal antibody directed against the viral protein gp48 then served to identify the coding gene *m06* (REUSCH et al. 1999).

3.1.2 Expression of a Genome Subset Library

There may be no strong interaction between viral and cellular proteins and therefore no formation of a complex. In this case, a positive screening of genome fragments may be used to identify the viral gene.

Genome fragments can be obtained by restriction enzyme digestion of the viral genome (Fig. 1, IA). The fragments can be expressed by co-injection into the nucleus of permissive cells together with DNA expressing the CMV genes *ie1* and *ie2* in HCMV (*ie3* in MCMV). The genes *ie1* and *ie2* (*ie3*) encode transactivators and permit the expression of viral genes of the early and perhaps the late phase. This approach was applied to identify the MCMV genes *m138* (Fc receptor homolog; THÄLE et al. 1994), *m152* (ZIEGLER et al. 1997) and *m06* (REUSCH et al. 1999). Alternatively, random fragments can be subcloned into a suitable expression vector and transiently expressed in cells. The subcloned fragments comprise only 2–5kb of viral DNA to be suitable for transfection into eukaryotic cells. This method was used to identify the HCMV exon 1 of the gene *UL37* as being responsible for inhibition of apoptosis (GOLDMACHER et al. 1999). Transfected cells expressing a certain subset of the genome are then screened for the specific phenotype. An example to screen for a phenotype was, for instance, the binding of the Fc receptor to IgG (THÄLE et al. 1994). To identify the MHC class I reactive MCMV gene *m152*, a screening for intracellular accumulation of MHC class I molecules in indirect immunofluorescence was applied (ZIEGLER et al. 1997) Likewise, indirect immunofluorescence with an antibody recognising gp48 (see Sect. 3.1.1) showed a distinct vesicular distribution of the viral protein and was applied to screen for genome fragments coding for gp48. To narrow down a possible ORF causing the infection phenotype, a combination of several restriction enzymes was used to obtain smaller genome fragments. With the use of certain restriction enzymes, the phenotype is lost because of the destruction of the ORF. The combination of these data with further information like predicted size of the viral protein and the DNA sequence will then allow the identification of the viral gene. The advantage of this approach is that a positive phenotype can be followed throughout the entire mapping procedure until the gene function is located.

Another approach involves the use of a cDNA library of CMV-infected cells (Fig. 1, IC). The mRNA of infected cells derives from both viral and cellular genes. By isolating the mRNA from these cells, genes which are upregulated after CMV infection can be identified (ZHU et al. 1997; REDPATH et al. 1999; BRESNAHAN and SHENK 2000). The reverse-transcribed mRNAs can then be subcloned into an expression vector as cDNAs and subsequently be transfected into cells and screened for a defined phenotype. So far, this method has not been successfully used to identify an immune-evasive gene.

3.1.3 Stable Expression of Single Candidate Genes

Provided that a certain knowledge of the region causing the phenotype of interest is already at hand, the specific subcloning of candidate genes into expression vectors and either stable or transient expression in cell lines is the method of choice (Fig. 1, IB).

This approach was followed to identify the gene *US6* of HCMV. Previous data had shown that HCMV infection causes an inhibition of the transporter associated with antigen processing (TAP). With the use of a temperature-sensitive HCMV deletion mutant, a region encompassing the genes *US1-US15* was identified to encode a gene product which inhibits peptide loading onto the heavy chain by TAP (HENGEL et al. 1996). An assay which measures peptide translocation was used to follow the transport of radioactively labelled peptides through the ER and Golgi. All candidate genes of the region *US1-US15* were subcloned and stably expressed (HENGEL et al. 1997). The transfectants were screened for TAP-mediated peptide transport, and only US6 transfectants showed a reduction of peptide translocation into the ER. A similar approach was applied by another group, and they also showed that US6 blocks peptide transport in HCMV-infected cells (AHN et al. 1997).

3.2 Negative Selection Procedures

A commonly used strategy for identification of viral gene(s) responsible for a specific phenotype is the investigation of virus mutants. In contrast to positive selection procedures, the infection of cells with mutants is a screen for loss of a phenotype in the natural context of viral infection. A prerequisite for this approach is that the specific gene function is not essential for viral replication. Because most genes with immunoregulatory function are not essential in vitro, they can be easily studied with deletion mutants. The location of numerous genes interfering with the immune system was found at the genome termini, which indicated possible genomic regions for testing (RAWLINSON et al. 1996; HENGEL et al. 1998).

The viral function can be identified best by testing mutants with large deletions first. Once a genome region is identified which causes a certain phenotype, further mutants with smaller deletions will lead to identification of the gene (Fig. 1, IIA). In case of previous knowledge of candidate genes, deletion of single genes is a possible method (Fig. 1, IIA). A random approach using transposon mutagenesis can be applied as a forward genetics approach and may be suitable, if no previous knowledge is available. The screening for the alteration of a given infection phenotype may lead to identification of new immune-evasive genes (Fig. 1, IIB).

MHC-reactive genes vary in their mode of action, but their phenotype is similar. To verify that indeed all viral genes with a similar function have been identified, a mutant with deletions of all those genes causing such a phenotype is the only possibility to exclude additional unknown viral genes with similar functions. Accordingly, an MCMV mutant was generated from which the genes *m04*, *m06* and *m152* were deleted. The lack of MHC class I reduction by this mutant confirmed that MCMV encodes only three MHC reactive properties, which is less than the number of MHC reactive genes in HCMV (M. Wagner, A. Gutermann, unpublished data).

3.2.1 Deletion Mutants Generated by Classic Site-Directed Mutagenesis

In the past, CMV mutants were produced by homologous recombination of mutated genome fragments with the CMV genome. This was only possible in eukaryotic cells, because the mutated genome was obtained by isolating the replicative virus mutant (MOCARSKI and KEMBLE 1996; BRUNE et al. 2000). An example for this reverse genetics approach is the search for HCMV genes involved in downregulation of MHC class I molecules from the cell surface by Jones and colleagues (JONES et al. 1995). They started the search for the gene(s) responsible for MHC class I downregulation by testing a mutant with a large deletion (comprising 18 genes – *UL33*, *IRS1*, *US1* to *US13*, *US27*, *US28* and *TRS1*) that did not show MHC class I downregulation any more. Cells infected with this mutant were tested for MHC class I surface expression by flow cytometry and immunoprecipitation analysis. Mutants with smaller deletions in this genome region led to identification of the gene *US11* which is responsible for MHC class I downregulation (Fig. 1, IIA). Further studies using deletion mutants revealed the genes *US2* (WIERTZ et al. 1996b; JONES and SUN 1997), *US3* (AHN et al. 1996; JONES et al. 1996) and *US6* (AHN et al. 1997; HENGEL et al. 1997) as causing MHC class I downregulation by various mechanisms.

Even when a specific virus mutant has been already attributed to a specific phenotype, it is not excluded that the same deleted gene or genomic region can be responsible for other phenotypes, which justifies the re-screening of pre-existing mutants. An example is the HCMV US2 deletion mutant which is involved in MHC class I downregulation. This gene was shown to be also responsible for MHC class II degradation (TOMAZIN et al. 1999).

3.2.2 Reverse Genetics with BAC Technology

Since the BAC (bacterial artificial chromosome) technology was introduced to the field of herpesvirology (MESSERLE et al. 1997), mutant virus genomes with gene interruptions at any position can be generated easily by mutagenesis of the viral BACs in *E. coli*. Both the MCMV smith strain and the HCMV AD169 strain genomes have been cloned as BACs (MESSERLE et al. 1997; BORST et al. 1999; WAGNER et al. 1999) and can be mutated by site-directed allelic exchange or randomly by transposon mutagenesis (reviewed in BRUNE et al. 2000).

Transposon mutagenesis allows fast generation of unlimited numbers of mutant CMV-BAC genomes, each with a random interruption of the genome by transposon insertion. The thus-generated BAC-based transposon mutant libraries of MCMV (C. Ménard, unpublished data) and HCMV (HOBOM et al. 2000) can be used for generation of a specific insertion virus mutant within a few weeks. This can be achieved by screening the CMV-BAC libraries for a specific gene interruption with a simple PCR screening method (HOBOM et al. 2000). The genomic distribution of 100 transposon insertions of an MCMV transposon library determined by sequencing confirmed the random distribution of the transposons in most areas of the genome (C. Ménard, U. Hobom, unpublished data). There are certain genomic "hot regions" with frequent but not identical insertions. The "hot region" in the *Hind*III-D fragment involves the genes *M54* (DNA polymerase) and *M55* (glycoprotein B). In the *Hind*III-E region the genes *m155–m164* are affected by multiple insertions. The function of these genes is so far unknown (Fig. 2). After identification of the desired mutant CMV-BAC clone within the library, the corresponding CMV-BAC DNA can be transfected into permissive cells for reconstitution of the deletion virus mutant. Mutants can then be screened for alteration of a certain phenotype. Recently, an indirect transposon mutagenesis approach for MCMV not using BACs has also been described (ZHAN et al. 2000). This approach combines subcloning of the MCMV genome fragments into plasmids and transposon mutagenesis of plasmids in *E. coli*. Generation of virus mutants then requires conventional insertion mutagenesis in cell culture. Fragments with transposon insertion are reintroduced into the MCMV wild-type genome by transfection of MCMV-infected cells and subsequent plaque purification of mutant viruses.

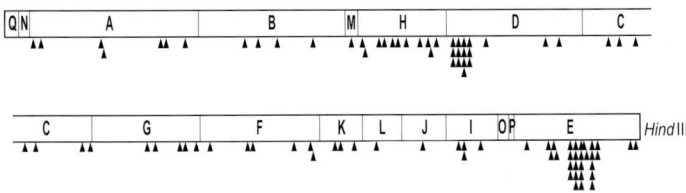

Fig. 2. Random distribution of Tn-insertions within the MCMV genome. Schematic map of the MCMV genome and transposon insertion sites of 100 mutant clones selected at random. The *open boxes* represent the different *Hind*III fragments of the MCMV genome and the *black arrows* represent the sequenced Tn-insertion mutants

Clearly, this multistep method, which was started before the BAC technology on full-length genomes became available, is much more laborious.

A more recent BAC-based approach is the one-step, PCR-based, site-directed mutagenesis of the CMV-BACs. This method allows generation of mutants with exact deletion of one or several genes or of larger sequences (up to several hundred kbp). Any other desired mutation (point mutation or insertion) within a genome region that possesses a potential candidate gene for the distinct phenotype can also be applied. For instance, MCMV single-, double and triple mutants with exact deletions of the MHC class I interacting genes *m04*, *m06* and *m152* have been generated recently (M. Wagner, unpublished data). Thereby, the relative importance of each gene for MHC class I retention can be investigated during the course of infection.

3.2.3 Forward Genetics with BAC Technology and Invasive Bacteria

Reverse genetics approaches allow candidate genes or genome regions to be tested by generation of the corresponding deletion mutant. But if no candidate gene(s) or gene regions are known, an unbiased approach is favourable. Unfortunately, a transposon library of the BAC-cloned CMV genome in *E. coli* is a library of genomes but not of viruses. Therefore, virus testing requires a prior decision on a particular genomic position. The newest addition to random herpesvirus mutagenesis is the usage of *E. coli* as vehicle to transfer the viral genome into eukaryotic cells. Introduction of bacterial invasion genes provides *E. coli* with the capacity to invade eukaryotic cells and to release an intact MCMV genome sufficient for virus replication (BRUNE et al. 2001). This new strategy is described in Fig 3 Briefly, a mixed pool of *E. coli* possessing MCMV-BAC plasmids with unknown random transposon insertions is generated. Into this pool of bacteria-harbouring mutant MCMV-BAC plasmids a second plasmid is introduced, which codes for the gene products invasin and listeriolysin. Thereby, the bacteria become invasive. After single bacterial clones possessing both plasmids are generated, these clones are transferred onto permissive cells for direct virus reconstitution without DNA isolation or transfection. This approach allows the random generation of viable MCMV mutants with unknown single transposon insertions in one step. By screening these virus mutants for cell type tropism, the gene *M45* was identified, which governs replication in endothelial cells. In principle, any assay can now be used to screen viable mutants of the transposon library for the alteration of a specific phenotype. We expect that this method will allow the identification of so-far unknown viral gene functions in the future by random screening approaches.

4 Functional Analysis of Immune-Evasive Genes

The identification of a gene with a role in virus-host interaction is only the first step in understanding virus-host interactions. The specific mode of action must be

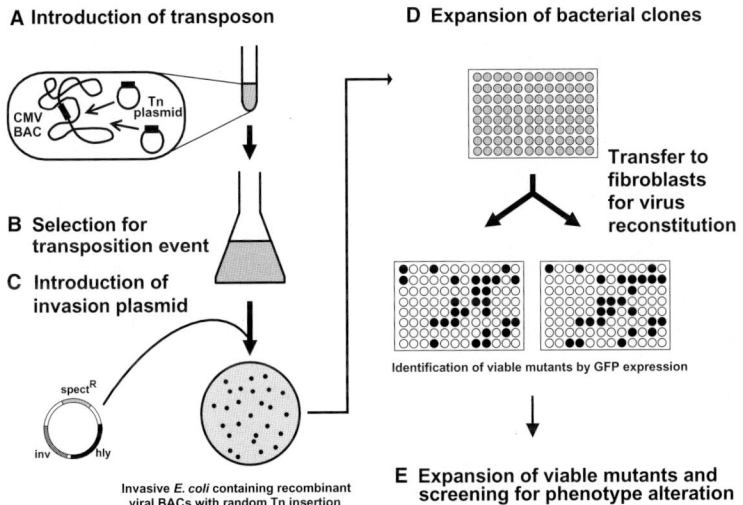

Fig. 3A–E. Strategy for creating a library of mutant CMV genomes by transposon mutagenesis and rapid conversion into a library of mutant viruses. **A** Random transposon mutagenesis is carried out in *E. coli* containing a pool of MCMV-BAC genomes with a chloramphenicol resistance mixed with the temperature-sensitive Tn donor plasmid. The Tn donor plasmid encodes an ampicillin resistance and carries a transposable element with a kanamycin resistance gene. Transposition occurs at the permissive temperature of 30°C. **B** To select for transposition events, an aliquot of bacteria is grown in the presence of chloramphenicol and kanamycin at the non-permissive temperature of 43°C at which the Tn donor plasmid cannot replicate. Subsequently, cells are made competent for transformation. **C** The competent bacteria are then transformed with a plasmid coding for the gene products listeriolysin (*hly*), invasin (*inv*) and a spectinomycin resistance (*spect^R*). Listeriolysin and invasin confer the ability to invade mammalian cells. Single bacterial clones are selected at 43°C on an agar plate with the antibiotics chloramphenicol, kanamycin and spectinomycin. **D** A library of invasive bacterial clones containing recombinant viral BACs is grown in microtitre plates. Aliquots of these bacteria, which are able to transfer the mutated viral genome to mammalian cells, are used to inoculate fibroblast cultures on replica plates. This results in virus reconstitution and viable mutant viruses are identified as green fluorescent plaques as the Tn contains a GFP gene. Replica plates are used because virus rescue is not always 100%. **E** Viral mutants are subsequently expanded and screened for phenotype alteration. (Modified with permission from BRUNE et al. 2001. Copyright 2001 American Association for the Advancement of Science)

elucidated at the molecular level to better understand the effect of immune-evasive genes on the host immune system. Here, we want to give two examples in the immune-evasive genes *m06* and *m152*.

The MCMV gene *m06* codes for the type I transmembrane glycoprotein gp48. MHC class I downregulation by gp48 results in a reduced level of MHC class I on the surface of stable m06 transfectants. The reduced antigen presentation leads to a protection of m06 transfectants from CD8$^+$ cytotoxic T-lymphocyte (CTL) lysis. The mechanism of MHC class I surface reduction by gp48 occurs at the post-translational level and reduces the half-life of MHC class I proteins. The *m06* gene product gp48 binds MHC class I molecules and re-routes them to the lysosome for degradation. The luminal domain of gp48 is sufficient to bind MHC class I but must be anchored in the membrane. The proximal di-leucine motif in the cytoplasmic tail of gp48 is responsible for the lysosomal targeting (U. Reusch, U.H. Koszinowski unpublished data, and REUSCH et al. 1999). A fusion protein of the cytoplasmic tail of *m06*/gp48 and the luminal domain of *m04*/gp34 shows a reduced half-life compared with gp34 (Fig. 4). A binding property of the fusion protein to MHC class I molecules is required for lysosomal targeting, as a fusion protein of CD4 and *m06*/gp48 does not show a reduced half-life (U. Reusch, U.H. Koszinowski, unpublished data). Thus the cytoplasmic domain of gp48 contains the information for the lysosomal targeting of the protein and it mediates surface reduction of MHC class I in combination with the MHC class I binding domain of gp34.

The MCMV gene *m152*/gp40 encodes a product which downregulates MHC class I on the cell surface by a different mechanism than *m06*/gp48. A direct interaction of gp40 and MHC class I has not been seen, so far. In fact, gp40 and MHC class I have different half-lives and different intracellular destinations. Even in the absence of gp40 the retained MHC class I molecules fail to reach the cell surface. MHC class I might be retained by gp40 after a transient interaction, and a so-far unknown biochemical modification might prevent the export of MHC class I from the ER-*cis*-Golgi intermediate compartment (ERGIC). Of course, this does not exclude the existence of further hitherto-unknown interaction partners (ZIEGLER et al. 1997, 2000). The functional domain of gp40 is the luminal domain. The transmembrane and cytoplasmic regions are dispensable for the retention of MHC class I molecules, but fusion of the CD4 transmembrane domain to the luminal domain of gp40 results in a more efficient retention of MHC class I (ZIEGLER et al. 2000).

MHC class I, the target of gp40, is retained because of properties in its luminal domain. A deletion mutant of murine H-2K^b which comprises the luminal domain of the heavy chain can be detected in the supernatant of transfectants and is retained in the ER in the presence of gp40 (Figs. 5, 6A). Thus the luminal part of MHC class I molecules is sufficient for recognition by gp40.

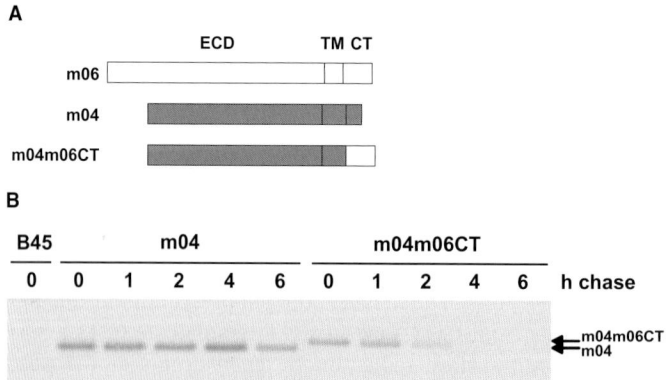

Fig. 4A,B. The lysosomal sorting motif of *m06*/gp48 can be transferred to MCMV *m04*/gp34. **A** Schematic presentation of MCMV *m06*, *m04* and the fusion protein *m04m06*CT. **B** Immunoprecipitation of the gene products of *m04* and *m04m06*CT expressed in NIH 3T3 cells. Whereas the gene product of *m04* is stable over a 6-h chase period, the gene product of *m04m06*CT is degraded after 4h. *ECD*, extracellular domain; *TM*, transmembrane region; *CT*, cytoplasmic tail

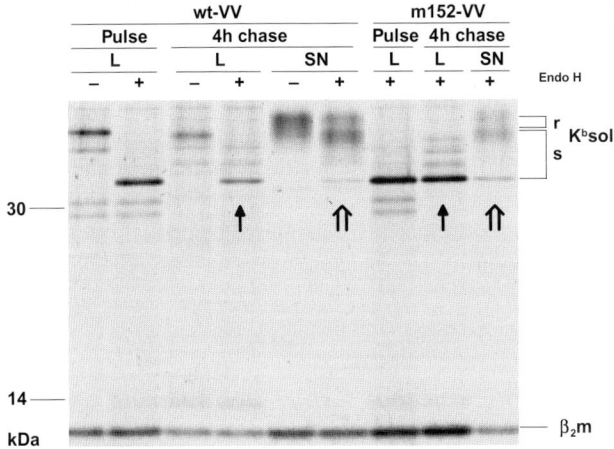

Fig. 5. The luminal domain of H-2K^b is retained by gp40. Immunoprecipitation of the luminal domain of H-2K^b ($K^b sol$). 1T-K^{b}sol stable transfectants were infected with vaccinia virus as indicated. After 4-h chase the cell lysate (*L*) of m152-VV-infected cells contains more K^{b}sol than wt-VV-infected cells (*black arrows*). Correspondingly, a lower amount of K^{b}sol is precipitated from the supernatant (*SN*) after 4-h chase in m152-VV-infected cells than from wt-VV-infected cells (*open arrows*). *L*, cell lysate; *SN*, supernatant; *r*, Endo H resistant; *s*, Endo H sensitive; *VV*, vaccinia virus

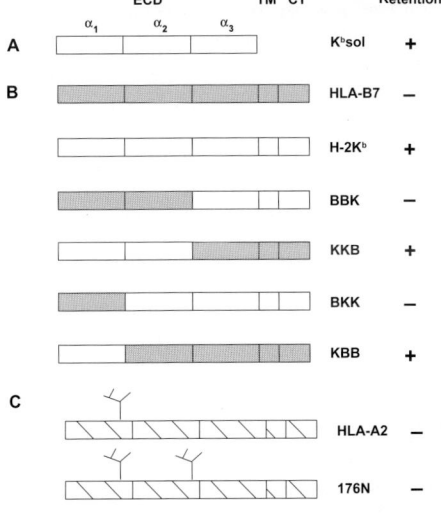

Fig. 6A–C. Modifications of the heavy chain and the effect on retention by gp40. Schematic representation of the heavy chain mutants. **A** Mutant H-2K^b heavy chain comprising only the luminal domain ($K^b sol$). **B** Wild-type and chimeric heavy chains of HLA-B7 (*grey*) and H-2K^b (*white*). **C** Representation of the HLA-A2 and mutant 176N heavy chain (*hatched*) with glycosylation sites indicated (*branched symbol*). The retention of the molecules in the presence of gp40 is indicated. *ECD*, extracellular domain; *TM*, transmembrane region; *CT*, cytoplasmic tail

An interesting property of gp40 is the species-specific retention of MHC class I molecules. In contrast to gp48, the human alleles HLA-A2 and HLA-B7 are not retained by gp40 (A. Gutermann, U. Reusch, unpublished data). This provides the chance to study the MHC class I domain(s) which are a target of gp40 by using chimeras of human and murine MHC class I molecules (Fig. 6B). Exchange of the

α_1 and α_2 domains of H-2K^b for the corresponding domains of HLA-B7 restores the transport of the molecule to the cell surface (Fig. 7A). Vice versa, the exchange of the α_1 and α_2 domains of HLA-B7 for the corresponding domains in H-2K^b molecules results in retention of the chimeric molecule (Fig. 7B). If only the α_1 domain of H-2K^b substitutes the corresponding domain of HLA-B7, retention by gp40 can still be observed (Fig. 7C). In conclusion, the α_1 domain of the heavy chain presents the area of interaction of MHC class I with gp40 or a third interaction partner, but MHC class I is more efficiently recognised by gp40 if both the α_1 and α_2 domains of H-2K^b are present.

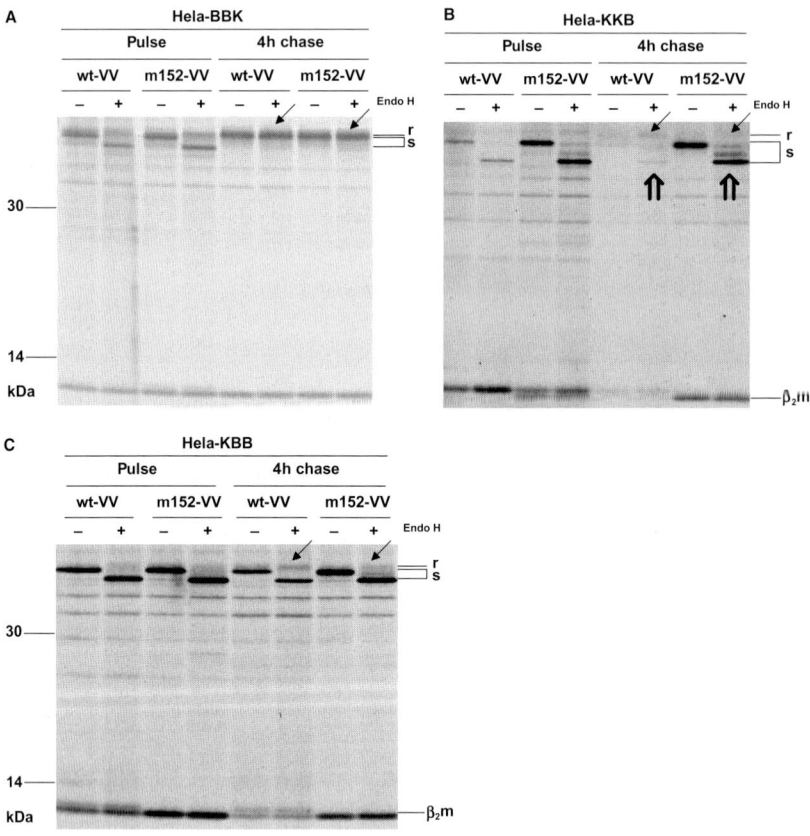

Fig. 7A–C. The α_1 domain of H-2K^b is still retained by gp40. Immunoprecipitation of chimeric MHC class I molecules expressed in HeLa cells after vaccinia virus infection. **A** The chimeric molecule BBK becomes Endo H resistant after 4-h chase in both wt- and m152-VV-infected cells (*black arrows*). **B** KKB molecules are retained by gp40 and remain Endo H sensitive in m152-VV-infected cells after 4-h chase (*open arrows*). Correspondingly, the faint band of Endo H-resistant molecules in wt-VV-infected cells is not detected in m152-VV-infected cells (*black arrows*). The Endo H-resistant molecule cannot be detected by the antibody as well as the Endo H-sensitive molecule. **C** The chimeric molecule KBB can still be retained by gp40 after 4-h chase. Even though most molecules do not pass the Golgi because of their altered structure, some of them are Endo H resistant in wt-VV-infected cells but remain Endo H sensitive in m152-VV-infected cells (*black arrows*). *r*, Endo H resistant; *s*, Endo H sensitive; *VV*, vaccinia virus

Human and murine MHC class I molecules differ in the number of glycosylation sites in the luminal domain. Whereas human MHC has one glycosylation site, H-2K^b has two. The additional glycosylation site of H-2K^b is not responsible for the species-specific retention by gp40. A mutant of HLA-A2 with a second glycosylation site corresponding to murine MHC class I is still transported to the cell surface in the presence of gp40 (Fig. 6C). Thus a role of the glycosylation pattern of MHC class I for the species-specific retention of gp40 can be excluded.

5 The Crucial Confirmation – In Vivo Studies

The relevance of an immune-evasive gene function during infection can only be determined by in vivo experiments. Infection of mice with MCMV is used as a model for HCMV, because the biological characteristics of the infection in the natural host are comparable. The study of the MCMV gene *m152*/gp40 in mice proved the relevance of the gene function which was studied in vitro. gp40 reduces MHC class I surface expression and protects cells from CTL-lysis in vitro (ZIEGLER et al. 1997). In neonatal mice, an MCMV mutant lacking the gene *m152* (Δ*m152*) replicates to lower titres than wild-type MCMV and leads to a reduced mortality in mice. Depletion of CD8$^+$ cytotoxic T-lymphocytes (CTL) in infected mice increases the virus titre of Δ*m152*-MCMV. Mice lacking the gene for β$_2$-microglobulin (β$_2$m$^{-/-}$) are deficient in the MHC class I antigen presentation pathway and cannot present peptides to CTLs. In β$_2$m$^{-/-}$ mice the Δ*m152* and wild-type MCMV both replicate to comparable titres (Fig. 8). Thus the MHC class I antigen presentation pathway is indeed modulated by the gene product of *m152*. This proved the role of *m152*/gp40 for MCMV to evade the CD8 T cell response in infected mice (KRMPOTIC et al. 1999).

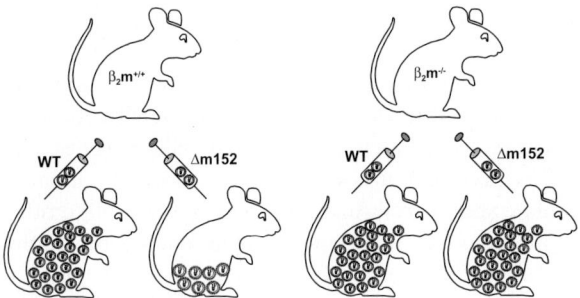

Fig. 8. MCMV *m152*/gp40 interferes with the MHC class I antigen presentation pathway in vivo. C57BL/6 mice were infected with wild-type (*WT*) or mutant MCMV lacking the gene *m152* (Δm152). Virus titres were determined in several organs, and titres are indicated by the number of symbolic virions in the mouse. In absence of MHC class I molecules – the cellular target of the *m152* gene function in vivo (β2m$^{-/-}$ knockout) – the deletion mutant has no phenotype

The role of an immune-evasive gene in vivo can only be proven for MCMV and RCMV as a model for HCMV. Even though the mode of action to evade the host immune system may be different in HCMV and MCMV with respect to mechanistic details, the use of an animal model can show the relevance of the interaction between the virus functions and the host immune system.

6 Future Aspects

Over the last few years a wealth of new insights into several individual mechanisms by which CMV interfere with the host immune system became available. The knowledge of the function of immune-evasive genes not only offers a better understanding of the virus pathogenesis but also has potential practical aspects. The information may be applied to use the interface between viral and cellular proteins as new targets of intervention with virus replication or to improve gene therapy vectors with genes of specific function (MESSERLE et al. 2000).

The immune-evasive genes provide a very important general lesson. After infection of a cell the CMV genome does not simply take command and stop the normal cellular gene expression program altogether to get the new viral blueprint to be realised but rather uses the cellular protein synthesis machinery to maintain functions that suit the virus life cycle. One simple explanation for this concept is the fact that CMV replication is slow and protracted. Perhaps an early host cell protein shut off in the infected cell would produce too few progeny to maintain this virus in nature. On the other hand, in a functionally competent cell the ongoing virus productivity is presented to the immune system unless the detection of the new antigens in infected cells is prevented. Thus immune-evasive genes serve to avoid the most dangerous consequences of the immune defence. The list of the known immune-evasive genes indicates the edge of major inhibitory defence mechanisms that threaten the virus and that need to be blunted. The list is most probably still incomplete, and new principles of virus-host interaction will be unravelled in the future.

An obvious question is whether this specific ability to modulate host functions is only used to avoid the consequences of the immune system. If the virus proteins are selected by evolution to redirect the function of normal cellular genes, this redirection of cellular functions need not necessarily be restricted to immune control. If the virus can modulate specific cell functions, why not modulate other aspects of cell biology of the infected cell as well? One example which points in this direction is the modulation of apoptosis. Certainly, the initiation of apoptosis is also one possibility to deploy an immune effector mechanism. On the other hand, apoptosis is used as a regulatory mechanism for various purposes in higher organisms. It is conceivable that events leading to apoptosis occur at different stages during virus replication. If such a lethal trigger is activated at various stages of the virus life cycle, virus replication is threatened unless inactivated by one or multiple viral functions. Therefore, if it is correct that the blockade of apoptosis has a general regulatory importance we expect to learn in the future that several viral genes have an antiapoptotic function.

It is a remarkable fact that the virus genes encoding proteins with functions in immunomodulation usually represent non-essential proteins. Does this mean that those functions must be non-essential because essential genes are used for the orchestra of essential virus proteins only? Probably not. It is more likely that the ease of the technical approach to study non-essential viral gene functions has focused attention on genes that can be eliminated from the genome. Given the potential for multifunctionality of a protein composed from different domains essential genes may also code for proteins that regulate cell functions as one of their tasks.

Given the size of CMV genomes, there are probably many more genes influencing the host immune response or other functions. It is of great importance to identify these genes to get a more comprehensive view of the virus-host interaction. Recently developed techniques will accelerate the search for unknown genes. Microarray techniques will allow the identification of genes, which are up- or downregulated during infection. The BAC-based transposon mutant libraries of human and murine CMV will allow a fast and extensive screening for phenotypes. Employing these methods will reveal novel genes which modify host functions and will give new insights on the interactions between virus and host.

Acknowledgements. This study was supported by grants of the Deutsche Forschungsgemeinschaft.

References

Ahn K, Angulo A, Ghazal P, Peterson PA, Yang Y, Fruh K (1996) Human cytomegalovirus inhibits antigen presentation by a sequential multistep process. Proc Natl Acad Sci USA 93:10990–10995

Ahn K, Gruhler A, Galocha B, Jones TR, Wiertz EJ, Ploegh HL, Peterson PA, Yang Y, Fruh K (1997) The ER-luminal domain of the HCMV glycoprotein US6 inhibits peptide translocation by TAP. Immunity 6:613–621

Alcami A, Koszinowski UH (2000) Viral mechanisms of immune evasion. Trends Microbiol 8:410–418

Beisser PS, Grauls G, Bruggeman CA, Vink C (1999) Deletion of the R78 G protein-coupled receptor gene from rat cytomegalovirus results in an attenuated, syncytium-inducing mutant strain. J Virol 73:7218–7230

Bodaghi B, Jones TR, Zipeto D, Vita C, Sun L, Laurent L, Arenzana-Seisdedos F, Virelizier JL, Michelson S (1999) Chemokine sequestration by viral chemoreceptors as a novel viral escape strategy: withdrawal of chemokines from the environment of cytomegalovirus-infected cells. J Exp Med 188:855–866

Borst EM, Hahn G, Koszinowski UH, Messerle M (1999) Cloning of the human cytomegalovirus (HCMV) genome as an infectious bacterial artificial chromosome in *Escherichia coli*: a new approach for construction of HCMV mutants. J Virol 73:8320–8329

Bresnahan WA, Shenk T (2000) A subset of viral transcripts packaged within human cytomegalovirus particles [see comments]. Science 288:2373–2376

Brune W, Messerle M, Koszinowski UH (2000) Forward with BACs: new tools for herpesvirus genomics. Trends Genet 16:254–259

Brune W, Menard C, Heesemann J, Koszinowski UH (2001) A ribonucleotide reductase homolog of cytomegalovirus and endothelial cell tropism. Science 291:303–305

Chambers J, Angulo A, Amaratunga D, Guo H, Jiang Y, Wan JS, Bittner A, Frueh K, Jackson MR, Peterson PA, Erlander MG, Ghazal P (1999) DNA microarrays of the complex human cytomegalovirus genome: profiling kinetic class with drug sensitivity of viral gene expression. J Virol 73:5757–5766

Chee MS, Bankier AT, Beck S, Bohni R, Brown CM, Cerny R, Horsnell T, Hutchison CA, Kouzarides T, Martignetti JA, et al. (1990a) Analysis of the protein-coding content of the sequence of human cytomegalovirus strain AD169. Curr Top Microbiol Immunol 154:125–169

Chee MS, Satchwell SC, Preddie E, Weston KM, Barrell BG (1990b) Human cytomegalovirus encodes three G protein-coupled receptor homologues [see comments]. Nature 344:774–777

Cosman D, Fanger N, Borges L (1999) Human cytomegalovirus, MHC class I and inhibitory signalling receptors: more questions than answers. Immunol Rev 168:177–185

Davis PN, Lynch DM, Vally H, Shellam GR, Rawlinson WD, Barrell BG, Farrell HE (1997) Identification and characterization of a G protein-coupled receptor homolog encoded by murine cytomegalovirus. J Virol 71:1521–1529

Farrell HE, Vally H, Lynch DM, Fleming P, Shellam GR, Scalzo AA, Davis PN (1997) Inhibition of natural killer cells by a cytomegalovirus MHC class I homologue in vivo [see comments]. Nature 386:510–514

Fleming P, Davis PN, Degli EM, Densley E, Papadimitriou J, Shellam G, Farrell H (1999) The murine cytomegalovirus chemokine homolog, m131/129, is a determinant of viral pathogenicity. J Virol 73:6800–6809

Goldmacher VS, Bartle LM, Skaletskaya A, Dionne CA, Kedersha NL, Vater CA, Han J, Lutz RJ, Watanabe S, McFarland ED, Kieff ED, Mocarski ES, Chittenden T (1999) A cytomegalovirus-encoded mitochondria-localised inhibitor of apoptosis structurally unrelated to Bcl-2. Proc Natl Acad Sci USA 96:12536–12541

Gompels UA, Macaulay HA (1995) Characterization of human telomeric repeat sequences from human herpesvirus 6 and relationship to replication. J Gen Virol 76:451–458

Goodbourn S, Didock L, Randall RE (2000) Interferons : cell signalling, immune modulation, antiviral responses and viral countermeasures. J Gen Virol 81:2341–2364

Hengel H, Flohr T, Hammerling GJ, Koszinowski UH, Momburg F (1996) Human cytomegalovirus inhibits peptide translocation into the endoplasmic reticulum for MHC class I assembly. J Gen Virol 77:2287–2296

Hengel H, Koopmann JO, Flohr T, Muranyi W, Goulmy E, Hammerling GJ, Koszinowski UH, Momburg F (1997) A viral ER-resident glycoprotein inactivates the MHC-encoded peptide transporter. Immunity 6:623–632

Hengel H, Brune W, Koszinowski UH (1998) Immune evasion by cytomegalovirus – survival strategies of a highly adapted opportunist [see comments]. Trends Microbiol 6:190–197

Hobom U, Brune W, Messerle M, Hahn G, Koszinowski UH (2000) Fast screening procedures for random transposon libraries of cloned herpesvirus genomes: mutational analysis of human cytomegalovirus envelope glycoprotein genes. J Virol 74:7720–7729

Jones TR, Hanson LK, Sun L, Slater JS, Stenberg RM, Campbell AE (1995) Multiple independent loci within the human cytomegalovirus unique short region down-regulate expression of major histocompatibility complex class I heavy chains. J Virol 69:4830–4841

Jones TR, Wiertz EJ, Sun L, Fish KN, Nelson JA, Ploegh HL (1996) Human cytomegalovirus US3 impairs transport and maturation of major histocompatibility complex class I heavy chains. Proc Natl Acad Sci USA 93:11327–11333

Jones TR, Sun L (1997) Human cytomegalovirus US2 destabilizes major histocompatibility complex class I heavy chains. J Virol 71:2970–2979

Kalvakolanu DV (1999) Virus interception of cytokine-regulated pathways. Trends Microbiol 7:166–171

Kleijnen MF, Huppa JB, Lucin P, Mukherjee S, Farrell H, Campbell AE, Koszinowski UH, Hill AB, Ploegh HL (1997) A mouse cytomegalovirus glycoprotein, gp34, forms a complex with folded class I MHC molecules in the ER which is not retained but is transported to the cell surface. EMBO J 16:685–694

Kotenko SV, Saccani S, Izotova LS, Mirochnitchenko OV, Pestka S (2000) Human cytomegalovirus harbors its own unique IL-10 homolog (cmvIL-10). Proc Natl Acad Sci USA 97:1695–1700

Krmpotic A, Messerle M, Crnkovic MI, Polic B, Jonjic S, Koszinowski UH (1999) The immunoevasive function encoded by the mouse cytomegalovirus gene m152 protects the virus against T cell control in vivo. J Exp Med 190:1285–1296

Lehner PJ, Karttunen JT, Wilkinson GW, Cresswell P (1997) The human cytomegalovirus US6 glycoprotein inhibits transporter associated with antigen processing-dependent peptide translocation. Proc Natl Acad Sci USA 94:6904–6909

Lockhart DJ, Winzeler EA (2000) Genomics, gene expression and DNA arrays. Nature 405:827–836

Lockridge KM, Zhou SS, Kravitz RH, Johnson JL, Sawai ET, Blewett EL, Barry PA (2000) Primate cytomegaloviruses encode and express an IL-10-like protein. Virology 268:272–280

MacDonald MR, Li XY, Virgin HW (1997) Late expression of a beta chemokine homolog by murine cytomegalovirus. J Virol 71:1671–1678

Messerle M, Crnkovic I, Hammerschmidt W, Ziegler H, Koszinowski UH (1997) Cloning and mutagenesis of a herpesvirus genome as an infectious bacterial artificial chromosome. Proc Natl Acad Sci USA 94:14759–14763

Messerle M, Hahn G, Brune W, Koszinowski UH (2000) Cytomegalovirus bacterial artificial chromosomes: a new herpesvirus vector approach. Adv Virus Res 55:463–478

Mocarski-ES J, Kemble GW (1996) Recombinant cytomegaloviruses for study of replication and pathogenesis. Intervirology 39:320–330

Pandey A, Mann M (2000) Proteomics to study genes and genomes. Nature 405:837–846

Penfold ME, Dairaghi DJ, Duke GM, Saederup N, Mocarski ES, Kemble GW, Schall TJ (1999) Cytomegalovirus encodes a potent alpha chemokine. Proc Natl Acad Sci USA 96:9839–9844

Rawlinson WD, Farrell HE, Barrell BG (1996) Analysis of the complete DNA sequence of murine cytomegalovirus. J Virol 70:8833–8849

Redpath S, Angulo A, Gascoigne NR, Ghazal P (1999) Murine cytomegalovirus infection down-regulates MHC class II expression on macrophages by induction of IL-10. J Immunol 162:6701–6707

Reusch U, Muranyi W, Lucin P, Burgert HG, Hengel H, Koszinowski UH (1999) A cytomegalovirus glycoprotein re-routes MHC class I complexes to lysosomes for degradation. EMBO J 18:1081–1091

Thäle R, Lucin P, Schneider K, Eggers M, Koszinowski UH (1994) Identification and expression of a murine cytomegalovirus early gene coding for an Fc receptor. J Virol 68:7757–7765

Thäle R, Szepan U, Hengel H, Geginat G, Lucin P, Koszinowski UH (1995) Identification of the mouse cytomegalovirus genomic region affecting major histocompatibility complex class I molecule transport. J Virol 69:6098–6105

Tomasec P, Braud VM, Rickards C, Powell MB, McSharry BP, Gadola S, Cerundolo V, Borysiewicz LK, McMichael AJ, Wilkinson GW (2000) Surface expression of HLA-E, an inhibitor of natural killer cells, enhanced by human cytomegalovirus gpUL40. Science 287:1031

Tomazin R, Boname J, Hegde NR, Lewinsohn DM, Altschuler Y, Jones TR, Cresswell P, Nelson JA, Riddell SR, Johnson DC (1999) Cytomegalovirus US2 destroys two components of the MHC class II pathway, preventing recognition by CD4+ T cells. Nat Med 5:1039–1043

Tortorella D, Gewurz B, Furman M, Schust D, Ploegh H (2000) Viral subversion of the immune system. Annu Rev Immunol 18:861–926

Ulbrecht M, Martinozzi S, Grzeschik M, Hengel H, Ellwart JW, Pla M, Weiss EH (2000) Cutting edge: the human cytomegalovirus UL40 gene product contains a ligand for HLA-E and prevents NK cell-mediated lysis. J Immunol 164:5019–5022

Vink C, Beuken E, Bruggeman CA (2000) Complete DNA sequence of the rat cytomegalovirus genome. J Virol 74:7656–7665

Wagner M, Jonjic S, Koszinowski UH, Messerle M (1999) Systematic excision of vector sequences from the BAC-cloned herpesvirus genome during virus reconstitution. J Virol 73:7056–7060

Wiertz EJ, Jones TR, Sun L, Bogyo M, Geuze HJ, Ploegh HL (1996a) The human cytomegalovirus US11 gene product dislocates MHC class I heavy chains from the endoplasmic reticulum to the cytosol. Cell 84:769–779

Wiertz EJ, Tortorella D, Bogyo M, Yu J, Mothes W, Jones TR, Rapoport TA, Ploegh HL (1996b) Sec61-mediated transfer of a membrane protein from the endoplasmic reticulum to the proteasome for destruction [see comments]. Nature 384:432–438

Zhan X, Lee M, Xiao J, Liu F (2000) Construction and characterization of murine cytomegaloviruses that contain transposon insertions at open reading frames m09 and M83. J Virol 74:7411–7421

Zhu H, Shen Y, Shenk T (1995) Human cytomegalovirus IE1 and IE2 proteins block apoptosis. J Virol 69:7960–7970

Zhu H, Cong JP, Shenk T (1997) Use of differential display analysis to assess the effect of human cytomegalovirus infection on the accumulation of cellular RNAs: induction of interferon-responsive RNAs. Proc Natl Acad Sci USA 94:13985–13990

Zhu H, Cong JP, Mamtora G, Gingeras T, Shenk T (1998) Cellular gene expression altered by human cytomegalovirus: global monitoring with oligonucleotide arrays. Proc Natl Acad Sci USA 95:14470–14475

Ziegler H, Thäle R, Lucin P, Muranyi W, Flohr T, Hengel H, Farrell H, Rawlinson W, Koszinowski UH (1997) A mouse cytomegalovirus glycoprotein retains MHC class I complexes in the ERGIC/cis-Golgi compartments. Immunity 6:57–66

Ziegler H, Muranyi W, Burgert HG, Kremmer E, Koszinowski UH (2000) The luminal part of the murine cytomegalovirus glycoprotein gp40 catalyzes the retention of MHC class I molecules. EMBO J 19:870–881

Avoiding Proteasomal Processing: The Case of EBNA1

N.P. Dantuma, A. Sharipo, and M.G. Masucci

Ubiquitin/proteasome-dependent proteolysis is involved in the regulation of a large variety of cellular processes including cell cycle progression, tissue development and atrophy, flux of substrates through metabolic pathways, selective elimination of abnormal proteins and processing of intracellular antigens for major histocompatibility complex (MHC) class I-restricted T-cell responses. Many viruses tamper with this proteolytic machinery by encoding proteins that interact with various components of the pathway. A particularly interesting example of a viral protein that interferes with proteasomal processing is the Epstein-Barr virus (EBV) nuclear antigen-1 (EBNA1). EBNA1 contains an internal repeat exclusively composed of glycines and alanines that inhibits *in cis* the presentation of MHC class I-restricted T-cell epitopes and prevents ubiquitin/proteasome-dependent proteolysis in vitro and in vivo. The glycine–alanine repeat acts as a transferable element on a variety of proteasomal substrates and may therefore provide a new approach to the modification of cellular proteins for therapeutic purposes.

1 Introduction	24
2 The Ubiquitin-Proteasome System for Regulated Proteolysis	25
3 Modulation of Proteolysis by the Glycine-Alanine Repeat	27
4 Implications for the Biology of EBV Infection	31
5 Concluding Remarks	32
References	33

Abbreviations

CTL	cytotoxic T lymphocyte
EBNA	EBV nuclear antigen
EBV	Epstein-Barr virus
GAr	glycine-alanine repeat
GFP	green fluorescent protein
LCL	lymphoblastoid cell line
LMP	latent membrane proteins
MHC	major histocompatibility complex
TAP	transporter associated with antigen presentation
UFD	ubiquitin fusion degradation

Microbiology and Tumor Biology Center, Karolinska Institutet, Stockholm, Sweden

1 Introduction

Major histocompatibility complex (MHC) class I-restricted T-cell responses play an important role in the control of virus infection by generating effectors that are able to recognise and kill infected target cells. Such recognition involves viral peptides derived from the processing of endogenously expressed viral proteins and presented at the target cell surface as complexes with MHC class I molecules (TOWNSEND et al. 1986). The presentation of antigenic peptides is intimately linked to the biosynthesis and intracellular trafficking of MHC molecules, and it is therefore not surprising that many viruses have evolved strategies that target different steps of the MHC class I maturation pathway, resulting in elimination of class I molecules from the cell surface (PLOEGH 1995). A few viruses appear to have chosen alternative strategies to selectively block the recognition of proteins that play important roles at discrete stages of viral infection. These strategies target the very first steps of antigen presentation by preventing the degradation of the antigenic protein by the ubiquitin/proteasome system. A fascinating example of this strategy is provided by the Epstein-Barr virus (EBV) nuclear antigen-1 (EBNA1).

EBV is a human γ-herpesvirus that in normal circumstances infects cells of the B lymphoid lineage and that can establish either latent (non-productive) or lytic (productive) infection in such cells (KIEFF 1996). Both forms of infection induce an array of $CD8^+$ cytotoxic T lymphocyte (CTL) responses that are apparent during primary EBV infection and persist in the memory of long-term virus carriers (RICKINSON and KIEFF 1996). The viral antigens expressed in latently infected cells are well characterised and comprise six nuclear proteins, the EBV-encoded nuclear antigens (EBNA 1-6), and two latent membrane proteins (LMP 1 and 2). When the $CD8^+$ T cell pool of virus carriers is challenged in vitro with cells from an autologous EBV latent antigen-expressing B lymphoblastoid cell line (LCL), the dominant CTL responses usually map to epitopes derived from the EBNA3, 4, 6 family of proteins, sometimes accompanied by subdominant responses to one of the other latent proteins but apparently never to EBNA1 (RICKINSON and Moss 1997). Such findings led to the discovery that endogenously expressed EBNA1 could not be presented to the $CD8^+$ T-cell repertoire because an internal glycine-alanine repeat (GAr) domain protected the protein from a key step in the MHC class I processing pathway, namely proteasomal degradation to peptides (LEVITSKAYA et al. 1995, 1997) This GAr-mediated protection proved to be extremely robust, holding firm even when the antigen was deliberately over-expressed in LCL cells, despite the fact that such latently infected B cells have an otherwise highly efficient antigen-presenting cell phenotype.

In this review we will summarise our current understanding of the mechanisms whereby the GAr domain of EBNA1 interferes with ubiquitin/proteasome-dependent proteolysis and speculate on how this viral strategy might contribute to the biology of EBV infection in healthy virus carriers and EBV-associated malignancies.

2 The Ubiquitin-Proteasome System for Regulated Proteolysis

Ubiquitin/proteasome-dependent proteolysis appears to be the major source of antigenic peptides recognised by MHC class I-restricted T-cell responses (ROCK and GOLDBERG 1999; ROCK et al. 1994). Dissection of the complex cascade of events leading to proteasomal destruction is therefore of paramount importance in identifying critical steps that may be targeted by viral escape mechanisms. Three major events have been identified and characterised in some details: (a) recognition of the substrate, (b) targeting of the substrate by covalent attachment of multiple ubiquitin molecules, and (c) degradation of the ubiquitin-tagged protein by the 26S proteasome (Fig. 1; HERSHKO and CIECHANOVER 1998). The ubiquitin-conjugation machinery is responsible for both the recognition and the targeting of substrates for proteasomal destruction. The recognition step involves specific degradation signals that may be either constitutive or generated by specific modifications of the target protein, such as phosphorylation or dephosphorylation (LANEY and HOCHSTRASSER 1999). A broad array of degradation signals has been identified, among which are PEST domains, the destruction box, the N-end rule degradation signal, and the ubiquitin fusion degradation (UFD) signals. These serve as binding sites for members of a large family of ubiquitin ligase enzymes called E3s (JOHNSON et al. 1995; RECHSTEINER and ROGERS 1996; VARSHAVSKY 1996). The E3s initiate the ubiquitination event by the linking an activated ubiquitin to the ε-NH$_2$ group of an internal lysine residue of the substrate or the free NH$_2$ terminus (HERSHKO and CIECHANOVER 1998). In successive reactions, a polyubiquitin chain is synthesised by consecutive transfer of ubiquitin moieties to an internal lysine residue of the previously conjugated ubiquitin.

The protease responsible for degradation of ubiquitinated proteins is the proteasome, a 26S multicatalytic endopeptidase complex that hydrolyses proteins

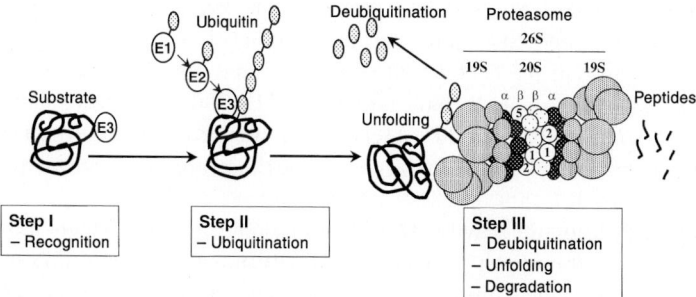

Fig. 1. The ubiquitin-proteasome system. Processing of substrates by the ubiquitin-proteasome pathway can be divided into three steps. *I* Substrates are recognised by ubiquitin ligases E3 by the presence of a degradation signal. *II* In a relay cascade involving a ubiquitin activase E1, a ubiquitin conjugase E2 and a ubiquitin ligase E3, a polyubiquitin tree is covalently linked to the substrate. *III* The polyubiquitinated substrate interacts with the 19S regulatory complex of the 26S proteasome, resulting in unfolding and degradation into small peptides by the 20S core. Before degradation, deubiquitination enzymes disassemble the polyubiquitin tree into reusable free ubiquitin monomers

to peptides of size ranging between 3 and 22 amino acids (KISSELEV et al. 1999). The proteasome is composed of a catalytic core, the 20S particle, and two types of regulatory subunits, the 19S or 11S caps (BOCHTLER et al. 1999). The 20S complex is a barrel-shaped structure consisting of four stacked rings, two outer α rings and two inner β rings. Each of the identical α or β rings is composed of seven distinct subunits, giving the general structure $\alpha_{1-7}\beta_{1-7}\beta_{1-7}\alpha_{1-7}$. The catalytic sites reside in the $\beta 5$ (X), $\beta 2$ (Z), and $\beta 1$ (Y) subunits and face the inner cavity of the proteasome. Thus proteins must be unfolded and tethered into the cavity to allow processing, which restricts proteolysis to substrates that are specifically targeted for destruction. The 19S regulatory particle mediates the recognition of polyubiquitinated substrates and their subsequent unfolding and transport into the cavity of the 20S (VOGES et al. 1999). Biochemical analysis in yeast revealed that the 19S can be resolved into two sub-complexes: the 'base', consisting of six ATPases of the AAA family and three additional proteins, and an eight-subunit 'lid' (GLICKMAN et al. 1998). The recruitment of the ubiquitinated substrates is likely to be mediated by subunits of the lid that recognise the polyubiquitin tree, such as the S5a subunit (FERRELL et al. 1996), or E3/substrate complexes (XIE and VARSHAVSKY 2000), whereas the ATPases of the base are probably responsible for substrate unfolding (VOGES et al. 1999).

Two structural features that are restricted to animals with an adaptive immune response highlight the involvement of the ubiquitin-proteasome pathway in the generation of antigenic peptides. First, interferon-γ induces the expression of three alternative proteolytic active β subunits: $\beta 5i$ (Lmp7), $\beta 2i$ (MECL1), and $\beta 1i$ (Lmp2), which are incorporated in the 20S proteasome instead of their constitutive counterparts (GACZYNSKA et al. 1993; HISAMATSU et al. 1996). The resulting 20S immunoproteasome has a slightly modified enzymatic activity, which enhances the production of peptides with hydrophobic or basic carboxy-termini having affinity for the peptide-binding pockets of MHC class I (GACZYNSKA et al. 1996). Second, a heptameric ring of PA28α and PA28β subunits, the 11S cap, can replace the 19S cap on the proteasome (KUEHN and DAHLMANN 1997). The PA28 subunits are also induced by interferon-γ and stimulate the production of antigenic peptides (GROETTRUP et al. 1996). The mode of action of the 11S cap is not well understood, but crystal structure analysis suggests that cavity of the 11S–20S complex is opened (WHITBY et al. 2000). On the basis of this observation it was proposed that the 11S cap might facilitate the exit of relatively large peptides that are better suited for loading onto MHC class I.

Peptides released from the cavity of the proteasome are translocated by transporter associated with antigen presentation (TAP) into the endoplasmic reticulum, where they are loaded onto MHC class I and transported to the cell surface (PAMER and CRESSWELL 1998). The recognition of complexes containing foreign or aberrant peptides triggers elimination of the affected cell by CD8$^+$ CTLs. This surveillance function requires that the blend of 'self' and 'non-self' peptides that reaches the cell surface is representative of the intracellular protein pool, a property that seems in conflict with the highly regulated and selective proteolytic function of the proteasome. A possible explanation was recently offered by the demonstration that a significant fraction of the peptides presented at MHC class I are derived from ubiquitin-dependent degradation of newly synthesised defective ribosomal products (REITS et al. 2000; SCHUBERT et al. 2000). Nevertheless, the presentation of viral epitopes is improved when the degradation of authentic substrates is accelerated by the introduction of known degradation signals (TOBERY and SILICIANO 1999), confirming the role of intact proteins as a source of antigenic peptides.

Although the efficiency of recognition and ubiquitination by specific E3s is considered to be the major determinant of the rate of turnover for many substrates, knowledge of additional factors involved in the regulation of this process is rapidly growing. For example, an new type of ubiquitination enzyme, originally called UFD2 (JOHNSON et al. 1995), was shown to be involved in extending the polyubiquitin tree in yeast, a process which is indispensable for the degradation of at least one known substrate (KOEGL et al. 1999). Another fascinating insight comes from the observation that protein degradation can also be regulated by deubiquitination enzymes (RORTH et al. 2000). The identification of a large family of highly diverse deubiquitinating enzymes suggests that the degradation of many substrates may be specifically modified downstream of the initial ubiquitination step (CHUNG and BAEK 1999). Finally, a variety of chaperones and co-factors appear to regulate the ubiquitin-dependent proteolysis of certain substrates (BERCOVICH et al. 1997; MEACHAM et al. 2001). This is especially intriguing in the light of the tight relationship between folding of proteins by chaperones and their destruction by proteolytic complexes both in prokaryotes and eukaryotes (BRAUN et al. 1999; PAK et al. 1999; STRICKLAND et al. 2000).

3 Modulation of Proteolysis by the Glycine-Alanine Repeat

Proteasomal processing of viral proteins is the first step in the antigen presentation cascade that can be manipulated by viruses in an attempt to avoid elimination of the host cell. This requires a high degree of selectivity because indiscriminate inactivation of the proteolytic pathway would affect the control of cell cycle progression and apoptosis, resulting in premature death of the infected cell. Although several viral proteins appear to modify the activity of the proteasome (BOYER et al. 1996; GRAND et al. 1999; HU et al. 1999; MANTOVANI and BANKS 1999; ROUSSET et al. 1996; SEEGER et al. 1997; TURNELL et al. 2000; ZHANG et al. 2000), only in two cases has this been shown to lead to blockade of T-cell responses. In human cytomegalovirus-infected cells, expression of the viral phosphoprotein pp65 inhibits the generation of virus-specific T-cell epitopes, probably by preventing their recognition and targeting for destruction (GILBERT et al. 1996). Another challenging example is the selective inhibition of proteasomal processing by the GAr domain of EBNA1.

As mentioned above, EBNA1 is the only viral protein expressed in latent EBV-infected B lymphocytes that is consistently unable to sensitise EBV-negative targets to recognition by virus-specific CTLs, even on overexpression through recombinant vaccinia or adenovirus vectors (BLAKE et al. 1997b; LEVITSKAYA et al. 1995). This was a particularly puzzling finding because EBNA1 is a relatively large protein of more than 600 amino acids and contains a large number of potential MHC class I-restricted epitopes, as defined by the presence of consensus motif for binding to a variety of class I alleles (STUBER et al. 1995). A first clue to the solution of this riddle came from a closer look to the amino acid sequence of EBNA1 and analysis of the known functional domains of the protein (Fig. 2A). EBNA1 is a phosphoprotein composed of unique N- and C-terminal domains joined by an internal domain exclusively composed of glycine and alanine residues that varies in length in different EBV isolates and is the major target of EBNA-specific antibody responses (DILLNER et al. 1984). The possibility that this peculiar repetitive sequence may

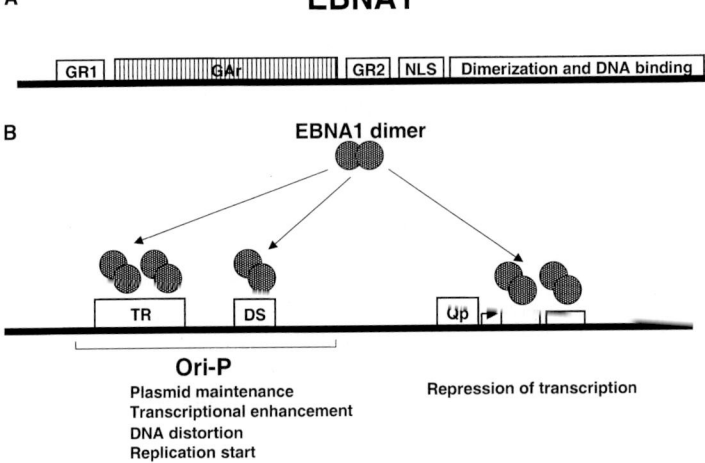

Fig. 2A,B. Schematic representation of EBNA1 and its known functional domains. **A** EBNA1 is a phosphoprotein composed of unique amino-terminal and carboxy-terminal domains (amino acids 1–89 and 327–607 in the B95.8 EBV sequence) that are joined by a repetitive sequence of glycine-arginine (GR)-containing motifs surrounding an internal glycine-alanine repeat (GA_R) of 239 amino acids. Nuclear localisation, DNA binding, and dimerisation domains are located in the carboxy-terminal domain. **B** EBNA1 binds to DNA via recognition of the partial palindrome TAGGATAGCATATGCTACCCAGATCCAG that is found in three sites in the EBV genome. The highest affinity sites consist of 20 tandem direct repeats of the cognate sequence and two cognate sequences in dyad symmetry and two in tandem. Binding of EBNA1 to the tandem repeat (*TR*) and dyad symmetry (*DS*) sites enables covalently closed circular DNA molecules to replicate and persist as episomes, and the region is therefore called OriP for origin of plasmid replication. In addition, the tandem-repeat component acts as an EBNA1 dependent enhancer. The third EBNA1 binding site is composed of two divergent tandem repeats located in the *Bam*HI Q fragment of the viral genome and is important for the negative regulation of an latent promoter (*Qp*) that drives the expression of EBNA1 in certain types of EBV-infected normal B lymphocytes and in EBV-associated malignancies. The N-terminal part of the dimerisation domain of EBNA1 is important for the formation of homopolymers after association with the DNA templates, although the C-terminal of EBNA1 determines its nuclear localisation by interacting with a specific chromosomal protein that is homogeneously distributed on chromosomes. This property is likely to be important for the segregation of the viral episomes into progeny nuclei during mitosis

influence the recognition of EBNA1 by CTLs was tested using recombinant vaccinia viruses encoding EBNA1 chimeras containing an immunodominant CTL epitope inserted in an intact or a GAr-deleted protein. Indeed, these experiments demonstrated that the GAr generates a *cis*-acting inhibitory signal that interferes with antigen presentation (LEVITSKAYA et al. 1995). Distinctive for this viral strategy of immunoescape appeared to be the selective effect on EBNA1, in the absence of detectable trans-inhibition on other viral antigens, even on huge overexpression. Furthermore, the GAr appeared to act as a transferable element because insertion of the domain in the highly immunogenic EBV latent antigen EBNA4 blocked the presentation of MHC class I-restricted CTL epitopes from this protein. Similar results were later obtained in a mouse model, in which presence of the repeats prevented the recognition of EBNA1 by CTLs generated by immunisation with a GAr-deleted mutant (MUKHERJEE et al. 1998), and by expression of EBNA1 in simian cell lines (BLAKE et al. 1997), confirming the general validity of the effect and its reproducibility across species. It is noteworthy that $CD8^+$ CTL precursors specific for MHC class I-restricted epitopes in EBNA1 were recently demonstrated in the memory pool of all EBV carriers (BLAKE et al. 1997). These CTLs kill only target cells that have internalised exogenous EBNA1 molecules, whereas cells that express EBNA1 as an endogenous antigen remain unaffected (BLAKE et al. 1997, 2000). Moreover, the presentation of exogenous EBNA1 is independent of the TAP transporter, further supporting the notion that it is accomplished through a mechanism of cross-priming. Cell-free degradation assays using in vitro translated substrates and semipurified components of the ubiquitin/ proteasome system confirmed that this proteolytic machinery is directly targeted by the inhibitory activity of the GAr (LEVITSKAYA et al. 1997). Whereas the highly immunogenic EBNA4 was efficiently degraded in an ATP/ubiquitin/proteasome-dependent fashion, EBNA1 was resistant to degradation. Processing of EBNA1 was accomplished by deletion of the GAr, whereas insertion of GAr domains of various lengths and in different positions prevented the degradation of EBNA4 without appreciable effect on ubiquitination. The involvement of the proteasome was further substantiated by the almost complete inhibition of ATP-dependent degradation on addition of the proteasome inhibitor MG132, which also inhibits the presentation of EBNA4 epitopes in fibroblasts infected with the relevant recombinant vaccinia virus. Of note, equal levels of inhibition were achieved by insertion of either the 239-amino acid-long repeat of the B95.8 EBNA1 or a shorter repeat containing only 17 glycine-alanine residues. This capacity to act across a wide size range is in line with the observation that the length of the repeats varies in natural EBV isolates, giving rise to EBNA1 polypeptides with characteristic patterns of migration by SDS-PAGE (FALK et al. 1995).

How does the GAr interfere with ubiquitin/proteasome-dependent proteolysis? In principle, any of the three identified steps in the pathway, recognition, targeting, or proteolysis, may be affected. However, the strictly *cis*-acting effect of the domain speaks against a direct interference with the enzymatic activity of the 20S particle. Indeed, addition of synthetic glycine-alanine peptides did not affect the hydrolysis of fluorogenic substrates by purified 20S proteasomes (A. Leonchiks, A. Sharipo,

M.G. Masucci, unpublished observation), pointing to a selective effect on the recognition or targeting steps. Two model systems were developed to further dissect the phenomenon. In the first model the effect of the viral GAr was investigated on the NF-κB inhibitor, IκBα. This is a well-known substrate of ubiquitin/proteasome-dependent proteolysis, and its targeting through recognition of phosphorylation-dependent and constitutive degradation signals as well as the E3 involved in specific ubiquitination have been characterised in great detail (KARIN and BEN-NERIAH 2000). In addition, we have constructed a set of artificial proteasomal substrates by inserting N-end rule and UFD degradation signals into the green fluorescent protein (GFP). Insertion of repeats of different length and amino acid composition in various positions of these model substrates has yielded some insight on the sequence and structure characteristics of the inhibitory domain and its mode of action in relation to various types of degradation signals.

Insertion of the GAr at the N-terminus or C-terminus of IκBα prevented its TNF-α-induced degradation and prolonged the half-life of the chimeric protein in transfected HeLa cells (SHARIPO et al. 1998). The GAr-containing chimeras were efficiently phosphorylated and ubiquitinated, suggesting that the repeats affect events lying downstream of substrate recognition and targeting. The ubiquitinated substrates appear to be unable to achieve an efficient interaction of the proteasome because, in contrast to wild-type IκBα, the chimeras were not co-precipitated by antibodies directed to α subunit of the proteasome. Somewhat surprisingly, an eight-amino acid-long GAr appears to be sufficient to block the degradation of IκBα, independent of its localisation and without affecting the folding properties of the substrate (LEONCHIKS et al. 1998). This suggests that the GAr may function as a recognition signal for events that could either alter the kinetics of interaction with the proteasome or sequester the substrate to a cellular compartment that is inaccessible to the activity of the proteasome. These possibilities were experimentally approached using the GFP reporters that allow visualisation and quantification of ubiquitin/proteasome-dependent proteolysis in living cells (DANTUMA et al. 2000a). Insertion of the repeats prevented the degradation of GFPs containing either N-end rule or the UFD signals without appreciable effect on cellular localisation (DANTUMA et al. 2000b). The comparable effect on different GFP reporters is especially interesting because detailed analysis in yeast has shown that N-end rule and UFD substrates depend differently on ubiquitination enzymes (JOHNSON et al. 1995; KOEGL et al. 1999; VARSHAVSKY 1996), AAA ATPase chaperones (GHISLAIN et al. 1996), and the S5a subunit of the 19S regulator (VAN NOCKER et al. 1996). This, together with the effect on signal-dependent and constitutive turnover of IκBα, which involve two additional sets of degradation signals and E3 ligases, support a broad effect of the GAr over a wide variety of substrates and targeting signals.

The easy quantification of the GFP reporters and their different rates of degradation revealed yet another puzzling feature of GAr. The protective effect appears to be inversely proportional to the strength of the degradation signal and directly proportional to the length of the repeat (DANTUMA et al. 2000b). Thus substrates that are efficiently targeted for proteasomal degradation are only partially rescued, whereas increasing the length of the repeat appears to enhance the protective effect. A clear illustration of this phenomenon is provided by the Ub-P-GFP and Ub^{G76V}-GFP reporters that are targeted for degradation with different efficiency by the same type of UFD signal. Insertion of a 25-amino acid-long GAr in Ub-P-GFP resulted in full protection, whereas a repeat of the same length had no measurable effect on the strongly destabilised Ub^{G76V}-GFP substrate. Furthermore, although the influence of the 25-amino acid repeat on UFD-harbouring substrates appeared to be minimal, a small but significant protection was accomplished in the strongly destabilised N-end rule substrate Ub-R-GFP. It is noteworthy that insertion of a strong degradation signal could overcome the stabilising effect of the GAr in the context of the native EBNA1 protein (DANTUMA et al. 2000b), suggesting that a relatively large size of the repeats may be required to counteract the endogenous degradation signal in the natural EBV isolates.

Very little is known on the mechanisms that influence the efficiency of degradation of a given proteasomal substrate. Conceivably, the rate of ubiquitination and/or deubiquitination could play a critical role (WILKINSON 2000). The affinity and avidity of interaction with docking and unfolding subunits of the 19S cap are also likely to be important because inappropriate acceleration or delay of these steps could result in premature release of the substrate. The demonstration that an eight-amino acid-long repeat is sufficient for full protection of IκBα has opened the possibility of investigating the structural requirement for the effect. With the use of a series of mutant peptides, it was shown that the inhibitory sequence must contain at least three alanine residues interspersed by one, two, or three glycine, an architecture that is strictly conserved in the EBNA1 GAr. The alanine residues can be substituted by small hydrophobic amino acids, such as valine, whereas glycine can be substituted by proline (LEVITSKAYA et al. 1997; SHARIPO et al. 2001). The minimal length of the active domain, together with the demonstration that the GAr lacks a stable conformation in solution (LEONCHIKS et al. 1998), is compatible with a model in which the repeat forms a hydrophobic surface on interaction with a putative partner, as described for SH3 domains. The observation that much longer repeats are required for stabilisation of certain substrates can be accommodated in a model in which multiple sequentially positioned recognition signals may tighten the interaction between the GAr and its cognate partner. An analogous situation is observed in the targeting of ubiquitinated substrate to the proteasome. Here, a large number of recognition signals, provided by the polyubiquitin tree, are required to establish an interaction which is sufficiently strong to initiate progressive degradation (THROWER et al. 2000).

4 Implications for the Biology of EBV Infection

The detection of GAr domains in the EBNA1 proteins of all EBV isolates analysed to date suggest that the repeat may be important for the function of this viral protein. This is further supported by the observation that the EBNA1 homologues of other Old World primate γ-herpesviruses also contain similar repeat domains, although these appear to be shorter in length and contain additional serine residues (BLAKE et al. 1999). The first and hitherto only example of an EBNA1 homologue lacking a GAr-like sequence comes from an EBV-related virus causing latent infection and B-cell lymphomas in a New World primate (CHO et al. 2001). However, the functional behaviour of this recently identified EBNA1 homologue awaits further clarification.

The significance of the stabilising activity of the GAr in the biology of EBV infection is still somewhat unclear. The major known function of EBNA1 is the maintenance of EBV episomes in proliferating virus-infected cells (Fig. 2B) (YATES et al. 1984). It also binds to downstream negative regulatory elements in the Qp promoter (WALLS and PERRICAUDET 1991) which is used for expression of EBNA1 in EBV-positive tumours (NONKWELO et al. 1997; SCHAEFER et al. 1991). Recent evidence demonstrates that Qp is also used for transcription of EBNA1 and in latently EBV-infected germinal centre B cells. In addition to EBNA1, these cells express the latent membrane proteins LMP1 and LMP2 but none of the remaining EBNAs which are believed to be important for B-cell immortalisation in vitro (BABCOCK and THORLEY-LAWSON 2000; BABCOCK et al. 2000). It should be stressed that the immortalisation-associated viral proteins that are expressed in EBV-infected B blasts are efficiently recognised by virus-specific immune responses, and it is therefore unlikely that failure to present epitopes from EBNA1 would significantly contribute to the escape of these infected cells from T-cell surveillance. Conceivably, inhibition of EBNA1 proteolysis by the GAr may be important in other types of virus–host relationships that are not recapitulated by the EBV-transformed B blasts. Although only a minority of freshly separated EBV carrying 'resting' B lymphocytes appear to express EBNA1 mRNA (CHEN et al. 1995; MIYASHITA et al. 1995), a stable EBNA1 protein could persist in these non-proliferating cells in the absence of transcription. Indeed, only few EBNA1 molecules per cell may be sufficient to repress transcription from the Qp promoter, allowing the establishment of a fully latent carrier state; failure to present EBNA1 epitopes could protect these cells from rejection. It is also possible that the stabilising effect of the GAr may serve a non-immunologic purpose. We have shown that an important effect of the repeats is the prolongation of protein half-life (SHARIPO et al. 1998, 2001). This is likely to have major consequences on protein expression and gene transcription in different types of virus-infected cells.

5 Concluding Remarks

The effect of the EBNA1 GAr domain on ubiquitin/proteasome-dependent proteolysis illustrates a new mechanism for the regulation of protein processing and degradation in vivo. Even though the GAr appears to induce a complete block of proteasomal degradation in the context of some proteins, such as EBNA1 and

IκBα, our studies with artificial proteasome substrates revealed that this repeat can delay proteasomal degradation without abrogating it. It is tempting to speculate that cellular substrates of the ubiquitin-proteasome system may harbour GAr-like sequences facilitating a fine-tuning of proteasomal degradation downstream of the ubiquitination step. The presence of such 'stabilisation signals' in natural proteasome substrates would add a new layer to the already complex regulation of proteasomal degradation.

Insertion of a GAr domain into relevant target proteins may provide a useful strategy to selectively modulate cellular processes that are regulated by the proteasome, such as the degradation of cyclins, transcription factors, or proto-oncogenes. This may have important medical applications (POWIS 1998). Because of the central part played by NF-κB and other Rel family members in immune and inflammatory responses, their inhibition by dominant-negative IκBα mutants generated by insertion of the GAr domain may provide a useful means to regulate these processes. Likewise, the generation of stable p53 mutants containing the GAr may provide a new tool for selective induction of apoptosis in a wide variety of human malignancies. There are now examples in which the efficacy of gene therapy may be hampered by the generation of immune responses to components of the transfer vector or the therapeutic product itself or by the rapid turnover of the therapeutic product in transduced cells (GLORIOSO et al. 2001; HIGH 2000). The demonstration that stabilising GAr domains can be inserted in various proteins while preserving their biological functions suggests a simple strategy to overcome these problems.

Acknowledgements. This work was supported by grants awarded by the Swedish Cancer Society, the Swedish Foundation of Strategy Research, the Swedish Research Council, and the Hedlund Foundation, Stockholm, Sweden. NPD is supported by a postdoctoral fellowship awarded by the European Commission Training and Mobility Program (ERBFMRXCT960026). AS is supported by the Visby program for exchange with Baltic States awarded by the Swedish Institute.

References

Babcock GJ, Thorley-Lawson DA (2000) Tonsillar memory B cells, latently infected with Epstein-Barr virus, express the restricted pattern of latent genes previously found only in Epstein-Barr virus-associated tumors. Proc Natl Acad Sci USA 97:12250–12255

Babcock JG, Hochberg D, Thorley-Lawson AD (2000) The expression pattern of Epstein-Barr virus latent genes in vivo is dependent upon the differentiation stage of the infected B cell. Immunity 13:497–506

Bercovich B, Stancovski I, Mayer A, Blumenfeld N, Laszlo A, Schwartz AL, Ciechanover A (1997) Ubiquitin-dependent degradation of certain protein substrates in vitro requires the molecular chaperone Hsc70. J Biol Chem 272:9002–9010

Blake N, Haigh T, Shaka'a G, Croom-Carter D, Rickinson A (2000) The importance of exogenous antigen in priming the human $CD8^+$ T cell response: lessons from the EBV nuclear antigen EBNA1. J Immunol 165:7078–7087

Blake N, Lee S, Redchenko I, Thomas W, Steven N, Leese A, Steigerwald-Mullen P, Kurilla MG, Frappier L, Rickinson A (1997) Human $CD8^+$ T cell responses to EBV EBNA1: HLA class I presentation of the (Gly-Ala)-containing protein requires exogenous processing. Immunity 7:791–802

Blake NW, Moghaddam A, Rao P, Kaur A, Glickman R, Cho YG, Marchini A, Haigh T, Johnson RP, Rickinson AB, Wang F (1999) Inhibition of antigen presentation by the glycine/alanine repeat domain is not conserved in simian homologues of Epstein-Barr virus nuclear antigen 1. J Virol 73: 7381–7389

Bochtler M, Ditzel L, Groll M, Hartmann C, Huber R (1999) The proteasome. Annu Rev Biophys Biomol Struct 28:295–317

Boyer SN, Wazer DE, Band V (1996) E7 protein of human papilloma virus-16 induces degradation of retinoblastoma protein through the ubiquitin-proteasome pathway. Cancer Res 56:4620–4624

Braun BC, Glickman M, Kraft R, Dahlmann B, Kloetzel PM, Finley D, Schmidt M (1999) The base of the proteasome regulatory particle exhibits chaperone-like activity. Nat Cell Biol 1:221–226

Chen F, Zou JZ, di Rienzo L, Wimberg G, Hu LF, Klein E, Klein G, Ernberg I (1995) A subpopulation of latently EBV infected normal B-cells resembles Burkitt lymphoma (BL) in expressing EBNA1 but not EBNA2 or LMP1. J Virol 1995:3752–3758

Cho YG, Ramer J, Rivailler P, Quink C, Garber RL, Beier DR, Wang F (2001) An Epstein-Barr-related herpesvirus from marmoset lymphomas. Proc Natl Acad Sci USA 98:1224–1229

Chung CH, Baek SH (1999) Deubiquitinating enzymes: their diversity and emerging roles. Biochem Biophys Res Commun 266:633–640

Dantuma NP, Lindsten K, Glas R, Jellne M, Masucci MG (2000a) Short-lived green fluorescent proteins for quantification of ubiquitin/proteasome-dependent proteolysis in living cells. Nat Biotech 18: 538–543

Dantuma NP, Heessen S, Lindsten K, Jellne M, Masucci MG (2000b) Inhibition of proteasomal degradation by the Gly-Ala repeat of Epstein-Barr virus is influenced by the length of the repeat and the strength of the degradation signal. Proc Natl Acad Sci USA 97:8381–8385

Dillner J, Sternås L, Kallin B, Alexander H, Ehlin-Henriksson B, Jörnvall J, Klein G, Lerner R (1984) Antibodies against a synthetic peptide identify the Epstein-Barr virus-determined nuclear antigen. Proc Natl Acad Sci USA 81:4652–4656

Falk K, Gratama JW, Rowe M, Zou JZ, Khanim F, Young LS, Oosterveer MAP, Ernberg I (1995) The role of repetitive DNA sequences in the size variation of Epstein-Barr virus (EBV) nuclear antigens, and the identification of different EBV isolates using RFLP and PCR analysis. J Gen Virol 76:779–790

Ferrell K, Deveraux Q, van Nocker S, Rechsteiner M (1996) Molecular cloning and expression of a multiubiquitin chain binding subunit of the human 26S protease. FEBS Lett 381:143–148

Gaczynska M, Goldberg AL, Tanaka K, Hendil KB, Rock KL (1996) Proteasome subunits X and Y alter peptidase activities in opposite ways to the interferon-γ-induced subunits LMP2 and LMP7. J Biol Chem 271:17275–17280

Gaczynska M, Rock KL, Goldberg AL (1993) γ-Interferon and expression of MHC genes regulate peptide hydrolysis by proteasomes. Nature 365:264–267

Ghislain M, Dohmen RJ, Levy F, Varshavsky A (1996) Cdc48p interacts with Ufd3p, a WD repeat protein required for ubiquitin-mediated proteolysis in *Saccharomyces cerevisiae*. EMBO J 15: 4884–4899

Gilbert MJ, Riddell SR, Plachter B, Greenberg PD (1996) Cytomegalovirus selectively blocks antigen processing and presentation of its immediate-early gene product. Nature 383:720–722

Glickman MH, Rubin DM, Coux O, Wefes I, Pfeifer G, Cjeka Z, Baumeister W, Fried VA, Finley D (1998) A subcomplex of the proteasome regulatory particle required for ubiquitin-conjugate degradation and related to the COP9-signalosome and eIF3. Cell 94:615–623

Glorioso JC, Naldini L, Kay MA (2001) Viral vectors for gene therapy: the art of turning infectious agents into vehicles of therapeutics. Nat Med 7:33–40

Grand RJA, Turnell AS, Mason GGF, Wnag W, Milner AE, Mymryk JS, Rookes SM, Rivett AJ, Gallimore PH (1999) Adenovirus early region 1A protein binds to mammalian SUG1-a regulatory component of the proteasome. Oncogene 18:449–458

Groettrup M, Soza A, Eggers M, Kuehn L, Dick TP, Schild H, Rammensee HG, Koszinowski UH, Kloetzel PM (1996) A role for the proteasome regulator PA28α in antigen presentation. Nature 381:166–168

Hershko A, Ciechanover A (1998) The ubiquitin system. Annu Rev Biochem 67:425–479

High KA (2000) Gene therapy in haematology and oncology. Lancet 356 Suppl:s8

Hisamatsu H, Shimbara N, Saito Y, Kristensen P, Hendil KB, Fujiwara T, Takahashi E, Tanahashi N, Tamura T, Ichihara A, Tanaka K (1996) Newly identified pair of proteasomal subunits regulated reciprocally by interferon γ. J Exp Med 183:1807–1816

Hu Z, Zhang Z, Doo E, Coux O, Goldberg AL, Liang TJ (1999) Hepatitis B virus X protein is both a substrate and a potential inhibitor of the proteasome complex. J Virol 73:7231–7240

Johnson ES, Ma PC, Ota IM, Varshavsky A (1995) A proteolytic pathway that recognizes ubiquitin as a degradation signal. J Biol Chem 270:17442–17456

Karin M, Ben-Neriah Y (2000) Phosphorylation meets ubiquitination: the control of NF-κB activity. Annu Rev Immunol 18:621–663

Kieff E. (1996) Epstein-Barr virus and its replication. In: Fields BN, Knipe DM, Howley PM (eds) Fields Virology, 3rd edition. Lippincott, Raven Publishers, Philadelphia, Vol 2, pp 2343–2396

Kisselev AF, Akopian TN, Woo KM, Goldberg AL (1999) The sizes of peptides generated from protein by mammalian 26 and 20S proteasomes. Implications for understanding the degradative mechanism and antigen presentation. J Biol Chem 274:3363–3371

Koegl M, Hoppe T, Schlenker S, Ulrich HD, Mayer TU, Jentsch S (1999) A novel ubiquitination factor, E4, is involved in multiubiquitin chain assembly. Cell 96:635–644

Kuehn L, Dahlmann B (1997) Structural and functional properties of proteasome activator PA28. Mol Biol Rep 24:89–93

Laney J, Hochstrasser M (1999) Substrate targeting in the ubiquitin system. Cell 97:427–430

Leonchiks A, Liepinsh E, Barishev M, Sharipo A, Masucci M, Otting G (1998) Random coil conformation of a Gly/Ala-rich insert in IκB α excludes structural stabilization as the mechanism for protection against proteasomal degradation. FEBS Lett 440:365–369

Levitskaya J, Coram M, Levitsky V, Imreh S, Stegerwald-Mullen PM, Klein G, Kurilla MG, Masucci MG (1995) Inhibition of antigen processing by the internal repeat region of the Epstein-Barr Virus nuclear antigen-1. Nature 375:685–688

Levitskaya J, Sharipo A, Leonchiks A, Ciechanover A, Masucci M (1997) Inhibition of ubiquitin/proteasome-dependent protein degradation by the Gly-Ala repeat domain of the Epstein-Barr virus nuclear antigen 1. Proc Natl Acad Sci USA 94:12616–12621

Mantovani F, Banks L (1999) Inhibition of E6 induced degradation of p53 is not sufficient for stabilization of p53 protein in cervical tumour derived cell lines. Oncogene 18:3309–3315

Meacham GC, Patterson C, Zhang W, Younger JM, Cyr DM (2001) The Hsc70 co-chaperone CHIP targets immature CFTR for proteasomal degradation. Nat Cell Biol 3:100–105

Miyashita EM, Yang B, Lam KM, Crawford DH, Thorley-Lawson DA (1995) A novel form of Epstein-Barr virus latency in normal B cells in vivo. Cell 80:593–601

Mukherjee S, Trivedi P, Dorfman DM, Klein G, Townsend A (1998) Murine cytotoxic T lymphocytes recognize an epitope in an EBNA-1 fragment, but fail to lyse EBNA-1-expressing mouse cells. J Exp Med 187:445–450

Nonkwelo C, Ruf IK, Sample J (1997) The Epstein-Barr virus EBNA-1 promoter Qp requires an initiator-like element. J Virol 71:354–361

Pak M, Hoskins JR, Singh SK, Maurizi MR, Wickner S (1999) Concurrent chaperone and protease activities of ClpAP and the requirement for the N-terminal ClpA ATP binding site for chaperone activity. J Biol Chem 274:19316–19322

Pamer E, Cresswell P (1998) Mechanisms of MHC class I-restricted antigen processing. Annu Rev Immunol 16:323–358

Ploegh HL (1995) Trafficking and assembly of MHC molecules: how viruses elude the immune system. Cold Spring Harbor Symp Quant Biol 60:263–266

Powis SH (1998) Lessons from an age-old war. Nat Med 4:887–888

Rechsteiner M, Rogers SW (1996) PEST sequences and regulation by proteolysis. Trends Biochem Sci 21:267–271

Reits EA, Vos JC, Gromme M, Neefjes J (2000) The major substrates for TAP in vivo are derived from newly synthesized proteins. Nature 404:774–778

Rickinson AB, Kieff E (1996) Epstein-Barr virus. In: Fields BN, Knipe DM, Howley PM (eds) Fields Virology, 3rd edition. Lippincott, Raven Publishers, Philadelphia, Vol 2, pp 2397–2446

Rickinson AB, Moss DJ (1997) Human cytotoxic T lymphocyte responses to Epstein-Barr virus infection. Annu Rev Immunol 15:405–431

Rock KL, Goldberg AL (1999) Degradation of cell proteins and the generation of MHC class I-presented peptides. Annu Rev Immunol 17:739–779

Rock KL, Gramm C, Rothstein L, Clark K, Stein R, Dick L, Hwang D, Goldberg AL (1994) Inhibitors of the proteasome block the degradation of most cell proteins and the generation of peptides presented on MHC class I molecules. Cell 78:761–771

Rorth P, Szabo K, Texido G (2000) The level of C/EBP protein is critical for cell migration during *Drosophila* oogenesis and is tightly controlled by regulated degradation. Mol Cell 6:23–30

Rousset R, Desbois C, Bantignies F, Jalinot P (1996) Effects on NF-κB1/p105 processing of the interaction between the HTLV-1 transactivator Tax and the proteasome. Nature 381:328–331

Schaefer BC, Woisetschlaeger M, Strominger JL, Speck SH (1991) Exclusive expression of Epstein-Barr virus nuclear antigen 1 in Burkitt lymphoma arises from a third promoter, distinct from the promoters used in latently infected lymphocytes. Proc Natl Acad Sci USA 88:6550–6554

Schubert U, Anton LC, Gibbs J, Norbury CC, Yewdell JW, Bennink JR (2000) Rapid degradation of a large fraction of newly synthesized proteins by proteasomes. Nature 404:770–774

Seeger M, Ferrell K, Frank R, Dubiel W (1997) HIV-1 tat inhibits the 20S proteasome and its 11S regulator-mediated activation. J Biol Chem 272:8145–8148

Sharipo A, Imreh M, Bränden CI, Masucci MG (2001) Cis-inhibition of proteasomal degradation by viral repeats: impact of length and amino acid composition. FEBS Lett 499:137–142

Sharipo A, Imreh M, Leonchiks A, Imreh S, Masucci M (1998) A minimal glycine-alanine repeat prevents the interaction of ubiquitinated IκB-α with the proteasome: a new mechanism for selective inhibition of proteolysis. Nat Med 4:939–944

Strickland E, Hakala K, Thomas PJ, DeMartino GN (2000) Recognition of misfolding proteins by PA700, the regulatory subcomplex of the 26S proteasome. J Biol Chem 275:5565–5572

Stuber G, Dillner J, Modrow S, Wolf H, Szekely L, Klein G, Klein E (1995) HLA-A0201 and HLA-B7 binding peptides in the EBV-encoded EBNA-1, EBNA-2 and BZLF-1 proteins detected in the MHC class I stabilization assay. Low proportion of binding motifs for several HLA class I alleles in EBNA-1. Int Immunol 7:653–663

Thrower JS, Hoffman L, Rechsteiner M, Pickart CM (2000) Recognition of the polyubiquitin proteolytic signal. EMBO J 19:94–102

Tobery T, Siliciano RF (1999) Cutting edge: induction of enhanced CTL-dependent protective immunity in vivo by N-end rule targeting of a model tumor antigen. J Immunol 162:639–642

Townsend A, Rothbard J, Gotch F, Bahadur B, Wraith D, McMichael A (1986) The epitopes of influenza nucleoprotein recognized by cytotoxic T lymphocytes can be defined with short synthetic peptides. Cell 44:959–968

Turnell AS, Grand RJ, Gorbea C, Zhang X, Wang W, Mymryk JS, Gallimore PH (2000) Regulation of the 26S proteasome by adenovirus E1A. EMBO J 19:4759–4773

van Nocker S, Sadis S, Rubin DM, Glickman M, Fu H, Coux O, Wefes I, Finley D, Vierstra RD (1996) The multiubiquitin-chain-binding protein Mcb1 is a component of the 26S proteasome in *Saccharomyces cerevisiae* and plays a nonessential, substrate-specific role in protein turnover. Mol Cell Biol 16:6020–6028

Varshavsky A (1996) The N-end rule: functions, mysteries, uses. Proc Natl Acad Sci USA 93:12142–12149

Voges D, Zwickl P, Baumeister W (1999) The 26S proteasome: a molecular machine designed for controlled proteolysis. Annu Rev Biochem 68:1015–1068

Walls D, Perricaudet M (1991) Novel downstream element upregulates transcription initiated from an Epstein-Barr virus latent promoter. EMBO J 10:143–151

Whitby FG, Masters EI, Kramer L, Knowlton JR, Yao Y, Wang CC, Hill CP (2000) Structural basis for the activation of 20S proteasomes by 11S regulators. Nature 408:115–120

Wilkinson KD (2000) Ubiquitination and deubiquitination: targeting of proteins for degradation by the proteasome. Semin Cell Dev Biol 11:141–148

Xie Y, Varshavsky A (2000) Physical association of ubiquitin ligases and the 26S proteasome. Proc Natl Acad Sci USA 97:2497–2502

Yates J, Warren N, Reisman D, Sugden B (1984) A *cis*-acting element from the Epstein-Barr viral genome that permits stable replication of recombinant plasmids in latently infected cells. Proc Natl Acad Sci USA 81:3806–3810

Zhang Z, Torii N, Furusaka A, Malayaman N, Hu Z, Liang TJ (2000) Structural and functional characterization of interaction between hepatitis B virus X protein and the proteasome complex. J Biol Chem 275:15157–15165

The HCMV Gene Products US2 and US11 Target MHC Class I Molecules for Degradation in the Cytosol

F.J. van der Wal, M. Kikkert, and E. Wiertz

Over millions of years of coevolution with their hosts, viruses have developed highly effective strategies to elude the host immune system. The degradation of major histocompatibility complex (MHC) class I heavy chains by human cytomegalovirus (HCMV) is an example of this. Two HCMV proteins, US2 and US11, target newly synthesized MHC class I heavy chains for destruction via a pathway that involves ubiquitin-dependent retrograde transport, or "dislocation", of the heavy chains from the ER to the cytosol, where the proteins are degraded by proteasomes. In this review, US2- and US11-mediated degradation of MHC class I heavy chains is discussed in relation to data concerning the degradation of other ER luminal proteins. A new, unified model for translocon-facilitated dislocation and degradation of MHC class I heavy chains is presented.

1	Introduction	38
2	Immune Evasion Strategies of Viruses	38
2.1	MHC Class I-Restricted Antigen Presentation	39
2.2	Herpesviruses Impede MHC Class I-Restricted Presentation of Peptides	39
3	HCMV Causes MHC Class I Heavy Chain Degradation	40
3.1	The HCMV Gene Products US2 and US11 Target MHC Class I Molecules for Destruction	40
3.2	US2 and US11 Bind to MHC Class I Heavy Chains	41
4	The Dislocation Pathway	42
4.1	The Translocon	43
4.2	Targeting of Degradation Substrates to the Translocon	44
4.3	Oxidoreductases	46
4.4	Ubiquitination and Protein Dislocation	47
4.5	Peptide:N-Glycanase	48
4.6	The Pulling Force	49
5	Concluding Remarks	50
	References	50

Abbreviations

β2m	β2-microglobulin
BiP	immunoglobulin heavy chain binding protein
CFTR	cystic fibrosis transmembrane conductance regulator

Department of Medical Microbiology, Leiden University Medical Center, P.O. Box 9600, 2300 RC Leiden, The Netherlands

CPY*	mutant yeast carboxypeptidase Y
ER	endoplasmic reticulum
HC	MHC class I heavy chain
HCMV	human cytomegalovirus
HLA	human leukocyte antigen
HMG-CoA	hydroxymethylglutaryl-coenzyme A reductase
MCMV	murine cytomegalovirus
MHC	major histocompatibility complex
PDI	protein disulfide isomerase
PNGase	Peptide:N-glycanase
TAP	transporter associated with antigen presentation

1 Introduction

The immune system has evolved so as to require only minute amounts of antigen for detection of a foreign invader. Herpesviruses have withstood the selective pressure exerted by the host through evasion of the host immune response and have had millions of years to bring these strategies to perfection. The fact that most, if not all, herpesviruses are capable of establishing persistent infections illustrates the effectiveness of the "stealth technology" employed by these viruses.

The diversity of immune escape mechanisms is astonishing: herpesviruses modify signal transduction and transcription factor activity, inhibit apoptosis, interfere with the function of cytokines and their receptors, interact with components of the complement system and modulate activation of T cells and NK cells, in addition to interference with normal functions of the cell (HENGEL et al. 1998; PLOEGH 1998; TORTORELLA et al. 2000; WIERTZ et al. 1997a,b; see also *Seminars in Virology* Vol. 8, No. 5, 1998). This review will focus on the interference with antigen presentation by two human cytomegalovirus proteins, US2 and US11, which target newly synthesized major histocompatibility complex class I molecules for destruction.

2 Immune Evasion Strategies of Viruses

Cytotoxic T cells play an important role in the detection and elimination of viruses. Activation of cytotoxic T cells is dependent on recognition of antigenic peptides derived from the pathogenic invader in the context of major histocompatibility complex (MHC) class I molecules (HEEMELS and PLOEGH 1995; ROCK and GOLDBERG 1999). Over millions of years of coevolution with their hosts, herpesviruses have developed highly effective strategies to elude the host immune system. Human cytomegalovirus (HCMV)-induced degradation of MHC class I heavy chains is an example of this.

2.1 MHC Class I-Restricted Antigen Presentation

MHC class I complexes consist of a heavy chain, which is a glycosylated type I membrane protein of about 45kDa, and a soluble light chain of 12kDa, β2-microglobulin (β2m). Heavy chain and β2m form a stable complex together with a peptide of eight or nine residues. These peptides are generated in the cytosol by proteasomal degradation of either cellular proteins or proteins from an invading pathogen (KISSELEV et al. 1999). A significant fraction of newly synthesized proteins is rapidly degraded by proteasomes (SCHUBERT et al. 2000). The resulting peptides are the principal substrates for the transporter associated with antigen presentation (TAP) (REITS et al. 2000). TAP transports the peptides into the ER, where the peptides are placed in a dedicated peptide-binding groove of the MHC class I heavy chain. It has been estimated that for every peptide ligand generated 10,000 proteins have to be degraded (YEWDELL 2001). The peptide-MHC complex then migrates to the cell surface through the secretory pathway and presents the antigenic peptide at the cell surface to cytotoxic T cells.

Herpesviruses have developed many different strategies to interfere with assembly and transport of MHC class I complexes. Although not formally proven, it is anticipated that these immune evasion mechanisms also play a role in the establishment of latency by herpesviruses.

2.2 Herpesviruses Impede MHC Class I-Restricted Presentation of Peptides

During the past decade, fascinating viral evasion strategies have been identified. Every single step in the assembly and transport of MHC class I complexes appears to present a potential target for viral evasion strategies. Viruses can compromise antigen processing at the earliest stage by preventing proteasomal degradation of potential antigens in the cytosol. The Epstein-Barr virus-encoded EBNA-1 serves as a paradigm for viral interference with proteasomal proteolysis. The presence of a Gly-Ala repeat within EBNA-1, varying in length between 70 and 600 residues, renders the protein refractory to proteasomal degradation (LEVITSKAYA et al. 1995, 1997).

Antigenic peptides that have been generated can be prevented from being transported into the ER. Herpes simplex virus does so by using a small cytosolic protein, ICP47, that competes with peptides for TAP (AHN et al. 1996; FRUH et al. 1995; HILL et al. 1995; TOMAZIN et al. 1996). HCMV employs an ER-resident membrane glycoprotein, US6, that inhibits peptide transport by preventing ATP binding to TAP (HEWITT et al. 2001; LEHNER et al. 1997).

If antigenic peptides do emerge in the ER, viruses can mediate retention of MHC class I molecules in intracellular compartments. Examples are reviewed elsewhere in this issue and include HCMV-encoded US3 and the murine cytomegalovirus (MCMV) m152 gene product gp40. The mechanism by which US3 binds to class I molecules and retains the complex in the ER, involves a retention signal in the luminal part of US3 (LEE et al. 2000). In contrast, the MCMV-encoded gp40 triggers retention of class I molecules in the *cis*-Golgi without forming a stable complex with the retained molecules (ZIEGLER et al. 2000).

Finally, herpesviruses produce proteins that target MHC class I heavy chains for destruction. MCMV induces degradation of MHC class I molecules in a lysosomal compartment (REUSCH et al. 1999). In HCMV-infected cells, MHC class I heavy chains are targeted for destruction to the proteasome via a novel degradation pathway that will be discussed below.

3 HCMV Causes MHC Class I Heavy Chain Degradation

HCMV infection results in downregulation of cell surface expression of MHC class I molecules (BROWNE et al. 1990; BARNES and GRUNDY 1992; YAMASHITA et al. 1993). This is not achieved by affecting expression levels of mRNA (BROWNE et al. 1990) but by rapid degradation of newly synthesized MHC class I heavy chains (BEERSMA et al. 1993; YAMASHITA et al. 1993). Downregulation of MHC class I surface expression is specifically due to degradation of MHC class I heavy chains, because steady-state levels of β2m and unrelated cellular proteins are not affected (BEERSMA et al. 1993; YAMASHITA et al. 1994). Whereas MCMV targets MHC class I molecules for destruction to the lysosomal compartment (REUSCH et al. 1999), in HCMV-infected cells this compartment was ruled out as the organelle where heavy chains are degraded (YAMASHITA et al. 1994).

3.1 The HCMV Gene Products US2 and US11 Target MHC Class I Molecules for Destruction

Two loci within the HCMV unique short region were identified as being responsible for the observed degradation of MHC class I heavy chains (JONES et al. 1995; JONES and SUN 1997). The first viral gene product identified, US11, encodes a 25-kDa protein that is predicted to be a type I membrane protein with a cleavable signal peptide and one consensus site for N-glycosylation in its luminal part. By immunogold labeling US11 was shown to reside in ER membranes exclusively (WIERTZ et al. 1996a).

The second viral protein involved, US2, is predicted to be a 23-kDa protein with a transmembrane domain in its C-terminal half (HOFMANN and STOFFEL 1993). The N-terminal part bears a resemblance to a signal peptide but is not cleaved off (GEWURZ et al. 2001a). US2 has three consensus sites for N-glycosylation, one of which is located toward the N-terminus of US2. The two other sites are in, or close to, the transmembrane domain. Although topology predictions strongly prefer an orientation in which the N-terminal part is not luminal but cytosolic, US2 does carry one single oligosaccharide (TORTORELLA et al. 1998;

WIERTZ et al. 1996b). Thus it is most likely that US2 occurs as a type I membrane protein of which the N-terminal site is glycosylated. Consistent with its function in the ER, US2 does not travel beyond the ER, because it never becomes endoH resistant in pulse-chase experiments (WIERTZ et al. 1996b).

Further studies were facilitated by transfection of human astrocytoma cells (U373) with US2 or US11. These cells show rapid degradation of endogenous MHC class I heavy chains, with half-lives of less than 1 min. Because degradation of many proteins is catalyzed by the proteasome (reviewed in VOGES et al. 1999), the effect of inhibiting its proteolytic activity has been investigated. In the presence of proteasome inhibitors, heavy chain degradation is inhibited and an intermediate with increased mobility by SDS-PAGE emerges. This intermediate appears to be a full-length class I heavy chain that has lost its N-linked glycan (WIERTZ et al. 1996a,b).

The involvement of the proteasome suggests that the MHC class I heavy chains migrate back to the cytosol. Indeed, a cytosolic disposition of the deglycosylated intermediate can be confirmed by subcellular fractionation experiments. Furthermore, this intermediate can be immunoprecipitated from material reactive with anti-proteasome antibodies. Apparently, US2 and US11 induce retrograde translocation, or "dislocation", of MHC class I heavy chains from the ER to the cytosol, where the heavy chains are subjected to proteolytic degradation by the proteasome (WIERTZ et al. 1996a,b).

3.2 US2 and US11 Bind to MHC Class I Heavy Chains

The mechanisms by which US2 and US11 force heavy chains into the degradation pathway seem to differ from each other. First of all, the viral proteins show an overlap in specificity, but there are differences as to the allelic products they can affect. US2 and US11 transfectants show degradation of heavy chains encoded by HLA-A and HLA-B alleles (BEERSMA et al. 1993; JONES et al. 1995; WIERTZ et al. 1996a,b) but not of HLA-C and HLA-G locus products (SCHUST et al. 1998). In mouse cells transfected with HLA-B27 only the human MHC class I heavy chains were degraded on HCMV infection, but not the murine class I products (BEERSMA et al. 1993). A study with human cells expressing murine MHC class I molecules showed that US2 induces degradation of H-2D^b and D^d, but not K^b or L^d. In US11-expressing cells K^b, D^d, D^b and L^d are degraded, whereas K^d is stable (MACHOLD et al. 1997). Studies with hybrid MHC class I heavy chains should reveal sequence requirements for interactions with the HCMV gene products.

Apart from a difference in specificity toward subtypes of MHC class I molecules, US2 and US11 seem to react differently with MHC class I heavy chains. In cells expressing US2 or US11 newly synthesized MHC class I heavy chains are rapidly degraded, regardless of whether they are correctly folded and assembled in a complex with β2m or occur as free heavy chains (WIERTZ et al. 1996a,b). In pulse-chase experiments, US2 coprecipitates with heavy chain-β2m complexes and with free heavy chains, whereas for US11 very little or no coprecipitation with class I heavy chains is observed (STORY et al. 1999). Importantly, deglycosylated US2 molecules exclusively coprecipitate with deglycosylated heavy chains (WIERTZ et al. 1996b). This suggests that US2 molecules accompany class I heavy chains on their way to the cytosol. In contrast, US11 remains glycosylated and associated with the ER membrane, implying a more transient interaction of US11 with MHC class I heavy chains (WIERTZ et al. 1996a). This might explain why these interactions are more difficult to visualize by immunoprecipitation techniques. Nevertheless, US11 can be caught interacting with heavy chains when degradation is impaired, for example, in cells overexpressing heavy chains devoid of their cytosolic tail (STORY et al. 1999), or when ubiquitination is prevented (KIKKERT et al. 2001; see below).

Recently, the crystal structure of US2 in a complex with class I heavy chain and β2m has been resolved (GEWURZ et al. 2001b). US2 appears to bind to the junction of the α3 domain and the region that binds peptide, and residues in HLA-A2 that are involved in recognition by US2 are identified. It is anticipated that these structures will contribute greatly to our understanding of how US2 mediates heavy chain degradation.

4 The Dislocation Pathway

Retrograde transport of MHC class I molecules, followed by proteasomal degradation, also occurs in the absence of US2 and US11 and is used to dispose of MHC class I heavy chains that fail to fold and assemble properly (HUGHES et al. 1997; WIERTZ et al. 1996b). MHC class I molecules misfold if antigenic peptides are lacking, as is the case in TAP-deficient cells, or if β2m is absent, as is the case in Daudi cells. As in US2 or US11-expressing cells, a deglycosylated heavy chain intermediate can be observed in the presence of proteasome inhibitors (HUGHES et al. 1997).

Whereas degradation of misfolded proteins was initially considered to take place in the ER (KLAUSNER and SITIA 1990), it has become clear now that degradation of many ER proteins is mediated by the proteasome in the cytosol. A quality control system in the ER lumen assists folding and maturation of newly synthesized proteins and serves to select misfolded proteins and unassembled subunits for degradation (ELLGAARD et al. 1999). A rapidly expanding list of proteins that are degraded by the proteasome includes yeast proteins, such as prepro-α factor (MCCRACKEN and BRODSKY 1996), mutant carboxypeptidase Y (CPY*) (HILLER et al. 1996), hydroxymethylglutaryl-coenzyme A reductase (HMG-CoA) (HAMPTON et al. 1996), and a mutant subunit of the translocon complex Sec61p (BIEDERER et al. 1996). In human cells, proteasomal degradation has been demonstrated for the cystic fibrosis transmembrane conductance regulator (CFTR) (JENSEN et al. 1995; WARD et al. 1995), α1-antitrypsin (LIU et al. 1999; QU et al. 1996), coagulation factor VIII (PIPE et al. 1998), secretory IgM (WINITZ et al. 1996), and apoprotein B (GINSBERG 1997). T cell receptor α chains, expressed in the absence of the other components of the TCR complex, are also transported back to the cytosol and degraded by the proteasome (HUPPA and PLOEGH 1997; YU et al. 1997). When the β chains of the MHC class II heterodimer are expressed in the absence of their counterparts, they undergo the same fate (DUSSELJEE et al. 1998). Apart from herpesviruses, this pathway has also been discovered by other viruses and is used by HIV-encoded Vpu to target CD4 for destruction (FUJITA et al. 1997; SCHUBERT et al. 1998).

From these examples, it is evident that protein dislocation is a common pathway used by cells to dispose of misfolded proteins. It is obvious that proteins that are located in the ER lumen have to travel across the ER membrane before they can be degraded by the cytosolic proteasome. The question arises whether retrograde translocation is mediated by the translocon or by a completely different translocation system dedicated to dislocating proteins (reviewed in KOPITO 1997).

4.1 The Translocon

Membrane and secretory proteins enter the mammalian ER via the translocon, which consists of the heterotrimeric Sec61 complex and TRAM (reviewed in JOHNSON and VAN WAES 1999). The observation that in US2-expressing cells deglycosylated degradation intermediates of MHC class I heavy chains coprecipitate with Sec61β initiated the hypothesis that the translocon is involved in retrograde transport of proteins from the ER lumen to the cytosol. In an experiment in which DTT was used to induce heavy chain degradation, heavy chains again could be seen interacting with Sec61β (WIERTZ et al. 1996b). This strongly suggested that, in addition to import of newly synthesized proteins, the translocon is involved in export of proteins destined for degradation by the proteasome.

Further support for the idea that retrograde transport is mediated by the translocon has been obtained in yeast. Sec61p is the yeast homologue of the mammalian Sec61α, i.e., the protein that forms the actual transmembrane pore of the translocon (JOHNSON and VAN WAES 1999). Several *sec61* mutants appeared to function in protein translocation but were impaired in dislocation. In such mutant strains misfolded proteins could be chemically cross-linked to Sec61p (PILON et al. 1997; PLEMPER et al. 1997; ZHOU and SCHEKMAN 1999). Recent studies suggest that the third transmembrane domain and the fourth luminal loop of Sec61p are required for retrograde protein transport (WILKINSON et al. 2000; ZHOU and SCHEKMAN 1999).

For MHC class I heavy chain and CFTR, both membrane proteins, it is clear that they coprecipitate with Sec61β in the course of degradation (WIERTZ et al. 1996b; BEBOK et al. 1998). Also, ribophorin I, which is an ER luminal protein, occurs in a complex with Sec61β during degradation (DEVIRGILIO et al. 1998). This subunit of the translocon might thus play an important role in protein dislocation. However, a deletion in *SBH1*, the yeast homologue of Sec61β, does not affect proteasomal degradation of the luminal protein CPY* (PLEMPER et al. 1997). This might reflect a difference in requirements for degradation per substrate or a difference between yeast and mammals in the mechanism of dislocation.

It is tempting to speculate as to which proteins are directly involved in handing over degradation substrates to the translocon. Kar2p, the yeast homologue of BiP, binds to Sec63p (MATLACK et al. 1999), and together they are involved in dislocation of proteins (BRODSKY et al. 1999; Plemper et al. 1997). It has been hypothesized that yeast strains mutated in Sec61p are perturbed for dislocation and posttranslational translocation because Sec63p cannot be recruited efficiently (WILKINSON et al. 2000). Recently, Sec62p and Sec63p homologues have been identified in mammalian cells (DAIMON et al. 1997; SKOWRONEK et al. 1999). Although these proteins interact directly with the mammalian Sec61α, they do not play a clear role in anterograde protein transport. On the basis of this observation, it has been suggested that in mammalian cells Sec62p and Sec63p may play a role in retrograde translocation of misfolded proteins (MEYER et al. 2000; TYEDMERS et al. 2000).

4.3 Oxidoreductases

It is likely that a protein has to be in a more or less unfolded state before it can be transported back from the ER lumen across the membrane to the cytosol. In US2- and US11-expressing cells the amount of heavy chain-β2m complexes decreases in favor of free heavy chains. These are then degraded by proteasomes or, in the presence of proteasome inhibitors, accumulate in the cytosol as soluble deglycosylated intermediates. Heavy chains have two disulfides of which the reduction occurs before deglycosylation (TORTORELLA et al. 1998). The conformation of a protein with intramolecular disulfide bridges may impose spatial restrictions on its dislocation. It therefore is conceivable that in dislocating proteins disulfide bonds are to be reduced before dislocation. When the translocon is occupied by a ribosome in action, the pore is wide open and can accommodate several transmembrane sequences (JOHNSON and VAN WAES 1999). Assuming that dislocating proteins engage an open pore, the requirement to "straighten" a protein before dislocation will depend on its conformation.

Pdi1p, the protein disulfide isomerase (PDI) homologue in yeast, is involved in retrograde translocation of misfolded proteins. It was shown that the efficiency of ER degradation decreases in strains expressing Pdi1p with deletions in its putative peptide binding domain (GILLECE et al. 1999). However, it is not clear yet how this chaperone would be linked to the dislocation machinery, because in Δpdi1 strains protein degradation remains unaffected when the essential *PDI1* gene is substituted for by one of its nonessential homologues (NORGAARD et al. 2001).

The mammalian ER luminal chaperone PDI is an oxidoreductase involved in the formation and reshuffling of disulfide bonds, although it is also capable of interactions with proteins that do not form disulfides (GILBERT 1997; NOIVA 1999; FERRARI and SOLING 1999). Because disulfides of dislocating MHC class I heavy chains are reduced just before the proteins are deglycosylated, it has been hypothesized that PDI may play a role in breaking up disulfide bonds in an early phase of dislocation (TORTORELLA et al. 1998). In addition to PDI, other disulfide isomerases could play a role in this process. A suitable contender is ERp57, because it is directly connected to the ER quality control apparatus via calnexin and calreticulin (OLIVER et al. 1997, 1999).

4.4 Ubiquitination and Protein Dislocation

The covalent attachment of multiple ubiquitin molecules to one or more lysines within a protein provides a well-established means of targeting the protein for destruction by the proteasome. Ubiquitination involves a set of enzymes that sequentially activate ubiquitin (E1 enzyme), conjugate it to the substrate (E2 or E3 enzyme), and elongate the multi-ubiquitin chain (E3 enzyme) (HERSHKO and CIECHANOVER 1998). Although E1 is a universal enzyme, E2 and particularly E3 enzymes usually have specificity for only one or a few degradation substrates.

Ubiquitinated MHC class I molecules could be detected upon expression of US11, whereas in the absence of US11 these were undetectable (SHAMU et al. 1999). Our recent study showed that a functional ubiquitin system is required for the US11 mediated breakdown of MHC class I molecules in a very early stage of the degradation pathway; disruption of the ubiquitin machinery in a mutant cell line causes retention of MHC class I molecules, bound to US11, in the ER membrane (KIKKERT et al. 2001). Dislocation of the MHC class I molecules across the ER membrane is apparently fully dependent on a functional ubiquitin system. Results from a recently published in vitro study using ubiquitin depleted cytosol are in complete agreement with this (SHAMU et al. 2001).

Ubiquitin dependent dislocation has also been observed for misfolded proteins or proteins that lack their oligomerization partners, including the T cell receptor α-chain (YU and KOPITO 1999), truncated ribophorin I (DEVIRGILIO et al. 1998), and CPY* (BORDALLO et al. 1998). Obviously, not only does dislocation entail ubiquitin, but subsequent proteasomal degradation requires ubiquitin as well, because proteins are usually targeted to the proteasome via the attachment of multiple ubiquitin molecules (COUX et al. 1996; HERSHKO and CIECHANOVER 1998). It is not clear, however, whether a single ubiquitination event accounts for both dislocation and degradation, or whether several distinct ubiquitination events are required. There is evidence that the lysines in the cytosolic tail of the heavy chain are not required for dislocation or degradation, indicating that their ubiquitination is not essential for retrograde transport (SHAMU et al. 1999). Despite this observation, a functional ubiquitination system is a prerequisite for heavy chain dislocation (KIKKERT et al. 2001). Because there is no evidence that US2 and US11 are ubiquitinated, a possibility might be that an auxiliary protein gets ubiquitinated to flag the US-HC complex for dislocation, analogous to ubiquitination of calnexin in degradation of α1-antitrypsin Z (QU et al. 1996). In the course of dislocation of the MHC class I heavy chain, the originally luminal lysines become exposed to the cytosol one by one, on which they are ubiquitinated. This would explain the occurrence of polyubiquitinated MHC class I heavy chains in cells in which proteasomal degradation is blocked. Recently it was proposed that polyubiquitin physically prevents dislocating heavy chains from sliding back into the ER lumen (SHAMU et al. 2001).

Several ubiquitinating enzymes that are involved in dislocation and degradation of ER glycoproteins have been identified in yeast. Ubc7 is a soluble E2 enzyme that is recruited from the cytoplasm to the ER membrane (BIEDERER et al. 1997). Ubc7 and two other E2 enzymes, Ubc1 and Ubc6, are involved in degradation of several yeast ER proteins (BIEDERER et al. 1996; HAMPTON and BHAKTA 1997; HILLER et al. 1996; WILHOVSKY et al. 2000; FRIEDLANDER et al. 2000). Together they may represent general E2 enzymes specialized for ER degradation substrates. Furthermore, a group of genes designated the "HRD genes" were found essential for regulated breakdown of yeast HMG-CoA (BORDALLO et al. 1998; BORDALLO and WOLF 1999). Hrd1p, also isolated as Der3p, is very likely to be an E3 enzyme and is part of a specific complex (GARDNER et al. 2000, 2001). There is experimental evidence indicating that these ubiquitinating enzymes have different substrate specificities (WILHOVSKY et al. 2000).

Recently, murine homologues of Ubc6 and Ubc7 have been identified that play a role in the initiation of retrograde transport of T cell receptor subunits (TIWARI and WEISSMAN 2001). Considering the data obtained for yeast, it may be possible that in mammalian cells ubiquitinating enzymes with a taste for MHC class I heavy chains are recruited to the site where heavy chains destined for degradation exit the ER.

4.5 Peptide:*N*-Glycanase

On the way from the ER to the proteasome the asparagine-linked glycan of MHC class I molecules is removed, resulting in the cytoplasmic degradation intermediates that are observed in the presence of proteasome inhibitors (WIERTZ et al. 1996a,b). Deglycosylation of the MHC class I molecules is accompanied by a charge shift, which can be explained by conversion of the asparagine into an aspartic acid residue. This conversion is a hallmark of deglycosylation by a peptide:*N*-glycanase (PNGase) (TAKAHASHI 1992). Recently, we showed that in US11-expressing cells MHC class I heavy chains accumulate in the ER when the ubiquitin system is inhibited (see above). The ER-retained MHC class I molecules are not deglycosylated. This suggests that deglycosylation takes place after dislocation, implying that the PNGase resides in the cytosol.

Further support for this hypothesis comes from studies on the export of glycosylated peptides. Glycosylated peptides are also exported from the ER to the cytosol through the translocon (GILLECE et al. 2000; KOOPMANN et al. 2000; ROMISCH and SCHEKMAN 1992). In a cell-free system, not only dislocation of glycosylated peptides from mammalian ER membranes, but also their deglycosylation is dependent on the addition of cytosol (ROMISCH and ALI 1997). This strongly suggests that mammalian cytosol contains a peptide:*N*-glycanase that deglycosylates glycosylated peptides.

Additional evidence that the pursued PNGase resides in the cytosol comes from studies in a glycosylation mutant of Chinese hamster ovary cells. In these cells, glucosylated oligosaccharides accumulate in the ER when protein synthesis is blocked. Under conditions in which protein synthesis occurs normally and degradation is induced, glucosylated oligosaccharides can be detected in the cytosol (DUVET et al. 1998). These sugar moieties must have been transported to the cytosol while attached to a glycoprotein. These results implicate that removal of oligosaccharides by PNGase takes place in the cytosol.

Soluble, nonlysosomal, intracellular peptide:N-glycanase activity has been found in several mammalian cell lines (SUZUKI et al. 1993). A PNGase enzyme was purified and characterized from mouse L929 cells (SUZUKI et al. 1994). The characteristics of these PNGases include maximum activity around pH 7 and the necessity of free sulfhydryls, consistent with a cytosolic location. Recently, a gene encoding a 42.5-kDa soluble PNGase was identified in yeast, which appears to have conserved homologues in many other eukaryotic species including human and mouse (SUZUKI et al. 2000). It is located in the nucleus as well as in the cytosol, has optimal activity around pH 7 and requires free sulfhydryls (SUZUKI et al. 2000). On the basis of these properties this enzyme forms a good candidate for the PNGase that catalyzes deglycosylation of dislocated degradation substrates. Very recently it was shown that yeast PNGase associates with Rad23p, which in turn associates with 26S proteasomes (SUZUKI et al. 2001). In addition, mouse PNGase has been found to associate with a regulatory subunit of the 19S proteasome (PARK et al. 2001). This clearly links PNGase activity to the proteasomal degradation pathway. It is, however, surprising that the degradation of the well-characterized mutant ER degradation substrate CPY* is only slightly delayed upon deletion of the characterized gene in yeast (SUZUKI et al. 2000). This may indicate that deglycosylation of substrates is not essential for their degradation by the proteasome, but merely accelerates the process.

4.6 The Pulling Force

At present, the actual extraction of glycoproteins from the ER is poorly understood. The pulling force that drives dislocation of the substrate may be provided by the proteasome (MAYER et al. 1998). Indeed, the observation that most degradation substrates accumulate in the ER when the proteolytic activity of the proteasome is inhibited with chemical compounds suggests a tight coupling between dislocation and degradation (KOPITO 1997; KOPITO and SITIA 2000). In fact, the soluble cytosolic MHC class I heavy chains and TCR α chains observed upon proteasome inhibition are exceptions (HUPPA and PLOEGH 1997; YU et al. 1997; STORY et al. 1999).

Cytosolic chaperones like Hsp90 (IMAMURA et al. 1998) or Hsp70 may also be involved in dislocation (BRODSKY et al. 1999). It has been hypothesized that the cytosolic tail of MHC class I heavy chains, which is required for dislocation, is used as a handle for extraction by cytosolic Hsp70 (STORY et al. 1999), similar to the action of yeast BiP in translocation (MATLACK et al. 1999). Although experimental data concerning the role of cytosolic Hsp70s in dislocations is not conclusive, it indicates that such chaperones could play a role in protein dislocation and degradation (BERCOVICH et al. 1997; BRODSKY et al. 1999; FISHER et al. 1997; GUSAROVA et al. 2001; MEACHAM et al. 2001; ZHANG et al. 2001).

5 Concluding Remarks

The degradation pathway used by HCMV to target MHC class I molecules for destruction represents a constitutive process that is generally used by the cell to dispose of misfolded proteins. Retrograde transport is anticipated to be a complex and tightly regulated process, involving a multitude of components that earmark potential substrates and target them to the cytosol. The HCMV-associated degradation of MHC class I molecules provides us with a unique model to decipher the molecular mechanisms underlying this intriguing process in more detail. The dislocation pathway also plays a role in diseases that involve degradation of cellular proteins, e.g., CFTR (JENSEN et al. 1995; WARD et al. 1995), and coagulation factor VIII (PIPE et al. 1998). A better understanding of the retrograde transport process might allow the development of therapies to treat such diseases.

Acknowledgements. We thank Gerco Hassink for comments on the manuscript.

References

Ahn K, Meyer TH, Uebel S, Sempe P, Djaballah H, Yang Y, Peterson PA, Fruh K, Tampe R (1996) Molecular mechanism and species specificity of TAP inhibition by herpes simplex virus ICP47. EMBO J 15:3247–3255
Barnes PD, Grundy JE (1992) Down-regulation of the class I HLA heterodimer and beta 2-microglobulin on the surface of cells infected with cytomegalovirus. J Gen Virol 73:2395–2403
Bebok Z, Mazzochi C, King SA, Hong JS, Sorscher EJ (1998) The mechanism underlying cystic fibrosis transmembrane conductance regulator transport from the endoplasmic reticulum to the proteasome includes Sec61beta and a cytosolic, deglycosylated intermediary. J Biol Chem 273:29873–29878
Beersma MF, Bijlmakers MJ, Ploegh HL (1993) Human cytomegalovirus down-regulates HLA class I expression by reducing the stability of class I H chains. J Immunol 151:4455–4464
Bercovich B, Stancovski I, Mayer A, Blumenfeld N, Laszlo A, Schwartz AL, Ciechanover A (1997) Ubiquitin-dependent degradation of certain protein substrates in vitro requires the molecular chaperone Hsc70. J Biol Chem 272:9002–9010
Biederer T, Volkwein C, Sommer T (1996) Degradation of subunits of the Sec61p complex, an integral component of the ER membrane, by the ubiquitin-proteasome pathway. EMBO J 15:2069–2076
Biederer T, Volkwein C, Sommer T (1997) Role of Cue1p in ubiquitination and degradation at the ER surface. Science 278:1806–1809
Bordallo J, Plemper RK, Finger A, Wolf DH (1998) De3p/Hrd1p is required for endoplasmic reticulum-associated degradation of misfolded lumenal and integral membrane proteins. Mol Biol Cell 9: 209–222

Bordallo J, Wolf DH (1999) A RING-H2 finger motif is essential for the function of Der3/Hrd1 in endoplasmic reticulum associated protein degradation in the yeast *Saccharomyces cerevisiae.* FEBS Lett 448:244–248

Brodsky JL, Werner ED, Dubas ME, Goeckeler JL, Kruse KB, McCracken AA (1999) The requirement for molecular chaperones during endoplasmic reticulum-associated protein degradation demonstrates that protein export and import are mechanistically distinct. J Biol Chem 274:3453–3460

Browne H, Smith G, Beck S, Minson T (1990) A complex between the MHC class I homologue encoded by human cytomegalovirus and beta 2 microglobulin. Nature 347:770–772

Coux O, Tanaka K, Goldberg AL (1996) Structure and functions of the 20S and 26S proteasomes. Annu Rev Biochem 65:801–847

Daimon M, Susa S, Suzuki K, Kato T, Yamatani K, Sasaki H (1997) Identification of a human cDNA homologue to the *Drosophila* translocation protein 1 (Dtrp1). Biochem Biophys Res Commun 230:100–104

deVirgilio M, Weninger H, Ivessa NE (1998) Ubiquitination is required for the retro-translocation of a short-lived luminal endoplasmic reticulum glycoprotein to the cytosol for degradation by the proteasome. J Biol Chem 273:9734–9743

Dusseljee S, Wubbolts R, Verwoerd D, Tulp A, Janssen H, Calafat J, Neefjes J (1998) Removal and degradation of the free MHC class II beta chain in the endoplasmic reticulum requires proteasomes and is accelerated by BFA. J Cell Sci 111:2217–2226

Duvet S, Labiau O, Mir AM, Kmiecik D, Krag SS, Verbert A, Cacan R (1998) Cytosolic deglycosylation process of newly synthesized glycoproteins generates oligomannosides possessing one GlcNAc residue at the reducing end. Biochem J 335:389–396

Ellgaard L, Molinari M, Helenius A (1999) Setting the standards: quality control in the secretory pathway. Science 286:1882–1888

Ferrari DM, Soling HD (1999) The protein disulphide-isomerase family: unravelling a string of folds. Biochem J 339:1–10

Fisher EA, Zhou M, Mitchell DM, Wu X, Omura S, Wang H, Goldberg AL, Ginsberg HN (1997) The degradation of apolipoprotein B100 is mediated by the ubiquitin-proteasome pathway and involves heat shock protein 70. J Biol Chem 272:20427–20434

Friedlander R, Jarosch E, Urban J, Volkwein C, Sommer T (2000) A regulatory link between ER-associated protein degradation and the unfolded-protein response. Nat Cell Biol 2:379–384

Fruh K, Ahn K, Djaballah H, Sempe P, van Endert PM, Tampe R, Peterson PA, Yang Y (1995) A viral inhibitor of peptide transporters for antigen presentation. Nature 375:415–418

Fujita K, Omura S, Silver J (1997) Rapid degradation of CD4 in cells expressing human immunodeficiency virus type 1 Env and Vpu is blocked by proteasome inhibitors. J Gen Virol 78:619–625

Gardner RG, Shearer AG, Hampton RY (2001) In vivo action of the HRD ubiquitin ligase complex: mechanisms of endoplasmic reticulum quality control and sterol regulation. Mol Cell Biol 21:4276–4291

Gardner RG, Swarbrick GM, Bays NW, Cronin SR, Wilhovsky S, Seelig L, Kim C, Hampton RY (2000) Endoplasmic reticulum degradation requires lumen to cytosol signaling. Transmembrane control of Hrd1p by Hrd3p. J Cell Biol 151:69–82

Gewurz BE, Wang EW, Tortorella D, Schust DJ, Ploegh HL (2001a) Human cytomegalovirus US2 endoplasmic reticulum-lumenal domain dictates association with major histocompatibility complex class I in a locus-specific manner. J Virol 75:5197–5204

Gewurz BE, Gaudet R, Tortorella D, Wang EW, Ploegh HL, Wiley DC (2001b) Antigen presentation subverted: Structure of the human cytomegalovirus protein US2 bound to the class I molecule HLA-A2. Proc Natl Acad Sci USA 98:6794–6799

Gilbert HF (1997) Protein disulfide isomerase and assisted protein folding. J Biol Chem 272:29399–29402

Gillece P, Luz JM, Lennarz WJ, de La Cruz FJ, Romisch K (1999) Export of a cysteine-free misfolded secretory protein from the endoplasmic reticulum for degradation requires interaction with protein disulfide isomerase. J Cell Biol 147:1443–1456

Gillece P, Pilon M, Romisch K (2000) The protein translocation channel mediates glycopeptide export across the endoplasmic reticulum membrane. Proc Natl Acad Sci USA 97:4609–4614

Ginsberg HN (1997) Role of lipid synthesis, chaperone proteins and proteasomes in the assembly and secretion of apoprotein B-containing lipoproteins from cultured liver cells. Clin Exp Pharmacol Physiol 24:A29–A32

Gusarova V, Caplan AJ, Brodsky JL, Fisher EA (2001) Apoprotein B degradation is promoted by the molecular chaperones hsp90 and hsp70. J Biol Chem 276:24891–24900

Hampton RY, Bhakta H (1997) Ubiquitin-mediated regulation of 3-hydroxy-3-methylglutaryl-CoA reductase. Proc Natl Acad Sci USA 94:12944–12948

Hampton RY, Gardner RG, Rine J (1996) Role of 26S proteasome and HRD genes in the degradation of 3-hydroxy-3-methylglutaryl-CoA reductase, an integral endoplasmic reticulum membrane protein. Mol Biol Cell 7:2029–2044

Heemels MT, Ploegh H (1995) Generation, translocation, and presentation of MHC class I-restricted peptides. Annu Rev Biochem 64:463–491

Hengel H, Brune W, Koszinowski UH (1998) Immune evasion by cytomegalovirus – survival strategies of a highly adapted opportunist. Trends Microbiol 6:190–197

Hershko A, Ciechanover A (1998) The ubiquitin system. Annu Rev Biochem 67:425–479

Hewitt EW, Gupta SS, Lehner PJ (2001) The human cytomegalovirus gene product US6 inhibits ATP binding by TAP. EMBO J 20:387–396

Hill A, Jugovic P, York I, Russ G, Bennink J, Yewdell J, Ploegh H, Johnson D (1995) Herpes simplex virus turns off the TAP to evade host immunity. Nature 375:411–415

Hiller MM, Finger A, Schweiger M, Wolf DH (1996) ER degradation of a misfolded luminal protein by the cytosolic ubiquitin-proteasome pathway. Science 273:1725–1728

Hofmann K, Stoffel W (1993) TMbase – a database of membrane spanning protein segments. Biol Chem Hoppe-Seyler 374:166

Hughes EA, Hammond C, Cresswell P (1997) Misfolded major histocompatibility complex class I heavy chains are translocated into the cytoplasm and degraded by the proteasome. Proc Natl Acad Sci USA 94:1896–1901

Huppa JB, Ploegh HL (1997) The alpha chain of the T cell antigen receptor is degraded in the cytosol. Immunity 7:113–122

Imamura T, Haruta T, Takata Y, Usui I, Iwata M, Ishihara H, Ishiki M, Ishibashi O, Ueno E, Sasaoka T, Kobayashi M (1998) Involvement of heat shock protein 90 in the degradation of mutant insulin receptors by the proteasome. J Biol Chem 273:11183–11188

Jensen TJ, Loo MA, Pind S, Williams DB, Goldberg AL, Riordan JR (1995) Multiple proteolytic systems, including the proteasome, contribute to CFTR processing. Cell 83:129–135

Johnson AE, van Waes MA (1999) The translocon: a dynamic gateway at the ER membrane. Annu Rev Cell Dev Biol 15:799–842

Jones TR, Hanson LK, Sun L, Slater JS, Stenberg RM, Campbell AE (1995) Multiple independent loci within the human cytomegalovirus unique short region down-regulate expression of major histocompatibility complex class I heavy chains. J Virol 69:4830–4841

Jones TR, Sun L (1997) Human cytomegalovirus US2 destabilizes major histocompatibility complex class I heavy chains. J Virol 71:2970–2979

Kikkert M, Hassink G, Barel M, Hirsch C, van der Wal FJ, Wiertz E (2001) Ubiquitination is essential for human cytomegalovirus US11-mediated dislocation of MHC class I molecules from the endoplasmic reticulum to the cytosol. Biochem J 358:369–377

Kisselev AF, Akopian TN, Woo KM, Goldberg AL (1999) The sizes of peptides generated from protein by mammalian 26 and 20S proteasomes. Implications for understanding the degradative mechanism and antigen presentation. J Biol Chem 274:3363–3371

Klausner RD, Sitia R (1990) Protein degradation in the endoplasmic reticulum. Cell 62:611–614

Koopmann JO, Albring J, Huter E, Bulbuc N, Spee P, Neefjes J, Hammerling GJ, Momburg F (2000) Export of antigenic peptides from the endoplasmic reticulum intersects with retrograde protein translocation through the Sec61p channel. Immunity 13:117–127

Kopito RR (1997) ER quality control: the cytoplasmic connection. Cell 88:427–430

Kopito RR, Sitia R (2000) Aggresomes and Russell bodies – symptoms of cellular indigestion? EMBO Rep 1:225–231

Lee S, Yoon J, Park B, Jun Y, Jin M, Sung HC, Kim IH, Kang S, Choi EJ, Ahn BY, Ahn K (2000) Structural and functional dissection of human cytomegalovirus US3 in binding major histocompatibility complex class I molecules. J Virol 74:11262–11269

Lehner PJ, Karttunen JT, Wilkinson GW, Cresswell P (1997) The human cytomegalovirus US6 glycoprotein inhibits transporter associated with antigen processing-dependent peptide translocation. Proc Natl Acad Sci USA 94:6904–6909

Levitskaya J, Coram M, Levitsky V, Imreh S, Steigerwald-Mullen PM, Klein G, Kurilla MG, Masucci MG (1995) Inhibition of antigen processing by the internal repeat region of the Epstein-Barr virus nuclear antigen-1. Nature 375:685–688

Levitskaya J, Sharipo A, Leonchiks A, Ciechanover A, Masucci MG (1997) Inhibition of ubiquitin/ proteasome-dependent protein degradation by the Gly-Ala repeat domain of the Epstein-Barr virus nuclear antigen 1. Proc Natl Acad Sci USA 94:12616–12621

Liu Y, Choudhury P, Cabral CM, Sifers RN (1997) Intracellular disposal of incompletely folded human alpha1-antitrypsin involves release from calnexin and post-translational trimming of asparagine-linked oligosaccharides. J Biol Chem 272:7946–7951

Liu Y, Choudhury P, Cabral CM, Sifers RN (1999) Oligosaccharide modification in the early secretory pathway directs the selection of a misfolded glycoprotein for degradation by the proteasome. J Biol Chem 274:5861–5867

Machold RP, Wiertz EJ, Jones TR, Ploegh HL (1997) The HCMV gene products US11 and US2 differ in their ability to attack allelic forms of murine major histocompatibility complex (MHC) class I heavy chains. J Exp Med 185:363–366

Matlack KE, Misselwitz B, Plath K, Rapoport TA (1999) BiP acts as a molecular ratchet during posttranslational transport of prepro-alpha factor across the ER membrane. Cell 97:553–564

Mayer TU, Braun T, Jentsch S (1998) Role of the proteasome in membrane extraction of a short-lived ER-transmembrane protein. EMBO J 17:3251–3257

McCracken AA, Brodsky JL (1996) Assembly of ER-associated protein degradation in vitro: dependence on cytosol, calnexin, and ATP. J Cell Biol 132:291–298

Meacham GC, Patterson C, Zhang W, Younger JM, Cyr DM (2001) The Hsc70 co-chaperone CHIP targets immature CFTR for proteasomal degradation. Nat Cell Biol 3:100–105

Meyer HA, Grau H, Kraft R, Kostka S, Prehn S, Kalies KU, Hartmann E (2000) Mammalian Sec61 is associated with Sec62 and Sec63. J Biol Chem 275:14550–14557

Noiva R (1999) Protein disulfide isomerase: the multifunctional redox chaperone of the endoplasmic reticulum. Semin Cell Dev Biol 10:481–493

Norgaard P, Westphal V, Tachibana C, Alsoe L, Holst B, Winther JR (2001) Functional differences in yeast protein disulfide isomerases. J Cell Biol 152:553–562

Oliver JD, Roderick HL, Llewellyn DH, High S (1999) ERp57 functions as a subunit of specific complexes formed with the ER lectins calreticulin and calnexin. Mol Biol Cell 10:2573–2582

Oliver JD, van der Wal FJ, Bulleid NJ, High S (1997) Interaction of the thiol-dependent reductase ERp57 with nascent glycoproteins. Science 275:86–88

Park H, Suzuki T, Lennarz WJ (2001) Identification of proteins that interact with mammalian peptide: N- glycanase and implicate this hydrolase in the proteasome-dependent pathway for protein degradation. Proc Natl Acad Sci USA

Pilon M, Schekman R, Romisch K (1997) Sec61p mediates export of a misfolded secretory protein from the endoplasmic reticulum to the cytosol for degradation. EMBO J 16:4540–4548

Pipe SW, Morris JA, Shah J, Kaufman RJ (1998) Differential interaction of coagulation factor VIII and factor V with protein chaperones calnexin and calreticulin. J Biol Chem 273:8537–8544

Plemper RK, Bohmler S, Bordallo J, Sommer T, Wolf DH (1997) Mutant analysis links the translocon and BiP to retrograde protein transport for ER degradation. Nature 388:891–895

Plemper RK, Deak PM, Otto RT, Wolf DH (1999) Re-entering the translocon from the lumenal side of the endoplasmic reticulum. Studies on mutated carboxypeptidase yscY species. FEBS Lett 443: 241–245

Ploegh HL (1998) Viral strategies of immune evasion. Science 280:248–253

Qu D, Teckman JH, Omura S, Perlmutter DH (1996) Degradation of a mutant secretory protein, alpha1-antitrypsin Z, in the endoplasmic reticulum requires proteasome activity. J Biol Chem 271: 22791–22795

Reits EA, Vos JC, Gromme M, Neefjes J (2000) The major substrates for TAP in vivo are derived from newly synthesized proteins. Nature 404:774–778

Reusch U, Muranyi W, Lucin P, Burgert HG, Hengel H, Koszinowski UH (1999) A cytomegalovirus glycoprotein re-routes MHC class I complexes to lysosomes for degradation. EMBO J 18:1081–1091

Rock KL, Goldberg AL (1999) Degradation of cell proteins and the generation of MHC class I-presented peptides. Annu Rev Immunol 17:739–779

Romisch K, Ali BR (1997) Similar processes mediate glycopeptide export from the endoplasmic reticulum in mammalian cells and *Saccharomyces cerevisiae*. Proc Natl Acad Sci USA 94:6730–6734

Romisch K, Schekman R (1992) Distinct processes mediate glycoprotein and glycopeptide export from the endoplasmic reticulum in *Saccharomyces cerevisiae*. Proc Natl Acad Sci USA 89:7227–7231

Schubert U, Anton LC, Bacik I, Cox JH, Bour S, Bennink JR, Orlowski M, Strebel K, Yewdell JW (1998) CD4 glycoprotein degradation induced by human immunodeficiency virus type 1 Vpu protein requires the function of proteasomes and the ubiquitin-conjugating pathway. J Virol 72:2280–2288

Schubert U, Anton LC, Gibbs J, Norbury CC, Yewdell JW, Bennink JR (2000) Rapid degradation of a large fraction of newly synthesized proteins by proteasomes. Nature 404:770–774

Schust DJ, Tortorella D, Seebach J, Phan C, Ploegh HL (1998) Trophoblast class I major histocompatibility complex (MHC) products are resistant to rapid degradation imposed by the human cytomegalovirus (HCMV) gene products US2 and US11. J Exp Med 188:497–503

Shamu CE, Flierman D, Ploegh HL, Rapoport TA, Chau V (2001) Polyubiquitination is required for US11-dependent movement of MHC class I heavy chain from endoplasmic reticulum into cytosol. Mol Biol Cell 12:2546–2555

Shamu CE, Story CM, Rapoport TA, Ploegh HL (1999) The pathway of US11-dependent degradation of MHC class I heavy chains involves a ubiquitin-conjugated intermediate. J Cell Biol 147:45–58

Skowronek MH, Rotter M, Haas IG (1999) Molecular characterization of a novel mammalian DnaJ-like Sec63p homolog. Biol Chem 380:1133–1138

Story CM, Furman MH, Ploegh HL (1999) The cytosolic tail of class I MHC heavy chain is required for its dislocation by the human cytomegalovirus US2 and US11 gene products. Proc Natl Acad Sci USA 96:8516–8521

Suzuki T, Park H, Hollingsworth NM, Sternglanz R, Lennarz WJ (2000) PNG1, a yeast gene encoding a highly conserved peptide:N-glycanase. J Cell Biol 149:1039–1052

Suzuki T, Park H, Kwofie MA, Lennarz WJ (2001) Rad23 provides a link between the Png1 deglycosylating enzyme and the 26S proteasome in yeast. J Biol Chem 276:21601–21607

Suzuki T, Seko A, Kitajima K, Inoue Y, Inoue S (1993) Identification of peptide:N-glycanase activity in mammalian-derived cultured cells. Biochem Biophys Res Commun 194:1124–1130

Suzuki T, Seko A, Kitajima K, Inoue Y, Inoue S (1994) Purification and enzymatic properties of peptide:N-glycanase from C3H mouse-derived L-929 fibroblast cells. Possible widespread occurrence of post-translational remodification of proteins by N-deglycosylation. J Biol Chem 269:17611–17618

Takahashi N (1992) Glycosamidases. In: Takahashi N, Muramatsu T (eds) CRC Handbook of endoglycosidases and glycoamidases. CRC Press Inc., Boca Raton, FL, pp 183–198

Tiwari S, Weissman AM (2001) Endoplasmic reticulum (ER)-associated degradation of T cell receptor subunits. Involvement of ER-associated ubiquitin-conjugating enzymes (E2s). J Biol Chem 276:16193–16200

Tomazin R, Hill AB, Jugovic P, York I, van Endert P, Ploegh HL, Andrews DW, Johnson DC (1996) Stable binding of the herpes simplex virus ICP47 protein to the peptide binding site of TAP. EMBO J 15:3256–3266

Tortorella D, Gewurz BE, Furman MH, Schust DJ, Ploegh HL (2000) Viral subversion of the immune system. Annu Rev Immunol 18:861–926

Tortorella D, Story CM, Huppa JB, Wiertz EJ, Jones TR, Bacik I, Bennink JR, Yewdell JW, Ploegh HL (1998) Dislocation of type I membrane proteins from the ER to the cytosol is sensitive to changes in redox potential. J Cell Biol 142:365–376

Tyedmers J, Lerner M, Bies C, Dudek J, Skowronek MH, Haas IG, Heim N, Nastainczyk W, Volkmer J, Zimmermann R (2000) Homologs of the yeast Sec complex subunits Sec62p and Sec63p are abundant proteins in dog pancreas microsomes. Proc Natl Acad Sci USA 97:7214–7219

Voges D, Zwickl P, Baumeister W (1999) The 26S proteasome: a molecular machine designed for controlled proteolysis. Annu Rev Biochem 68:1015–1068

Ward CL, Omura S, Kopito RR (1995) Degradation of CFTR by the ubiquitin-proteasome pathway. Cell 83:121–127

Wiertz E, Hill A, Tortorella D, Ploegh H (1997a) Cytomegaloviruses use multiple mechanisms to elude the host immune response. Immunol Lett 57:213–216

Wiertz EJ, Jones TR, Sun L, Bogyo M, Geuze HJ, Ploegh HL (1996a) The human cytomegalovirus US11 gene product dislocates MHC class I heavy chains from the endoplasmic reticulum to the cytosol. Cell 84:769–779

Wiertz EJ, Mukherjee S, Ploegh HL (1997b) Viruses use stealth technology to escape from the host immune system. Mol Med Today 3:116–123

Wiertz EJHJ, Tortorella D, Bogyo M, Yu J, Mothes W, Jones TR, Rapoport TA, Ploegh HL (1996b) Sec61-mediated transfer of a membrane protein from the endoplasmic reticulum to the proteasome for destruction. Nature 384:432–438

Wilhovsky S, Gardner R, Hampton R (2000) HRD gene dependence of endoplasmic reticulum-associated degradation. Mol Biol Cell 11:1697–1708

Wilkinson BM, Tyson JR, Reid PJ, Stirling CJ (2000) Distinct domains within yeast Sec61p involved in post-translational translocation and protein dislocation. J Biol Chem 275:521–529

Winitz D, Shachar I, Elkabetz Y, Amitay R, Samuelov M, Bar-Nun S (1996) Degradation of distinct assembly forms of immunoglobulin M occurs in multiple sites in permeabilized B cells. J Biol Chem 271:27645–27651

Yamashita Y, Shimokata K, Mizuno S, Yamaguchi H, Nishiyama Y (1993) Down-regulation of the surface expression of class I MHC antigens by human cytomegalovirus. Virology 193:727–736

Yamashita Y, Shimokata K, Saga S, Mizuno S, Tsurumi T, Nishiyama Y (1994) Rapid degradation of the heavy chain of class I major histocompatibility complex antigens in the endoplasmic reticulum of human cytomegalovirus-infected cells. J Virol 68:7933–7943

Yewdell JW (2001) Not such a dismal science: the economics of protein synthesis, folding, degradation and antigen processing. Trends Cell Biol 11:294–297

Yu H, Kaung G, Kobayashi S, Kopito RR (1997) Cytosolic degradation of T-cell receptor alpha chains by the proteasome. J Biol Chem 272:20800–20804

Yu H, Kopito RR (1999) The role of multiubiquitination in dislocation and degradation of the alpha subunit of the T cell antigen receptor. J Biol Chem 274:36852–36858

Zhang Y, Nijbroek G, Sullivan ML, McCracken AA, Watkins SC, Michaelis S, Brodsky JL (2001) Hsp70 molecular chaperone facilitates endoplasmic reticulum-associated protein degradation of cystic fibrosis transmembrane conductance regulator in yeast. Mol Biol Cell 12:1303–1314

Zhou MY, Schekman R (1999) The engagement of Sec61p in the ER dislocation process. Mol Cell 4:925–934

Ziegler H, Muranyi W, Burgert HG, Kremmer E, Koszinowski UH (2000) The luminal part of the murine cytomegalovirus glycoprotein gp40 catalyzes the retention of MHC class I molecules. EMBO J 19:870–881

Corking the Bottleneck: The Transporter Associated with Antigen Processing as a Target for Immune Subversion by Viruses

F. Momburg[1] and H. Hengel[2]

In this chapter, mechanisms are reviewed that viruses use to inhibit the function of the peptide transporter associated with antigen processing (TAP), which translocates cytosolic peptides into the endoplasmic reticulum (ER) for binding to MHC class I molecules. Although some DNA viruses, such as adenovirus or EBV, downmodulate TAP expression on the transcriptional level, members of the alpha and beta subfamily of herpesviruses, such as herpes simplex virus (HSV) and human cytomegalovirus (HCMV), express proteins that bind to TAP and interfere with peptide translocation. The modes of action of the HSV-encoded cytosolic TAP inhibitor ICP47 and the HCMV-encoded ER-resident TAP inhibitor gpUS6 are discussed in detail. Viral interference with antigen presentation through TAP inhibition is not only relevant for the immunobiology of persistent viral infections but also contributes to the understanding of the translocation mechanism utilized by the ATP-binding cassette transporter TAP.

1	Introduction	57
2	TAP Peptide Transporters: Biological Role, Structure and Molecular Mechanism	58
3	Control of TAP Gene Expression in Virus-Induced Tumors	60
3.1	Inhibition of TAP Synthesis by Virus-Encoded IL-10	62
4	Viruses Affecting Peptide Translocation	62
4.1	TAP Inhibition by HSV ICP47	63
4.2	TAP Function in CMV-Infected Cells	63
4.2.1	Molecular Function of HCMV gpUS6	64
4.2.2	TAP-Independent Peptide Transport of an HCMV-Encoded HLA-E Ligand, gpUL40	68
4.2.3	Altered Phosphorylation of TAP in HCMV-Infected Cells	68
4.3	Adenovirus E3/19K Protein	68
4.4	Augmented Peptide Transport in Flavivirus-Infected Cells	69
5	Perspectives	69
References		70

1 Introduction

CD8[+] T lymphocytes represent deadly weapons of the adapted immune system destined to track down virus-infected cells and to direct a lethal hit before the virus has multiplied and spread. To guide CD8[+] T cell recognition, MHC class I

---

[1] Deutsches Krebsforschungszentrum, Department of Molecular Immunology, 69120 Heidelberg, Germany
[2] Robert Koch-Institut, Division of Viral Infections, 13353 Berlin, Germany

(MHC-I) molecules sample the cytosolic compartment and present antigenic (e.g., viral) peptides to antigen-specific T cell receptors on $CD8^+$ T lymphocytes. The membrane-anchored MHC-I heavy and soluble light (β_2-microglobulin) chains are biosynthesized into the endoplasmic reticulum (ER). Assembly of stable ternary MHC-I complexes in the ER requires the binding of short antigenic peptides that are supplied by the transporter associated with antigen processing, TAP. Without its peptide cargo, class I molecules are unstable and dissociate. Peptide-loaded MHC-I complexes are allowed to exit from the ER and migrate to the cell surface. Modulation of MHC-I surface expression is an efficient strategy to escape the adaptive immune response by $CD8^+$ T lymphocytes. In particular, those viruses that establish persistent infections and replicate in the face of a primed T cell response, e.g., herpesviruses, adenoviruses or HIV, have found elegant means to manipulate the expression and function of MHC-I molecules. The identification of the responsible viral inhibitors gave information about their molecular function and revealed that their physical target is not necessarily the MHC-I complex itself, but components of the cellular machinery for antigen processing and presentation.

As delineated below, TAP plays a dual and critical role within the MHC-I pathway. First, TAP represents the principal route of entry into the MHC-I-restricted pathway for peptide ligands which are generated in the cytosol. Second, TAP constitutes the center of a transient "loading complex" containing peptide-free MHC-I molecules and ER-resident chaperones. These features make TAP a prime target for viral interference.

2 TAP Peptide Transporters: Biological Role, Structure and Molecular Mechanism

The vast majority of antigenic peptides destined to associate with MHC-I molecules must be translocated into the ER lumen by the TAP peptide transporter. This can be concluded from severely reduced levels of MHC-I surface expression in TAP-deficient cells and from the general failure of such mutant cells to present peptides derived from viral and other antigens expressed in the cytosol (reviewed in ELLIOTT 1997; MOMBURG and HÄMMERLING 1998). Peptides imported by TAP mostly result from the degradation of proteins by the proteasome, an abundant protease complex exerting multiple proteolytic specificities (reviewed in PAMER and CRESSWELL 1998; ROCK and GOLDBERG 1999). Cytosolic (amino)peptidases can, however, contribute to the generation of TAP substrates (reviewed in ROCK and GOLDBERG 1999; FRÜH and YANG 1999). Moreover, some MHC-I-binding peptides can be generated in the ER lumen from cleaved signal sequences or even through the proteolytic processing of membrane-bound or soluble ER proteins (reviewed in MOMBURG and HÄMMERLING 1998; ROCK and GOLDBERG 1999). Finally, loading of recycling MHC-I molecules in endosomal compartments has been reported (reviewed in WATTS 1997; YEWDELL et al. 1999), but such alternative pathways clearly appear to be of minor biological importance.

TAP belongs to the large family of ATP-binding cassette (ABC) transporters. The noncovalently associated half-transporters TAP1 and TAP2 each contain an N-terminal domain that spans the ER membrane several times and one nucleotide-binding domain (NBD) that is oriented toward the cytoplasm (reviewed in ELLIOTT 1997; MOMBURG and HÄMMERLING 1998). Analysis of truncation mutants revealed eight transmembrane segments (TMS) in the TAP1 chain and seven TMS in the TAP2 chain (Vos et al. 1999a). Experimental evidence suggests that a pore-like structure may be formed by TMS1–6 of TAP1 and TMS1–5 of TAP2 in a head-head/tail-tail orientation (Vos et al. 1999b). From this topology it can be concluded that only minor parts (<10%) of the linear sequences of TAP1 and TAP2 are exposed to the ER lumen whereas long loops extend into the cytoplasm. Biochemical and genetic analyses have implicated loops in the central portions of TAP1/TAP2 heterodimers in mediating peptide binding and control of substrate specificity (reviewed in MOMBURG and HÄMMERLING 1998). TAP preferentially translocates peptides of 8- to 12-amino acid length, but peptides of up to 40 residues can also serve as substrates (reviewed in ELLIOTT 1997; MOMBURG and HÄMMERLING 1998). Concerning the peptide sequence, it has been worked out that the N-terminal three residues as well as the C-terminal residue is of special importance. Allelic rat TAPs, or human and mouse TAP, show striking differences regarding the preferred C-terminal amino acid (exclusively hydrophobic vs. nonselective) of translocated peptides (reviewed in ELLIOTT 1997; MOMBURG and HÄMMERLING 1998). With regard to lengths and sequences of translocated peptides, the specificities of different TAP variants appear to be in good concordance with the peptide-binding preferences of MHC-I molecules that are expressed *in trans* in the respective species or in rat inbred strains. On the other hand, the relatively promiscuous substrate specificity of TAP helps to supply various co-expressed MHC-I molecules with sufficient amounts of potential ligands.

Constraints in the sequence specificity of TAP predict that some MHC-I-binding peptides are translocated as slightly extended precursor products that subsequently undergo final trimming in the ER lumen, and this has indeed been demonstrated for several virus-derived peptides (reviewed in MOMBURG and HÄMMERLING 1998). Trimming is restricted to the amino terminus of the precursor and may be directed by the MHC-I molecule binding the precursor (PAZ et al. 1999).

Peptide translocation into the ER strictly requires the presence of hydrolyzable trinucleotides (reviewed in ELLIOTT 1997; MOMBURG and HÄMMERLING 1998) and the functionality of both NBDs (KNITTLER et al. 1999). ATP-dependent peptide binding induces conformational changes in the TAP1/TAP2 heterodimer (NEUMANN and TAMPÉ 1999; REITS et al. 2000) and seems to trigger ATP hydrolysis (ABELE and TAMPÉ 1999). The exact order of events that lead to the completion of the translocation cycle and restoration of the initial peptide-receptive conformation awaits detailed investigation. Furthermore, cooperative and alternating roles of NBDs in TAP1 and TAP2, which can putatively be inferred from mechanistically better understood ABC transporters, need to be elucidated.

TAP is an integrated part of the pathway that leads to the generation of properly peptide-loaded MHC-I molecules. In conjunction with the ER chaperone calreticulin and the oxidoreductase ERp57, MHC-I heavy chain/β_2-microglobulin dimers transiently bind, facilitated by the adaptor molecule tapasin, to TAP1/2 heterodimers (reviewed in PAMER and CRESSWELL 1998; VAN ENDERT 1999a). Release of MHC-I molecules from this loading complex requires the TAP-mediated translocation of appropriate peptides that are believed to induce a conformational change in MHC-I (reviewed in PAMER and CRESSWELL 1998; VAN ENDERT 1999a). The recruitment of MHC-I into the TAP-linked complex by tapasin is essential for the loading of an optimized spectrum of peptides that is a prerequisite for the formation of long-lived MHC-I heterotrimers (BARNDEN et al. 2000; GARBI et al. 2000). Independent of this function, tapasin has a chaperoning function for TAP because it increases the steady-state levels of TAP1/2 heterodimers (BANGIA et al. 1999).

3 Control of TAP Gene Expression in Virus-Induced Tumors

A major mechanism to achieve a reduction of TAP-mediated peptide translocation of peptides employs the suppression of transcription of *TAP* genes, either *TAP1* or *TAP2* or both. Deficient *TAP* gene expression is frequently found in certain virus-transformed cell lines and tumor tissue expressing viral genes. Highly oncogenic rodent cells transformed by a group A adenovirus, Ad12, exhibit a very low level of MHC-I transcription and surface expression due to the expression of a viral early protein, E1A (SCHRIER et al. 1983; EAGER et al. 1985), that decreases the binding of NF-κB to the enhancer element of MHC-I genes (LIU et al. 1996; KUSHNER et al. 1996). Analysis of steady-state mRNA of the *TAP* transporter genes revealed a 100-fold reduction of TAP2 and 5- to 10-fold reduction of TAP1 mRNA (ROTEM-YEHUDAR et al. 1994). In contrast to the full reconstitution by interferon-γ, experiments using stable transfection of TAP genes or recombinant vaccinia viruses restored murine MHC-I surface expression only partially, suggesting that further factors involved in the MHC-I pathway are rate-limiting in Ad12-transformed cells (ROTEM-YEHUDAR et al. 1996).

Papillomaviruses form a large family of oncogenic DNA viruses. Infections of the genital tract with specific types of human papillomaviruses are etiologically related with squamous cell carcinomas, most notably of the uterine cervix (reviewed in ZUR HAUSEN 2000). In a majority of cervical carcinomas HLA-A and -B locus products are downregulated, and in a substantial proportion of tumors the TAP expression is found very low (CROMME et. 1994a; KEATING et al. 1995). This downregulation was even more pronounced in cervical cancer lymph node metastases (CROMME et al. 1994b). Recurrent respiratory papillomatosis is caused by human papillomavirus (HPV) infection and characterized by a variable clinical course from disease recurrence to severe morbidity and occasional mortality. TAP1 and MHC-I synthesis is found concomitantly downregulated in laryngeal papilloma tissue biopsies and cell culture of primary explants (VAMBUTAS et al. 2000). Notably, reduction of TAP1 expression in biopsy tissue correlates with rapid recurrence of disease, and expression of TAP1 could be used as a prognostic marker of the further course of disease (VAMBUTAS et al. 2000).

Human herpesvirus 8 (HHV8) is a recently identified human γ_2-herpesvirus associated with various tumors such as Kaposi's sarcoma, primary effusion lymphoma, and Castleman's disease. B cell lines latently infected with HHV8 are impaired in presentation of endogenous peptides to $CD8^+$ cytotoxic T lymphocytes (CTL) (BRANDER et al. 2000). Incubation of HHV8-infected cell lines with high concentrations of exogenously added peptides did not restore the deficient MHC-I surface expression, suggesting deficient transport of peptides into the ER. This notion was supported by the finding that TAP1 expression was significantly diminished in HHV8-infected cell lines whereas TAP2 expression was not affected (BRANDER et al. 2000).

Another human γ-herpesvirus, Epstein-Barr virus (EBV), is associated with various malignancies arising from latently infected cells, including Burkitt's lymphoma (BL), nasopharyngeal carcinoma (NPC), and certain subtypes of Hodgkin's Disease (HD) (reviewed in RICKINSON and KIEFF 1996). Remarkably, the efficiency of antigen presentation within the MHC-I pathway differs considerably between these tumor types (reviewed in Moss et al. 1999). BL cells display a drastically downregulated expression of MHC-I, TAP1, and TAP2 compared with EBV-transformed lymphoblastoid cell lines because of a greatly diminished transcription of *TAP1* and *TAP2* genes (KHANNA et al. 1994). In constitutively expressing gene transfectants, the EBV-encoded latent membrane protein-1 (LMP1) has been demonstrated to induce upregulation of both TAP and MHC-I synthesis comparable to the effects mediated by treatment of BL cells with interferonγ (ROWE et al. 1995). Unlike BL, Hodgkin/Reed-Sternberg (HRS) cells of EBV-associated HD consistently express LMP1 and LMP2 as well as high levels of TAP1 and TAP2 (MURRAY et al. 1998), consistent with the finding that HRS cell lines are readily recognized by EBV-specific $CD8^+$ CTL (LEE et al. 1998a). Likewise, analysis of NPC cells has revealed normal expression of TAP1 and TAP2, together with high levels of MHC-I on the cell surface (KHANNA et al. 1998; Moss et al. 1999). Together, reduced or absent gene expression of *TAP1*, *TAP2*, or both, has been documented in a large number of virus-associated malignancies. So far, little is known about the responsible viral genes and the molecular mechanisms used by their products to shut down *TAP* gene expression. TAP-deficient phenotypes are also prevalent in many nonvirally induced tumors (reviewed in SELIGER et al. 1997 and in SELIGER et al. 2000), suggesting that several states of malignant cell transformation are accompanied by the inhibition of *TAP* gene transcription.

3.1 Inhibition of TAP Synthesis by Virus-Encoded IL-10

IL-10 is a pleiotropic immunomodulatory cytokine produced by T lymphocytes, monocytes, macrophages, B cells, and keratinocytes. IL-10 is a potent immunosuppressor because it blocks the synthesis of proinflammatory cytokines and MHC class II molecules and impairs the costimulatory function of professional antigen-presenting cells such as macrophages (reviewed in MOORE et al. 1993). Remarkably, a number of large DNA viruses harbor viral homologs of IL-10, e.g., EBV (MOORE et al. 1990; HSU et al. 1990), equine herpesvirus type 2 (RODE et al. 1993), human (KOTENKO et al. 2000) and primate (LOCKRIDGE et al. 2000) CMV, and orf poxvirus (FLEMING et al. 1997). EBV-encoded IL-10 shares many but not all biological activities of human IL-10. Both factors have been shown to downregulate steady-state levels of TAP1- and LMP2-specific mRNAs but not TAP2 and LMP7 (ZEIDLER et al. 1997). Interestingly, the TAP1 and LMP2 genes share a bidirectional promotor element (WRIGHT et al. 1995). The EBV-IL-10 mediated effect moderately affects the peptide translocation function of exposed primary B cells and their surface expression of MHC-I molecules (ZEIDLER et al. 1997). On the other hand, immortalized EBV-infected B lymphocyte cell lines express LMP1, a viral factor that is suspected to upregulate TAP synthesis via the NF-κB/Rel pathway (ROWE et al. 1995) and thus compensates for the EBV-IL-10 mediated inhibitory effect.

4 Viruses Affecting Peptide Translocation

Distinct members of the *Alphaherpesvirinae* and *Betaherpesvirinae* subfamilies express specific polypeptides designed to interfere with the peptide translocation function of TAP, whereas other related herpesviruses of the same subfamily apparently do not (HENGEL et al. 1996). Analysis of viral deletion mutants allowed mapping of the gene regions of the viral genomes required for TAP inhibition in human herpes simplex virus (HSV) type 1-infected cells (YORK et al. 1994) and human cytomegalovirus (HCMV)-infected cells (HENGEL et al. 1996). Exhibiting vastly distinct mechanisms of TAP inactivation, the viral proteins of HSV and HCMV proved to be valuable molecular tools for unraveling distinct molecular requirements of TAP function. Notably, both TAP inhibitors ICP47 of HSV as well as gpUS6 of HCMV inhibit TAP in a species-specific manner (see below). This common theme reflects their precise molecular design as natural inhibitors of TAP and represents a remarkable example for the close adaptation of herpesviruses to their natural hosts. The suspected viral factors that inhibit TAP in cells infected with bovine herpesvirus 1 (HINKLEY et al. 1998; KOPPERS-LALIC et al. 2001) or pseudorabies virus (AMBAGALA et al. 2000) await their identification.

4.1 TAP Inhibition by HSV ICP47

HSV types 1 and 2 express an immediate-early-phase gene product that leads to retention of class I molecules in the ER and renders infected fibroblasts resistant to CTL lysis (York et al. 1994; Hill et al. 1994). The responsible gene product of HSV type 1 was identified as the 9-kDa cytosolic protein ICP47 encoded by the *US12* immediate-early gene (York et al. 1994). ICP47 inhibits peptide transport into the ER by stably interacting with the cytosolic face of TAP (Früh et al. 1995; Hill et al. 1995). ICP47, which itself is not translocated into the ER, binds to TAP with higher affinity (~50 nM) than usual peptide substrates (Ahn et al. 1996a; Tomazin et al. 1996). High concentrations of peptide can compete with the TAP association of ICP47, indicating that it binds to a site that overlaps with the peptide binding site of the transporter (Tomazin et al. 1996). Experiments with truncated ICP47 peptides have shown that the active inhibitory domain is contained in an N-terminal fragment comprising residues 2–35 (Galocha et al. 1997, Neumann et al. 1997). Photocross-linking studies suggest an asymmetrical binding to TAP1 and TAP2 (Galocha et al. 1997). ICP47 does not inhibit labeling of TAP1 or TAP2 with 8-azido-ATP, indicating that ICP47 does not obstruct the NBDs of the half-transporters (Ahn et al. 1996a; Tomazin et al. 1996). In addition to its role as potent blocker of peptide binding to TAP, incubation with ICP47 at 37°C was found to completely abrogate chemical cross-linking of TAP1/2 heterodimer (Lacaille and Androlewicz 1998), whereas regular peptide substrates of high affinity (Lacaille and Androlewicz 1998; van Endert 1999b) as well as various trinucleotides or ADP (van Endert 1999b) rescue a temperature-induced destabilization of the TAP heterodimer. Thus binding of ICP47 seems to disturb a stable conformation of the TAP1/2 complex. Significant sequence homology with the TAP inhibitor of HSV1, ICP47-1, guided the identification of the responsible gene product of HSV type 2, ICP47-2, which exhibits a similar potency of TAP inhibition (Galocha et al. 1997; Tomazin et al. 1998). TAP inhibition by ICP47-1 or ICP47-2 is species-specific because TAP molecules in human, pig, cow, and monkey cells are inhibited, whereas mouse, rat, guinea pig, and rabbit TAP molecules are insensitive to ICP47 (Ahn et al. 1996a; Tomazin et al. 1996, 1998; Jugovic et al. 1998). The affinity to mouse TAP is ~100-fold reduced compared with human TAP (Ahn et al. 1996a). The ability of ICP47 to suppress surface expression of HLA-G molecules that serve as inhibitors of NK cells in placental trophoblast may explain the epidemiological association between HSV infection and spontaneous abortion (Schust et al. 1996; Robb et al. 1986). Furthermore, it was demonstrated that HSV-infected cells become susceptible to lysis by NK cells because NK-inhibitory HLA-C molecules are downregulated by ICP 47 (Huard and Früh 2000).

4.2 TAP Function in CMV-Infected Cells

Presentation of endogenous peptides to $CD8^+$ CTL is completely abrogated in HCMV-infected cells (Hengel et al. 1995), the MHC-I assembly and surface expression of which is drastically reduced (BEERSMA et al. 1993, YAMASHITA et al. 1993). Reminiscent of the situation in TAP-deficient cell mutants (SALTER and CRESSWELL 1986; LJUNGGREN et al. 1990), MHC-I complexes are unstable in cells infected with HCMV for 72 h whereas the abundance of free MHC-I heavy chains is significantly less affected (HENGEL et al. 1996). In parallel, TAP-dependent peptide translocation into the ER of HCMV-infected fibroblasts declines continuously over the HCMV replication cycle although synthesis and formation of TAP1/2 heterodimers is significantly increased compared with mock-infected cells (HENGEL et al. 1996). Analysis of HCMV deletion mutants lacking genes in the short component of the HCMV genome has revealed four independent and consecutively expressed genes of the HCMV-specific *US2/US6* gene family coding for type I transmembrane glycoproteins, all of which downregulate polymorphic MHC-I molecules in a selective manner (reviewed in HENGEL et al. 1997b, PLOEGH 1998). One of these factors, gpUS6, turned out to be a specific inhibitor of peptide translocation by TAP (HENGEL et al. 1997a; AHN et al. 1997; LEHNER et al. 1997), whereas the *US2-*, *US3-*, and *US11*-encoded glycoproteins directly interact with MHC class I (WIERTZ et al. 1996a,b; JONES et al. 1996; AHN et al. 1996b). By preventing the delivery of TAP-dependent peptides into the ER, gpUS6 affects assembly of both classic MHC-I and nonclassic MHC-I molecules such as HLA-G (JUN et al. 2000).

4.2.1 Molecular Function of HCMV gpUS6

The putative amino acid (a.a.) sequence of the *US6* open reading frame of HCMV strain AD169 codes for a 21-kDa type I transmembrane protein comprising 183 amino acids (CHEE et al. 1990) with a predicted signal sequence (a.a. 7–21), a transmembrane domain (a.a. 150–166), a short cytoplasmic tail (a.a. 167–183), and a single N-linked glycosylation site (a.a. 52). Transcription of *US6* is detected few hours after infection, but gpUS6 synthesis is most abundant in the late phase of the HCMV replication cycle correlating with the continuously decreasing transport of peptide throughout the HCMV replication cycle (JONES and MUZITHRAS 1992; HENGEL et al. 1997a). In IFN-γ-treated cells, synthesis of gpUS6 is drastically suppressed after HCMV infection (BENZ and HENGEL 2000). Considering the induction of TAP gene expression by this cytokine, IFN-γ may be a potent counterregulator able to restore peptide translocation in HCMV-infected tissues. Analysis of the intracellular distribution of gpUS6 revealed a typical ER-like staining pattern superimposed with the intracellular localization of TAP1/2 (HENGEL et al. 1997a). The complete sensitivity of gpUS6 to endoglycosidase H digestion indicated efficient retention in the ER. In *US6*-transfected cells the viral protein is found associated with the transient multimeric assembly complex composed of TAP1, TAP2, tapasin, MHC-I, β_2-microglobulin, and calreticulin as well as with calnexin (HENGEL et al. 1997a). In a minimal model, the inactivation of peptide transport could be explained by the physical association of gpUS6 with TAP without further factors being involved. Alternatively, gpUS6 could require a cellular cofactor to shut down peptide release into the ER (Fig. 1). A number of independent experimental approaches have been pursued to unravel direct and indirect interactions of gpUS6

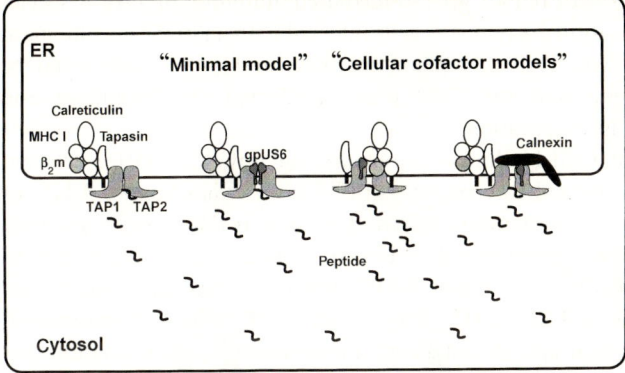

Fig. 1. Alternative models of HCMV gpUS6 interaction with the transient TAP complex containing peptide-free MHC I molecules and ER-resident chaperones. According to a "minimal model", gpUS6 binds directly to TAP itself (*left*). The "cofactor models" depict scenarios in which gpUS6 does not bind physically to TAP but to a protein present in the TAP complex (*center*) or gpUS6 recruits an unrelated molecule, e.g., calnexin, to TAP (*right*), resulting in functional inactivation of peptide transport. *ER*, endoplasmic reticulum; $\beta_2 m$, β_2-microglobulin. See text for details

with ER proteins and to determine the cellular factors required for gpUS6 function. In human cell lines deficient for MHC-I, β_2-microglobulin, tapasin, or calnexin, the inhibition of peptide transport by gpUS6 is maintained, indicating that these molecules are dispensable for gpUS6 function (HENGEL et al. 1997a; D. Bauer and H. Hengel, manuscript in preparation). Given the significant sequence homology of *US6* with the MHC-I binding proteins encoded by the *US2/6* gene family members *US2*, *US3*, and *US11*, it is noteworthy that in TAP-deficient T2 cells gpUS6 expression fails to downregulate MHC-I, indicating that the MHC-I subversive function of gpUS6 is exclusively dependent on TAP1/2 expression (D. Bauer and H. Hengel, manuscript in preparation). Comparative analysis of proteins associating with gpUS6 in human and in murine cells revealed characteristic differences. Although the gpUS6-associated protein complex consisting of TAP1, TAP2, tapasin, MHC-I, β_2-microglobulin, and calreticulin could consistently be retrieved from lysates of several human gp*US6*-transfected cell lines (including the calnexin-deficient line CEM-NKR), in lysates of gp*US6*-transfected murine cell lines only calnexin, but not TAP1 or TAP2, was found to coprecipitate with gpUS6 (D. Bauer and H. Hengel, manuscript in preparation). Together, the data suggested that gpUS6 binds directly but independently to both TAP and the lectin-like ER chaperone calnexin.

Several findings indicate that gpUS6 regulates the translocation of peptides by interactions with the luminal surface of the TAP1/TAP2 heterodimer. Unlike ICP47, gpUS6 does not interfere with peptide binding to the cytosolic face of TAP, whereas ICP47 prevents peptide binding to TAP in the presence of gpUS6 (AHN et al. 1997; HENGEL et al. 1997a). Moreover, expression of a truncated mutant of gpUS6 lacking the transmembrane and cytoplasmic domains resulted in a MHC-I-downregulated phenotype (AHN et al. 1997).

gpUS6, whereas TAP2 labeling was still intact (HEWITT et al. 2001). Thus gpUS6 may cause a marked change in the behavior of the NBD of TAP1.

4.2.2 TAP-Independent Peptide Transport of an HCMV-Encoded HLA-E Ligand, gpUL40

Expression of viral factors that interfere with the fate of MHC-I molecules results in a decreased MHC-I cell-surface expression that can be sensed by killer inhibitory receptors (KIR) on NK cells. In consequence, this strategy may render HCMV-infected cells susceptible to NK-mediated lysis. The inhibitory receptor CD94/NKG2A is broadly distributed on NK cells and recognizes the ubiquitously expressed MHC class Ib molecule HLA-E. Surface expression of HLA-E requires the TAP-dependent binding of conserved peptides corresponding to residues 3–11 of the signal sequences from various HLA-A, HLA-C, and some HLA-B heavy chains as well as from HLA-G (BRAUD et al. 1998a,b; BORREGO et al. 1998; LEE et al. 1998b). Notably, the motif of HLA-E ligands is also present in the leader sequence of the HCMV *UL40* open reading frame (CHEE et al. 1990), which is expressed during the early and late phases of infection (TOMASEC et al. 2000; ULBRECHT et al. 2000). Expression of *UL40* conferred resistance to NK cell lysis via the CD94/NKG2A receptor by inducing membrane expression of HLA-E (TOMASEC et al. 2000; ULBRECHT et al. 2000). Exploiting the HLA-E-mediated inhibition of NK cells appears to be a perfect strategy to counteract NK attack of MHC-I-deficient target cells, provided that the gpUL40-derived HLA-E ligand is able to bypass the TAP block established by the simultaneously expressed gpUS6. Expression of gp*UL40* in TAP-deficient RMA-S cells efficiently induced surface expression of cotransfected HLA-E molecules, indicating that binding of the gpUL40-derived peptide ligand is indeed TAP independent (ULBRECHT et al. 2000).

4.2.3 Altered Phosphorylation of TAP in HCMV-Infected Cells

Recently, evidence was provided for a regulation of TAP by phosphorylation. Both TAP1 and TAP2 as well as tapasin were shown to be phosphorylated under physiological conditions (LI et al. 2000). Phosphorylated TAP transporters were still able to bind peptides and ATP but lost the capacity to transport peptides into the ER. The data by the Yang group further suggest that HCMV may induce increased TAP phosphorylation in human fibroblasts during the late phase of infection. The enhanced TAP phosphorylation in the presence of HCMV gene expression is accompanied by a reduced peptide transport rate.

4.3 Adenovirus E3/19K Protein

The first intensively studied viral protein that was demonstrated to interfere with the MHC-I pathway of antigen presentation is the adenovirus E3/19K glycoprotein. The type I transmembrane glycoprotein E3/19K binds via its luminal domain to a large number of human MHC-I allelic variants in the ER (BURGERT and KVIST 1985; ANDERSSON et al. 1985). E3/19K contains a di-lysine ER retention motif in its cytoplasmic tail. Recently, an additional mechanism of E3/19K was elucidated through which the molecule may inhibit MHC-I expression and which does not require the ER retention signal (BENNETT et al. 1999). By coprecipitation with an E3/19K-specific antibody, E3/19K was shown to associate with MHC-I and TAP2. TAP2 was also coprecipitated from lysates of MHC-I- or tapasin-deficient cells, suggesting that these molecules are dispensable for E3/19K binding to TAP. E3/19K expression strongly reduced the steady-state association of MHC-I heavy chains with TAP1. Through independent binding to both MHC-I and TAP, E3/19K may prevent tapasin from forming a bridge between MHC-I and TAP, leading to defective peptide loading of MHC-I.

4.4 Augmented Peptide Transport in Flavivirus-Infected Cells

In clear contrast to many other viruses that interfere with a MHC-I-restricted immune response by downregulating components of the MHC-I pathway, flaviviruses are known to upregulate MHC-I surface expression in an interferon-independent manner (LIU et al. 1989; KING et al. 1989). In a previous report flavivirus-mediated augmentation of MHC-I surface expression and antigen presentation was shown to occur in TAP2-deficient RMA-S cells, suggesting TAP independence of this phenomenon (MÜLLBACHER and LOBIGS 1995). Further investigation, however, revealed that flavivirus-induced upregulation of peptide supply to the ER is fully dependent on the presence of functional peptide transporters and is associated with a significantly enhanced protection from recognition by NK cells (MOMBURG et al. 2001). During the early phase of infection with flavivirus, TAP-mediated peptide transport was consistently increased, whereas during later stages, or when high virus doses were used for infection, peptide transport was reduced below control levels. Because biosynthesis and steady-state levels of TAP were not significantly altered during infection, these data suggest that flavivirus gene products may interfere with the regulation of TAP activity itself. The mechanism(s) through which flavivirus up- or downregulates TAP activity awaits further experimentation.

5 Perspectives

The heterodimeric ABC transporter TAP is a complex molecular machine that combines membrane translocation of a diverse array of peptide substrates with conformational changes entailed by catalytic cycles of the two interdependent NBDs. Herpesviral proteins like ICP47 or gpUS6, which alter substrate or ATP binding to TAP, proved to be instrumental to delineate the order of molecular events that eventually result in peptide release into the ER. However, both proteins impose profound conformational changes on TAP1/TAP2 that make it difficult to differentiate between the initial TAP association of the inhibitor and TAP-inactivating *trans*-effects. New molecular approaches are required for further insight into where and at which step of the translocation cycle viral proteins interfere with the intramolecular communication of TAP domains, e.g., NBDs and luminal parts of the transporter finally forming the pore.

First experiments using intrachain hybrids of human and rat TAP1 as well as human and rat TAP2 served to delineate those portions of human TAP1/2 that mediate binding and inhibition of transport by the human-specific TAP inhibitor gpUS6 (H. Hengel and F. Momburg, unpublished results). Because gpUS6 seems to merely interact with luminal TAP sequences, our preliminary findings would call for a previously undetected pair of TMS in each TAP1 and TAP2. This contention is supported by TMS prediction programs as well as sequence alignments with other ABC transporters. Thus studies using gpUS6 will help to further characterize the topology of TAP1 and TAP2 polypeptides in the ER membrane, which, because of the weak hydrophobicity and poor separation of some TMS, is not easily amenable to structural analysis.

Likewise, the use of interspecies TAP1/2 hybrids in peptide translocation assays has led to the observation that the TAP2 subunit must be human for the inhibitory effect of HSV1 ICP47 whereas the TAP1 subunit can also be of rodent origin (F. Momburg and K. Früh, unpublished results). Investigation of rat-human or mouse-human TAP2 chimeras will help to pinpoint the interaction site of the peptidomimetic inhibitor ICP47 with human TAP2. This could in turn provide structural information about the peptide binding site in the TAP1/2 heterodimer that is blocked by ICP47.

Acknowledgements. Our work is supported by the DFG through project He2526/3-2, SFB 455 A6, SFB 352 B6, B12, and by the Bundesministerium für Bildung und Forschung.

References

Abele R, Tampé R (1999) Function of the transport complex TAP in cellular immune recognition. Biochim Biophys Acta 1461:405–419
Ahn K, Meyer TH, Uebel S, Sempe P, Djaballah H, Yang Y, Peterson PA, Früh K, Tampe R (1996a) Molecular mechanism and species specificity of TAP inhibition by herpes simplex virus ICP47. EMBO J 15:3247–3255
Ahn K, Angulo A, Ghazal P, Peterson PA, Yang Y, Früh K (1996b) Human cytomegalovirus inhibits antigen presentation by a sequential multistep process. Proc Natl Acad Sci USA 93:10990–10995
Ahn K, Gruhler A, Galocha B, Jones TR, Wiertz EJ, Ploegh HL, Peterson PA, Yang Y, Früh K (1997) The ER-luminal domain of the HCMV glycoprotein US6 inhibits peptide translocation by TAP. Immunity 6:613–621
Ambagala AP, Hinkley S, Srikumaran S (2000) An early pseudorabies virus protein down-regulates porcine MHC class I expression by inhibition of transporter associated with antigen processing (TAP). J Immunol 164:93–99
Andersson M, Pääbo S, Nilsson T, Peterson PA (1985) Impaired intracellular transport of class I MHC antigens as a possible means for adenoviruses to evade immune surveillance. Cell 43:215–222
Bangia N, Lehner PJ, Hughes EA, Surman M, Cresswell P (1999) The N-terminal region of tapasin is required to stabilize the MHC class I loading complex. Eur J Immunol 29:1858–1870

Barnden MJ, Purcell AW, Gorman JJ, McCluskey J (2000) Tapasin-mediated retention and optimization of peptide ligands during the assembly of class I molecules. J Immunol 165:322–30

Beersma MF, Bijlmakers MJ, Ploegh HL (1993) Human cytomegalovirus down-regulates HLA class I expression by reducing the stability of class I H chains. J Immunol 151:4455–4464

Bennett EM, Bennink JR, Yewdell JW, Brodsky FM (1999) Cutting edge: adenovirus E19 has two mechanisms for affecting class I MHC expression. J Immunol 162:5049–5052

Benz C, Hengel H (2000) MHC class I-subversive gene functions of cytomegalovirus and their regulation by interferons – an intricate balance. Virus Genes 21:39–47

Borrego F, Ulbrecht M, Weiss EH, Coligan JE, Brooks AG (1998) Recognition of human histocompatibility leukocyte antigen (HLA)-E complexed with HLA class I signal sequence-derived peptides by CD94/NKG2 confers protection from natural killer cell-mediated lysis. J Exp Med 187: 813–818

Brander C, Suscovich T, Lee Y, Nguyen PT, O'Connor P, Seebach J, Jones NG, van Gorder M, Walker BD, Scadden DT (2000) Impaired CTL recognition of cells latently infected with Kaposi's sarcoma-associated herpes virus. J Immunol 165:2077–2083

Braud VM, Allan DS, Wilson D, McMichael AJ (1998a) TAP- and tapasin-dependent HLA-E surface expression correlates with the binding of an MHC class I leader peptide. Curr Biol 8:1–10

Braud VM, Allan DS, O'Callaghan CA, Soderstrom K, D'Andrea A, Ogg GS, Lazetic S, Young NT, Bell JI, Phillips JH, Lanier LL, McMichael AJ (1998b) HLA-E binds to natural killer cell receptors CD94/NKG2 A, B and C. Nature 391:795–799

Burgert HG, Kvist S (1985) An adenovirus type 2 glycoprotein blocks cell surface expression of human histocompatibility class I antigens. Cell 41:987–997

Chee MS, Bankier AT, Beck S, Bohni R, Brown CM, Cerny R, Horsnell T, Hutchison CA 3d, Kouzarides T, Martignetti JA, Preddi E, Satchwell SC, Tomlinson P, Weston KM, Barrell BG (1990) Analysis of the protein-coding content of the sequence of human cytomegalovirus strain AD169. Curr Top Microbiol Immunol 154:125–169

Cromme FV, Airey J, Heemels MT, Ploegh HL, Keating PJ, Stern PL, Meijer CJ, Walboomers JM (1994a) Loss of transporter protein, encoded by the TAP-1 gene, is highly correlated with loss of HLA expression in cervical carcinomas. J Exp Med 179:335–340

Cromme FV, van Bommel PF, Walboomers JM, Gallee MP, Stern PL, Kenemans P, Helmerhorst TJ, Stukart MJ, Meijer CJ (1994b) Differences in MHC and TAP-1 expression in cervical cancer lymph node metastases as compared with the primary tumours. Br J Cancer 69:1176–1181

Eager KB, Williams J, Breiding D, Pan S, Knowles B, Appella E, Ricciardi RP (1985) Expression of histocompatibility antigens H-2K, -D, and -L is reduced in adenovirus-12-transformed mouse cells and is restored by interferon gamma. Proc Natl Acad Sci USA 82:5525–5529

Elliott T (1997) Transporter associated with antigen processing. Adv Immunol 65:47–109

Fleming SB, McCaughan CA, Andrews AE, Nash AD, Mercer AA (1997) A homolog of interleukin-10 is encoded by the poxvirus orf virus. J Virol 71:4857–4861

Früh K, Ahn K, Djaballah H, Sempe P, van Endert PM, Tampe R, Peterson PA, Yang Y (1995) A viral inhibitor of peptide transporters for antigen presentation. Nature 375:415–418

Früh K, Yang Y (1999) Antigen presentation by MHC class I and its regulation by interferon γ. Curr Opin Immunol 11:76–81

Galocha B, Hill A, Barnett BC, Dolan A, Raimondi A, Cook RF, Brunner J, McGeoch DJ, Ploegh HL (1997) The active site of ICP47, a herpes simplex virus-encoded inhibitor of the major histocompatibility complex (MHC)-encoded peptide transporter associated with antigen processing (TAP), maps to the NH_2-terminal 35 residues. J Exp Med 185:1565–1572

Garbi N, Tan P, Diehl AD, Chambers BJ, Ljunggren HG, Momburg F, Hämmerling GJ (2000) Impaired immune responses and altered peptide repertoire in tapsin-deficient mice. Nature Immunol 1:234–238

Hengel H, Eßlinger C, Pool J, Goulmy E, Koszinowski UH (1995) Cytokines restore MHC class I complex formation and control antigen presentation in human cytomegalovirus-infected cells. J Gen Virol 76:2987–2997

Hengel H, Flohr T, Hämmerling GJ, Koszinowski UH, Momburg F (1996) Human cytomegalovirus inhibits peptide translocation into the endoplasmic reticulum for MHC class I assembly. J Gen Virol 77:2287–2296

Hengel H, Koopmann JO, Flohr T, Muranyi W, Goulmy E, Hämmerling GJ, Koszinowski UH, Momburg F (1997a) A viral ER-resident glycoprotein inactivates the MHC-encoded peptide transporter. Immunity 6:623–632

Hengel H, Koszinowski UH (1997b) Interference with antigen processing by viruses. Curr Opin Immunol 9:470–476

Hewitt EW, Gupta SS, Lehner PJ (2001) The human cytomegalovirus gene product US6 inhibits ATP binding by TAP. EMBO J 20:387–396

Hill AB, Barnett BC, McMichael AJ, McGeoch DJ (1994) HLA class I molecules are not transported to the cell surface in cells infected with herpes simplex virus types 1 and 2. J Immunol 152:2736–2741

Hill A, Jugovic P, York I, Russ G, Bennink J, Yewdell J, Ploegh H, Johnson D (1995) Herpes simplex virus turns off the TAP to evade host immunity. Nature 375:411–415

Hinkley S, Hill AB, Srikumaran S (1998) Bovine herpesvirus-1 infection affects the peptide transport activity in bovine cells. Virus Res 53:91–96

Hsu DH, de Waal Malefyt R, Fiorentino DF, Dang MN, Vieira P, de Vries J, Spits H, Mosmann TR, Moore KW (1990) Expression of interleukin-10 activity by Epstein-Barr virus protein BCRF1. Science 250:830–832

Huard B, Früh K (2000) A role for MHC class I down-regulation in NK cell lysis of herpes virus-infected cells. Eur J Immunol 30:509–515

Jones TR, Muzithras VP (1992) A cluster of dispensable genes within the human cytomegalovirus genome short component: IRS1, US1 through US5, and the US6 family. J Virol 66:2541–2546

Jones TR, Wiertz EJ, Sun L, Fish KN, Nelson JA, Ploegh HL (1996) Human cytomegalovirus US3 impairs transport and maturation of major histocompatibility complex class I heavy chains. Proc Natl Acad Sci USA 93:11327–11333

Jugovic P, Hill AM, Tomazin R, Ploegh H, Johnson DC (1998) Inhibition of major histocompatibility complex class I antigen presentation in pig and primate cells by herpes simplex virus type 1 and 2 ICP47. J Virol 72:5076–5084

Jun Y, Kim E, Jin M, Sung HC, Han H, Geraghty DE, Ahn K (2000) Human cytomegalovirus gene products US3 and US6 down-regulate trophoblast class I MHC molecules. J Immunol 164:805–811

Keating PJ, Cromme FV, Duggan-Keen M, Snijders PJ, Walboomers JM, Hunter RD, Dyer PA, Stern PL (1995) Frequency of down-regulation of individual HLA-A and -B alleles in cervical carcinomas in relation to TAP-1 expression. Br J Cancer 72:405–411

Khanna R, Burrows SR, Argaet V, Moss DJ (1994) Endoplasmic reticulum signal sequence facilitated transport of peptide epitopes restores immunogenicity of an antigen processing defective tumour cell line. Int Immunol 6:639–645

Khanna R, Busson P, Burrows SR, Raffoux C, Moss DJ, Nicholls JM, Cooper L (1998) Molecular characterization of antigen-processing function in nasopharyngeal carcinoma (NPC): evidence for efficient presentation of Epstein-Barr virus cytotoxic T-cell epitopes by NPC cells. Cancer Res 58:310–314

King NJ, Maxwell LE, Kesson AM (1989) Induction of class I major histocompatibility complex antigen expression by West Nile virus on γ interferon-refractory early murine trophoblast cells. Proc Natl Acad Sci USA 86:911–915

Knittler MR, Alberts P, Deverson EV, Howard JC (1999) Nucleotide binding by TAP mediates association with peptide and release of assembled MHC class I molecules. Curr Biol 9:999–1008

Koppers-Lalic D, Rijsewijk FAM, Verschuren SBE, van Gaans-van den Brink JAM, Neisig A, Ressing ME, Neefjes J, Wiertz EJHJ (2001) The UL41-encoded virion host shutoff (vhs) protein and vhs-independent mechanisms are responsible for down-regulation of MHC class I molecules by bovine herpesvirus 1. J Gen Virol 82:2071–2081

Kotenko SV, Saccani S, Izotova LS, Mirochnitchenko OV, Pestka S (2000) Human cytomegalovirus harbors its own unique IL-10 homolog (cmvIL-10). Proc Natl Acad Sci USA 97:1695–1700

Kushner DB, Pereira DS, Liu X, Graham FL, Ricciardi RP (1996) The first exon of Ad12 E1A excluding the transactivation domain mediates differential binding of COUP-TF and NF-κB to the MHC class I enhancer in transformed cells. Oncogene 12:143–151

Lacaille VG, Androlewicz MJ (1998) Herpes simplex virus inhibitor ICP47 destabilizes the transporter associated with antigen processing (TAP) heterodimer. J Biol Chem 273:17386–17390

Lee SP, Constandinou CM, Thomas WA, Croom-Carter D, Blake NW, Murray PG, Crocker J, Rickinson AB (1998a) Antigen presenting phenotype of Hodgkin Reed-Sternberg cells: analysis of the HLA class I processing pathway and the effects of interleukin-10 on epstein-barr virus-specific cytotoxic T-cell recognition. Blood 92:1020–1030

Lee N, Goodlett DR, Ishitani A, Marquardt H, Geraghty DE (1998b) HLA-E surface expression depends on binding of TAP-dependent peptides derived from certain HLA class I signal sequences. J Immunol 160:4951–4960

Lehner PJ, Karttunen JT, Wilkinson GW, Cresswell P (1997) The human cytomegalovirus US6 glycoprotein inhibits transporter associated with antigen processing-dependent peptide translocation. Proc Natl Acad Sci USA 94:6904–6909

Li Y, Salter-Cid L, Vitiello A, Preckel T, Lee JD, Angulo A, Cai Z, Peterson PA, Yang Y (2000) Regulation of transporter associated with antigen processing by phosphorylation. J Biol Chem 275:24130–24135

Liu Y, King N, Kesson A, Blanden RV, Mullbacher A (1989) Flavivirus infection up-regulates the expression of class I and class II major histocompatibility antigens on and enhances T cell recognition of astrocytes in vitro. J Neuroimmunol 21:157–168

Liu Y, Kitsis RN (1996) Induction of DNA synthesis and apoptosis in cardiac myocytes by E1A oncoprotein. J Cell Biol 133:325–334

Ljunggren HG, Stam NJ, Ohlen C, Neefjes JJ, Hoglund P, Heemels MT, Bastin J, Schumacher TN, Townsend A, Kärre K, Ploegh HL (1990) Empty MHC class I molecules come out in the cold. Nature 346:476–480

Lockridge KM, Zhou SS, Kravitz RH, Johnson JL, Sawai ET, Blewett EL, Barry PA. (2000) Primate cytomegaloviruses encode and express an IL-10-like protein. Virology 268:272–280

Momburg F, Hämmerling GJ (1998) Generation and TAP-mediated transport of peptides for major histocompatibility complex class I molecules. Adv Immunol 68:191–256

Momburg F, Müllbacher A, Lobigs M (2001) Modulation of transporter associated with antigen processing (TAP)-mediated peptide import into the endoplasmic reticulum by flavivirus infection. J Virol 75:5663–5671

Momburg F, Roelse J, Hämmerling GJ, Neefjes JJ (1994) Peptide size selection by the major histocompatibility complex-encoded peptide transporter. J Exp Med 179:1613–1623

Moore KW, Vieira P, Fiorentino DF, Trounstine ML, Khan TA, Mosmann TR (1990) Homology of cytokine synthesis inhibitory factor (IL-10) to the Epstein-Barr virus gene BCRFI. Science 248:1230–1234

Moore KW, O'Garra A, de Waal Malefyt R, Vieira P, Mosmann TR (1993) Interleukin-10. Annu Rev Immunol 11:165–190

Moss DJ, Khanna R, Sherritt M, Elliott SL, Burrows SR (1999) Developing immunotherapeutic strategies for the control of Epstein-Barr virus-associated malignancies. J Acquir Immune Defic Syndr 21:S80–S83

Müllbacher A, Lobigs M (1995) Up-regulation of MHC class I by flavivirus-induced peptide translocation into the endoplasmic reticulum. Immunity 3:207–214

Murray PG, Constandinou CM, Crocker J, Young LS, Ambinder RF (1998) Analysis of major histocompatibility complex class I, TAP expression, and LMP2 epitope sequence in Epstein-Barr virus-positive Hodgkin's disease. Blood 92:2477–2483

Neumann L, Kraas W, Uebel S, Jung G, Tampé R (1997) The active domain of the herpes simplex virus protein ICP47: a potent inhibitor of the transporter associated with antigen processing. J Mol Biol 272:484–492

Neumann L, Tampe R (1999) Kinetic analysis of peptide binding to the TAP transport complex: evidence for structural rearrangements induced by substrate binding. J Mol Biol 294:1203–1213

Pamer E, Cresswell P (1998) Mechanisms of MHC class I-restricted antigen processing. Annu Rev Immunol 16:323–358

Paz P, Brouwenstijn N, Perry R, Shastri N (1999) Discrete proteolytic intermediates in the MHC class I antigen processing pathway and MHC I-dependent peptide trimming in the ER. Immunity 11:241–251

Ploegh HL (1998) Viral strategies of immune evasion. Science 280:248–253

Reits EA, Vos JC, Gromme M, Neefjes J (2000) The major substrates for TAP in vivo are derived from newly synthesized proteins. Nature 404:774–778

Rickinson AB, Kieff E (1996) Epstein-Barr Virus. In: Fields BN, Knipe DM, Howley PM (eds) Fields Virology. Lippincott-Raven, Philadelphia, pp 2397–2445

Robb JA, Benirschke K, Barmeyer R (1986) Intrauterine latent herpes simplex virus infection: I. Spontaneous abortion. Hum Pathol 17:1196–1209

Rock KL, Goldberg AL (1999) Degradation of cell proteins and the generation of MHC class I-presented peptides. Annu Rev Immunol 17:739–779

Rode HJ, Janssen W, Rosen-Wolff A, Bugert JJ, Thein P, Becker Y, Darai G (1993) The genome of equine herpesvirus type 2 harbors an interleukin 10 (IL10)-like gene. Virus Genes 7:111–116

Rotem-Yehudar R, Winograd S, Sela S, Coligan JE, Ehrlich R (1994) Downregulation of peptide transporter genes in cell lines transformed with the highly oncogenic adenovirus 12. J Exp Med 180:477–488

Rotem-Yehudar R, Groettrup M, Soza A, Kloetzel PM, Ehrlich R (1996) LMP-associated proteolytic activities and TAP-dependent peptide transport for class 1 MHC molecules are suppressed in cell lines transformed by the highly oncogenic adenovirus 12. J Exp Med 183:499–514

Rowe M, Khanna R, Jacob CA, Argaet V, Kelly A, Powis S, Belich M, Croom-Carter D, Lee S, Burrows SR, Trowsdale J, Moss, DJ, Rickinson, AB (1995) Restoration of endogenous antigen processing in Burkitt's lymphoma cells by Epstein-Barr virus latent membrane protein-1: coordinate up-regulation of peptide transporters and HLA-class I antigen expression. Eur J Immunol 25:1374–1384

Salter RD, Cresswell P (1986) Impaired assembly and transport of HLA-A and -B antigens in a mutant TxB cell hybrid. EMBO J 5:943–949

Schrier PI, Bernards R, Vaessen RT, Houweling A, van der Eb AJ (1983) Expression of class I major histocompatibility antigens switched off by highly oncogenic adenovirus 12 in transformed rat cells. Nature 305:771–775

Schust DJ, Hill AB, Ploegh HL (1996) Herpes simplex virus blocks intracellular transport of HLA-G in placentally derived human cells. J Immunol 157:3375–3380

Seliger B, Maeurer MJ, Ferrone S (1997) TAP off – tumors on. Immunol Today 18:292–299

Seliger B, Maeurer MJ, Ferrone S (2000) Antigen-processing machinery breakdown and tumor growth. Immunol Today 21:455–464

Tomasec P, Braud VM, Rickards C, Powell MB, McSharry BP, Gadola S, Cerundolo V, Borysiewicz LK, McMichael AJ, Wilkinson GW (2000). Surface expression of HLA-E, an inhibitor of natural killer cells, enhanced by human cytomegalovirus gpUL40. Science 287:1031–1033

Tomazin R, Hill AB, Jugovic P, York I, van Endert P, Ploegh HL, Andrews DW, Johnson DC (1996) Stable binding of the herpes simplex virus ICP47 protein to the peptide binding site of TAP. EMBO J 15:3256–3266

Tomazin R, van Schoot NE, Goldsmith K, Jugovic P, Sempe P, Früh K, Johnson DC (1998) Herpes simplex virus type 2 ICP47 inhibits human TAP but not mouse TAP. J Virol 72:2560–2563

Ulbrecht M, Martinozzi S, Grzeschik M, Hengel H, Ellwart JW, Pla M, Weiss EH (2000) Cutting edge: the human cytomegalovirus UL40 gene product contains a ligand for HLA-E and prevents NK cell-mediated lysis. J Immunol 164:5019–5022

van Endert PM (1999a) Genes regulating MHC class I processing of antigen. Curr Opin Immunol 11:82–88

van Endert PM (1999b) Role of nucleotides and peptide substrate for stability and functional state of the human ABC family transporters associated with antigen processing. J Biol Chem 274:14632–14638

Vambutas A, Bonagura VR, Steinberg BM (2000) Altered expression of TAP-1 and major histocompatibility complex class I in laryngeal papillomatosis: correlation of TAP-1 with disease. Clin Diagn Lab Immunol 7:79–85

Vos JC, Spee P, Momburg F, Neefjes J (1999a) Membrane topology and dimerization of the two subunits of the transporter associated with antigen processing reveal a three-domain structure. J Immunol 163:6679–6685

Vos JC, Reits EA, Wojcik-Jacobs E, Neefjes J (1999b) Head-head/tail-tail relative orientation of the pore-forming domains of the heterodimeric ABC transporter TAP. Curr Biol 10:1–7

Watts C (1997) Capture and processing of exogenous antigens for presentation on MHC molecules. Annu Rev Immunol 15:821–50

Wiertz EJ, Jones TR, Sun L, Bogyo M, Geuze HJ, Ploegh HL (1996a) The human cytomegalovirus US11 gene product dislocates MHC class I heavy chains from the endoplasmic reticulum to the cytosol. Cell 84:769–779

Wiertz EJ, Tortorella D, Bogyo M, Yu J, Mothes W, Jones TR, Rapoport TA, Ploegh HL (1996b) Sec61-mediated transfer of a membrane protein from the endoplasmic reticulum to the proteasome for destruction. Nature 384:432

Wright KL, White LC, Kelly A, Beck S, Trowsdale J, Ting JP (1995) Coordinate regulation of the human TAP1 and LMP2 genes from a shared bidirectional promoter. J Exp Med 181:1459–1471

Yamashita Y, Shimokata K, Mizuno S, Yamaguchi H, Nishiyama Y (1993) Down-regulation of the surface expression of class I MHC antigens by human cytomegalovirus. Virology 193:727–736

Yewdell JW, Norbury CC, Bennink JR (1999) Mechanisms of exogenous antigen presentation by MHC class I molecules in vitro and in vivo: implications for generating CD8+ T cell responses to infectious agents, tumors, transplants, and vaccines. Adv Immunol 73:1–77

York IA, Roop C, Andrews DW, Riddell SR, Graham FL, Johnson DC (1994) A cytosolic herpes simplex virus protein inhibits antigen presentation to CD8+ T lymphocytes. Cell 77:525–535

Zeidler R, Eissner G, Meissner P, Uebel S, Tampe R, Lazis S, Hammerschmidt W (1997) Downregulation of TAP1 in B lymphocytes by cellular and Epstein-Barr virus-encoded interleukin-10. Blood 90:2390–2397

zur Hausen H (2000) Papillomaviruses causing cancer: evasion from host-cell control in early events in carcinogenesis. J Natl Cancer Inst 92:690–698

Herpes Viral Proteins Manipulating the Peptide Transporter TAP

E. REITS, A. GRIEKSPOOR, and J. NEEFJES

The peptide transporter associated with antigen processing (TAP) is crucial for class I-restricted antigen presentation because it transfers cytosolic peptides into the endoplasmic reticulum (ER) lumen for class I binding. It is therefore not surprising that TAP is targeted for inactivation by many viruses. Herpesviruses have been very successful in designing various proteins that inactivate TAP. We summarise current knowledge on the class I antigen presentation pathway and the function, structure and action of TAP and its viral inhibitors.

1 Introduction	75
2 Antigen Presentation by MHC Class I Molecules: Degradation, Transport and Binding	76
3 Function and Structure of the Peptide Transporter TAP	77
4 Viral Inhibitors of TAP and Their Action	79
References	81

1 Introduction

Antigen presentation by MHC class I molecules is Nature's solution to the problem of visualising the intracellular protein content to the outside surveying immune system. This is achieved through a number of steps involving protein breakdown into peptides, transfer of the breakdown products over the endoplasmic reticulum (ER) membrane and binding to MHC class I molecules present in the ER. The MHC class I molecules ultimately transport a blueprint of the intracellular protein content – in the form of small peptides – to the plasma membrane. A critical step in this cascade of events is the translocation of peptides over the ER membrane, which is performed by a dedicated peptide transporter associated with antigen processing (TAP). Because both viruses and their hosts battle in the struggle for survival, some viruses have developed unique and specific tools to block their presentation by MHC class I molecules to the immune system. One of the targets is the peptide transporter TAP, for which several viral inhibitors have now been defined. Here we describe the current model of action of TAP and discuss how viruses can block the

---

Division of Tumor Biology, Plesmanlaan 121, 1066 CX Amsterdam, The Netherlands

transporter. This information may be used to design both other inhibitors for TAP, which may be useful in immune suppression, and inhibitors of TAP related pumps like those expressed by *M. tuberculosis*.

2 Antigen Presentation by MHC Class I Molecules: Degradation, Transport and Binding

Peptides that are presented by MHC class I molecules are almost all generated by the action of the proteasome from nuclear, cytosolic and most ER proteins (reviewed by Coux et al. 1996; Pamer and Cresswell 1998). The proteasome is a large multicatalytic protease, abundantly present in the nucleus and the cytoplasm. Proteasomes diffuse in these two compartments and probably find most of their substrate proteins by simple collision (Reits et al. 1997). As a result, a wide range of differently sized proteasomal degradation products are deposited in the cytosol and the nucleus. This implies that peptides have to undergo at least one translocation step over a membrane before they can contact the peptide-binding groove of MHC class I molecules in the ER.

Biochemical experiments have shown that MHC class I molecules have to associate with a peptide in the ER lumen to be transported to the cell surface. Consequently, if MHC class I molecules fail to acquire a peptide, they are retained in the ER, resulting in MHC class I depletion from the plasma membrane (Townsend et al. 1989). Cells deficient in one of the steps in the MHC class I pathway can therefore be isolated through their resistance against MHC class I antibody and complement or because they are ignored after transplantation by the immune system of other mice (DeMars et al. 1985; Salter and Cresswell 1986). Many of these mutant cell lines appeared to be deficient in one of the two subunits of a transporter encoded by two closely linked genes in the MHC locus. These genes were named *TAP1* and *TAP2*, and the proteins form a heterodimeric complex called TAP (Kelly et al. 1992). Inactivation of one of the two subunits of TAP is sufficient to abrogate antigen presentation by MHC class I molecules, implying that TAP is crucial and that cells do not have an alternative system for TAP. Some peptides are still presented by MHC class I molecules in TAP-deficient cells, which appear to be derived from signal sequences that enter the ER during translocation of novel nascent proteins and are generated by ER-located signal peptidases (Wei and Cresswell 1992; Henderson et al. 1992). Even so, TAP is the bottleneck through which most antigenic peptides have to travel before binding to and presentation by MHC class I molecules. Because TAP is crucial for antigen presentation but not for other cellular processes, it is an ideal target for viral inhibition.

TAP belongs to a large family of ATP binding cassette (ABC)-containing transporters that translocate many different substrates including lipids, drugs and glutathione conjugates (reviewed by Higgins 1992). TAP is a rather unique ABC transporter, not only because of its localisation in the ER membrane but also because it acts as a platform for assembly and loading of nascent MHC class I molecules. In the MHC locus another gene is located encoding for a protein called tapasin, which associates with TAP and recruits MHC class I molecules to TAP to facilitate efficient peptide capture (ORTMANN et al. 1997). Tapasin appears to act as a dedicated chaperone that stabilises unloaded MHC class I molecules (which are also stabilised by the more common chaperones calnexin and Erp57) and only releases them when MHC class I has captured a peptide. Recruitment to the TAP-tapasin complex is, however, not a prerequisite for successful loading of class I molecule, because certain MHC class I molecules can be found free in the ER lumen (NEISIG et al. 1996). TAP appears to be central in a dynamic complex consisting of TAP, tapasin, MHC class I, Erp57 and calnexin. After peptide binding, the MHC class I molecules are released from the various chaperones and allowed to leave the ER to present their cargo at the plasma membrane. In this dynamic process (Fig. 1) three steps have been major targets for viral inhibition: inhibition of TAP, degradation of nascent MHC class I in the ER and removal of MHC class I molecules from the plasma membrane. We will elaborate on the inhibition of TAP by viruses and will first describe the function and the composition of TAP in detail.

3 Function and Structure of the Peptide Transporter TAP

ABC transporters, including TAP, have a common architecture. They are constructed of series of closely packed transmembrane segments that form a pore in the membrane and contain two ATP-binding cassettes that drive the conformational changes inducing unidirectional flow of substrates. The peptide-binding domain of

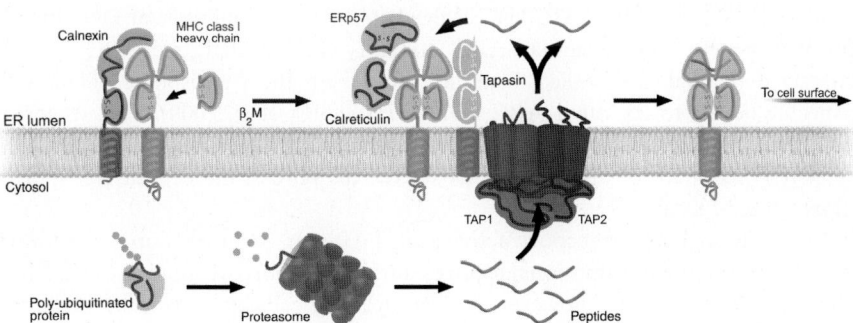

Fig. 1. Schematic representation of the process of antigen presentation by MHC class I molecules. Endogenous proteins are degraded by the proteasome present in the cytoplasm and the nucleus. The resulting peptides are transported by TAP into the ER lumen. Most MHC class I molecules are bound to TAP in an antigen loading complex including tapasin as a bridging molecule. When loaded with peptide, MHC class I molecules are transported to the cell surface for presentation to the immune system

TAP is located on the cytoplasmic side between the pore and the ABC domains (reviewed by (REITS et al. 2000a). Unlike MHC class I molecules, TAP is not polymorphic. Because TAP has to supply a large variety of MHC class I molecules with peptides, TAP cannot have a very restricted peptide sequence specificity. Indeed, experiments with systematically varied peptide substrates (MOMBURG et al. 1994a; NEEFJES et al. 1995) or peptide libraries (UEBEL et al. 1997) revealed that human TAP has broad substrate specificity. Further experiments revealed that TAP also had broader size selectivity than MHC class I molecules. Whereas the latter mainly binds peptides of 8–10 amino acids, TAP has a lower size limit of 7 amino acids and an optimum for binding peptides of around 9–12 amino acids. Longer peptides are also translocated albeit with an efficiency decreasing with size (MOMBURG et al. 1994b; SCHUMACHER et al. 1994). Peptide binding by TAP appears to be similar to that of MHC class I molecules, where most binding energy is generated by interaction between MHC class I and the peptide backbone as well as the peptide amino- and carboxy- termini. With exception of a proline residue at position 2 or 3, side chains are usually not important and can easily be substituted for unnatural side chains (ANDROLEWICZ and CRESSWELL 1994), phosphorylated amino acids (ANDERSEN et al. 1999) or elongated side chains. The latter allows the synthesis of TAP inhibitors. By systematically increasing the size of the side chain, a peptide can be generated which is able to bind TAP but not able to be translocated (GROMME et al. 1997).

Although the arguments above suggest that TAP may have a peptide-binding domain similar to MHC class I molecules, no structural data are available to validate this. Photoaffinity labelling (NIJENHUIS and HAMMERLING 1996) and mutational studies (DEVERSON et al. 1998) have defined the peptide binding domain, which is composed of segments of both TAP1 and TAP2 and which immediately follows the pore domain (VOS et al. 1999). The two subunits of TAP are orientated in a head-head tail-tail topology, suggesting that the peptide-binding domain is located to one side of the pore followed by the two ATP binding sites (Fig. 2) (REITS et al. 2000b). It is unclear how peptide and ATP binding actually trigger peptide translocation by TAP. The resolved structure of a number of DNA-repair enzymes which contain similar ABC domains showed that one ABC domain interacted through a so-called signature motif with the γ-phosphate of an ATP molecule bound to the other ABC domain (HOPFNER et al. 2000). This sensor of ATP binding may also support the arrangement of alternating cycles of ATP hydrolysis required for opening and closure of the pore of the peptide transporter (REITS et al. 2000a).

Even though the transport activity of TAP can be directly measured with model substrates, conformational changes during transport are difficult to visualise. The activity of TAP can, however, be read in living cells by measuring the lateral mobility of TAP over the ER membrane (REITS et al. 2000b). With the use of bleaching protocols and fluorescent GFP-tagged TAP complexes, it turned out that the mobility of inactive TAP is higher compared with TAP actively pumping peptides. The activity of TAP can thus be visualised through the diffusion rate of the complex in the ER membrane. The decrease in mobility is not due to added

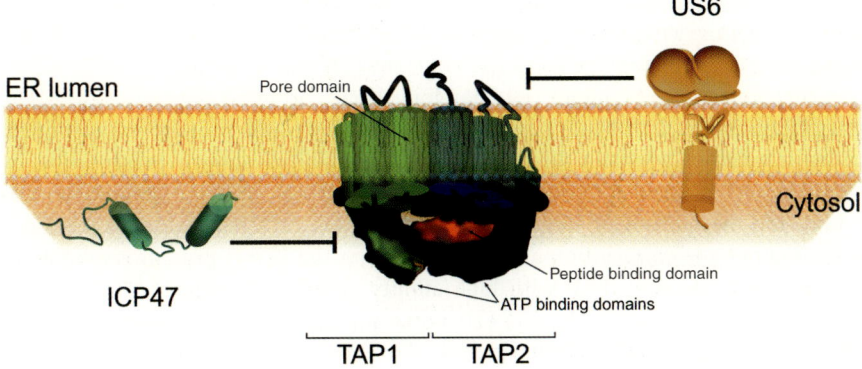

Fig. 2. Model of TAP. TAP forms a transmembrane pore in the ER membrane. The pore is followed by a peptide-binding domain located to one side of the pore at the cytoplasmic side. The structure is concluded by the two ATP-binding domains. In this model, both peptides and ICP47 approach the binding site of TAP from the cytosolic side, whereas US6 interacts with TAP from the ER luminal side

mass of the peptide substrate but is most likely the result of conformational changes within the TAP complex during pore opening and peptide transfer. This observation has been used to study the origin of peptides in vivo. It turned out that most peptides generated by the proteasome are derived from newly translated proteins, which was confirmed in biochemical experiments (SCHUBERT et al. 2000). In addition, the effect of chemical and viral TAP inhibitors can be visualised as long as they affect conformational changes within the TAP complex. For example, a chemical inhibitor acting as a peptide with a long side chain appears to block TAP in an opened conformation, which is suggestive for a transition-state trap (REITS et al. 2000b). Viruses may use similar ways to trap and thereby inhibit TAP activity and thereby antigen presentation.

4 Viral Inhibitors of TAP and Their Action

It is obvious that viruses (but not their hosts) would prosper when they could survive and propagate without detection by the host immune system. Many viruses indeed downregulate MHC class I expression in infected cells, but only in a few cases is this achieved by inhibiting the peptide transporter TAP. Alternatively, MHC class I molecules are retained and degraded in the ER or routed to lysosomes for degradation. As yet, only two viral products that inhibit TAP have been defined, ICP47 expressed by herpes simplex virus (HSV) and US6 expressed by human and mouse cytomegalovirus (CMV). HSV and CMV belong to different subfamilies of herpesviruses, which are large DNA-containing enveloped viruses. Interestingly, each virus has found its own solution to the challenge of inhibiting TAP. Whereas ICP47 is a cytosolic protein expressed early during infection, US6 is expressed at late times after infection and blocks TAP from the ER luminal side, thus interacting with different domains of TAP and probably using a different mechanism as well (Fig. 2).

HSV infection of human cells has long been known to downregulate MHC class I expression. In 1994, a small HSV-encoded cytosolic protein was identified, called infected-cell protein 47 (ICP47), which inhibited antigen presentation by MHC class I molecules (YORK et al. 1994). Because no cellular homologue was found, the mechanism of inhibition by ICP47 was rather obscure. Experiments including peptide competition assays showed that the ICP47 protein is binding to the peptide domain of TAP with high affinity, thereby blocking TAP-dependent peptide translocation into the ER (TOMAZIN et al. 1996; AHN et al. 1996). ICP47 affects the structure of TAP in vitro, as the TAP heterodimer is destabilised by ICP47 in detergent extracts (LACAILLE and ANDROLEWICZ 1998). ICP47 is 88 amino acids in length, which is considerably larger than a usual peptide, and although it has no effect on ATP binding by TAP, it remains on the cytosolic side of the TAP complex. Apparently, ICP47 acts as a pseudo-substrate inhibitor, but how is not well understood. The N-terminal residues 2–35 are sufficient for inhibiting peptide binding to TAP, and this region is conserved between ICP47 proteins from HSV-1 and HSV-2 (NEUMANN et al. 1997). It contains several regions required for interactions with the peptide-binding domain and possibly surrounding areas to increase stable binding with TAP (GALOCHA et al. 1997). Spectroscopy studies suggest that ICP47 is mainly unstructured in solution, but on interaction with membranes an α-helical structure in the protein is induced (BEINERT et al. 1997). Apparently the N-terminal region undergoes a conformational change required for membrane anchoring, which is, however, much weaker than the association with TAP. It remains unclear whether the conformation is altered on TAP binding and how binding and inhibition of TAP by ICP47 occur.

The HCMV-encoded US6 protein disables TAP through an entirely different mechanism. Human CMV can use several unique short (US) region-encoded proteins to block the MHC class I antigen presentation pathway (reviewed by TORTORELLA et al. 2000). Various HCMV proteins were already known to be involved in early degradation of MHC class I when two groups reported in 1997 the identification of a viral ER-resident glycoprotein that inhibited TAP (AHN et al. 1997; HENGEL et al. 1997). Surprisingly, US6 does not prevent the binding of peptides to TAP but apparently inhibits the actual translocation while associating to TAP from the ER luminal side. US6 is a small 21-kDa type I transmembrane glycoprotein with cytosolic ER retention signals. The soluble ER part of US6 suffices to inhibit TAP, and because it is enriched in cysteine residues (8 of 129 amino acids) it may form many internal disulfide bridges resulting in a compact structure. Although US6 does not have significant homology to any human protein, the cysteine content suggests that it may be forming a so-called double knot structure, which is also seen in various growth factors (SUN and DAVIES 1995). How US6 inhibits TAP function is unclear. TAP mobility experiments using bleaching protocols showed that US6 traps TAP in a defined conformation, which is not responsive to the absence or presence of ATP or peptide substrates (REITS et al.

2000b). As a result, ATP hydrolysis by TAP is also blocked (HEWITT et al. 2001). Although the exact site of interaction is unclear at the moment, it is likely that US6 is interacting with the pore of TAP, thereby preventing further conformational changes required for opening and closure of the TAP pore. The unique mechanism of action of US6 may be highly relevant for the generation of inhibitors of TAP and other ABC transporters in vivo.

It is likely that other unique viral inhibitors of TAP will be defined, including early proteins derived from bovine herpes virus BHV1 (HINKLEY et al. 1998) and pseudorabies virus (AMBAGALA et al. 2000). These will not only result in recognition of the strategies viruses use to hide for the immune system but will also provide a flow of information on the mechanism of action of the peptide transporter TAP.

References

Ahn K, Gruhler A, Galocha B, Jones TR, Wiertz EJ, Ploegh HL, Peterson PA, Yang Y, Fruh K (1997) The ER-luminal domain of the HCMV glycoprotein US6 inhibits peptide translocation by TAP. Immunity 6:613–621

Ahn K, Meyer TH, Uebel S, Sempe P, Djaballah H, Yang Y, Peterson PA, Fruh K, Tampe R (1996) Molecular mechanism and species specificity of TAP inhibition by herpes simplex virus ICP47. EMBO J 15:3247–3255

Ambagala AP, Hinkley S, Srikumaran S (2000) An early pseudorabies virus protein down-regulates porcine MHC class I expression by inhibition of transporter associated with antigen processing (TAP). J Immunol 164:93–99

Andersen MH, Bonfill JE, Neisig A, Arsequell G, Sondergaard I, Valencia G, Neefjes J, Zeuthen J, Elliott T, Haurum JS (1999) Phosphorylated peptides can be transported by TAP molecules, presented by class I MHC molecules, and recognized by phosphopeptide-specific CTL. J Immunol 163:3812–3818

Androlewicz MJ and Cresswell P (1994) Human transporters associated with antigen processing possess a promiscuous peptide-binding site. Immunity 1:7–14

Beinert D, Neumann L, Uebel S, Tampe R (1997) Structure of the viral TAP-inhibitor ICP47 induced by membrane association. Biochemistry 36:4694–4700

Coux O, Tanaka K, Goldberg AL (1996) Structure and functions of the 20S and 26S proteasomes. Annu Rev Biochem 65:801–847

DeMars R, Rudersdorf R, Chang C, Petersen J, Strandtmann J, Korn N, Sidwell B, Orr HT (1985) Mutations that impair a posttranscriptional step in expression of HLA-A and -B antigens. Proc Natl Acad Sci USA 82:8183–8187

Deverson EV, Leong L, Seelig A, Coadwell WJ, Tredgett EM, Butcher GW, Howard JC (1998) Functional analysis by site-directed mutagenesis of the complex polymorphism in rat transporter associated with antigen processing. J Immunol 160:2767–2779

Galocha B, Hill A, Barnett BC, Dolan A, Raimondi A, Cook RF, Brunner J, McGeoch DJ, Ploegh HL (1997) The active site of ICP47, a herpes simplex virus-encoded inhibitor of the major histocompatibility complex (MHC)-encoded peptide transporter associated with antigen processing (TAP), maps to the NH2-terminal 35 residues. J Exp Med 185:1565–1572

Gromme M, van der Valk R, Sliedregt K, Vernie L, Liskamp R, Hammerling G, Koopmann JO, Momburg F, Neefjes J (1997) The rational design of TAP inhibitors using peptide substrate modifications and peptidomimetics. Eur J Immunol 27:898–904

Henderson RA, Michel H, Sakaguchi K, Shabanowitz J, Appella E, Hunt DF, Engelhard VH (1992) HLA-A2.1-associated peptides from a mutant cell line: a second pathway of antigen presentation. Science 255:1264–1266

Hengel H, Koopmann JO, Flohr T, Muranyi W, Goulmy E, Hammerling GJ, Koszinowski UH, Momburg F (1997) A viral ER-resident glycoprotein inactivates the MHC-encoded peptide transporter. Immunity 6:623–632

Hewitt EW, Gupta SS, Lehner PJ (2001) The human cytomegalovirus gene product US6 inhibits ATP binding by TAP. EMBO J 20:387–396

Higgins CF (1992) ABC transporters: from microorganisms to man. Annu Rev Cell Biol 8:67–113

Hinkley S, Hill AB, Srikumaran S (1998) Bovine herpesvirus-1 infection affects the peptide transport activity in bovine cells. Virus Res 53:91–96

Hopfner KP, Karcher A, Shin DS, Craig L, Arthur LM, Carney JP, Tainer JA (2000) Structural biology of Rad50 ATPase: ATP-driven conformational control in DNA double-strand break repair and the ABC-ATPase superfamily. Cell 101:789–800

Kelly A, Powis SH, Kerr LA, Mockridge I, Elliott T, Bastin J, Uchanska-Ziegler B, Ziegler A, Trowsdale J, Townsend A (1992) Assembly and function of the two ABC transporter proteins encoded in the human major histocompatibility complex. Nature 355:641–644

Lacaille VG and Androlewicz MJ (1998) Herpes simplex virus inhibitor ICP47 destabilizes the transporter associated with antigen processing (TAP) heterodimer. J Biol Chem 273:17386–17390

Momburg F, Roelse J, Howard JC, Butcher GW, Hammerling GJ, Neefjes JJ (1994a) Selectivity of MHC-encoded peptide transporters from human, mouse and rat. Nature 367:648–651

Momburg F, Neefjes JJ, Hammerling GJ (1994b) Peptide selection by MHC-encoded TAP transporters. Current Opinion in Immunology 6:32–37

Neefjes J, Gottfried E, Roelse J, Gromme M, Obst R, Hammerling GJ, Momburg F (1995) Analysis of the fine specificity of rat, mouse and human TAP peptide transporters. Eur J Immunol 25:1133–1136

Neisig A, Wubbolts R, Zang X, Melief C, Neefjes J (1996) Allele-specific differences in the interaction of MHC class I molecules with transporters associated with antigen processing. J Immunol 156:3196–3206

Neumann L, Kraas W, Uebel S, Jung G, Tampe R (1997) The active domain of the herpes simplex virus protein ICP47: a potent inhibitor of the transporter associated with antigen processing. J Mol Biol 272:484–492

Nijenhuis M and Hammerling GJ (1996) Multiple regions of the transporter associated with antigen processing (TAP) contribute to its peptide binding site. J Immunol 157:5467–5477

Ortmann B, Copeman J, Lehner PJ, Sadasivan B, Herberg JA, Grandea AG, Riddell SR, Tampe R, Spies T, Trowsdale J, Cresswell P (1997) A critical role for tapasin in the assembly and function of multimeric MHC class I-TAP complexes. Science 277:1306–1309

Pamer E and Cresswell P (1998) Mechanisms of MHC class I-restricted antigen processing. Annu Rev Immunol 16:323–358

Reits EA, Griekspoor AC, Neefjes J (2000a) How does TAP pump peptides? Insights from DNA repair and traffic ATPases. Immunol Today 21:598–600

Reits EA, Vos JC, Gromme M, Neefjes J (2000b) The major substrates for TAP in vivo are derived from newly synthesized proteins. Nature 404:774–778

Reits EA, Benham AM, Plougastel B, Neefjes J, Trowsdale J (1997) Dynamics of proteasome distribution in living cells. EMBO J 16:6087–6094

Salter RD and Cresswell P (1986) Impaired assembly and transport of HLA-A and -B antigens in a mutant TxB cell hybrid. EMBO J 5:943–949

Schubert U, Anton LC, Gibbs J, Norbury CC, Yewdell JW, Bennink JR (2000) Rapid degradation of a large fraction of newly synthesized proteins by proteasomes. Nature 404:770–774

Schumacher TN, Kantesaria DV, Heemels MT, Ashton-Rickardt PG, Shepherd JC, Fruh K, Yang Y, Peterson PA, Tonegawa S, Ploegh HL (1994) Peptide length and sequence specificity of the mouse TAP1/TAP2 translocator. J Exp Med 179:533–540

Sun PD and Davies DR (1995) The cystine-knot growth-factor superfamily. Annu Rev Biophys Biomol Struct 24:269–291

Tomazin R, Hill AB, Jugovic P, York I, van Endert P, Ploegh HL, Andrews DW, Johnson DC (1996) Stable binding of the herpes simplex virus ICP47 protein to the peptide binding site of TAP. EMBO J 15:3256–3266

Tortorella D, Gewurz BE, Furman MH, Schust DJ, Ploegh HL (2000) Viral subversion of the immune system. Annu Rev Immunol 18:861–926

Townsend A, Ohlen C, Bastin J, Ljunggren HG, Foster L, Karre K (1989) Association of class I major histocompatibility heavy and light chains induced by viral peptides. Nature 340:443–448

Uebel S, Kraas W, Kienle S, Wiesmuller KH, Jung G, Tampe R (1997) Recognition principle of the TAP transporter disclosed by combinatorial peptide libraries. Proc Natl Acad Sci USA 94:8976–8981

Vos JC, Spee P, Momburg F, Neefjes J (1999) Membrane topology and dimerization of the two subunits of the transporter associated with antigen processing reveal a three-domain structure. J Immunol 163:6679–6685

Wei ML and Cresswell P (1992) HLA-A2 molecules in an antigen-processing mutant cell contain signal sequence-derived peptides. Nature 356:443–446

York IA, Roop C, Andrews DW, Riddell SR, Graham FL, Johnson DC (1994) A cytosolic herpes simplex virus protein inhibits antigen presentation to CD8+ T lymphocytes. Cell 77:525–535

Herpes Viral Proteins Blocking the Transporter Associated with Antigen Processing TAP – From Genes to Function and Structure

D. BAUER and R. TAMPÉ

In adaptation to the immune system, viruses have developed manifold mechanisms to evade the immune response, causing lifelong persistence in the host. Several members of the herpesvirus family are known to interfere with antigen presentation via MHC class I molecules. Here we compare the mechanistic and structural aspects of two unrelated herpesviral proteins, both of which have selected the transporter associated with antigen processing (TAP) as target for immune evasion. However, ICP47 (IE12) encoded by the herpes simplex virus and US6 from human cytomegalovirus utilize entirely different strategies to block TAP function. Detailed knowledge of the function and structure of these viral factors will help to understand TAP function and to design novel immune suppressors or vectors for gene transfer.

1	Introduction	85
2	Pathway of Antigen Processing via MHC Class I	86
2.1	TAP – the Transporter Associated with Antigen Processing	87
3	The Immune Evasion of HSV: Identification of ICP47	89
3.1	ICP47: From Gene to Function and Structure	90
4	The Immune Evasion of HCMV: Identification of US6	93
4.1	US6: From Gene to Function and Structure	93
5	Different Strategies, One Target	95
6	Perspectives	96
	References	97

1 Introduction

During evolution vertebrates have developed an adaptive immune system to defend themselves against viral and bacterial encounters. The executors of this system are macrophages as well as B and T lymphocytes, which are able to eliminate foreign invaders or to lyse infected cells. In the process of antigen processing (reviewed by PAMER and CRESSWELL 1998; YEWDELL et al. 1999), short fragments (8–10 amino acids in length) derived from endogenously synthesized proteins are presented on

---

Institute of Biochemistry, Biozentrum Frankfurt, Goethe-University Frankfurt, Marie-Curie-Str. 9, 60439 Frankfurt, Germany
E-mail: tampe@em.uni-frankfurt.de

the cell surface in complex with MHC class I molecules. Peptides are mainly generated by proteasomal cleavage in the cytosol and have to be translocated into the endoplasmic reticulum (ER) to reach the MHC class I molecules. This intracellular peptide transport is mediated by the transporter associated with antigen processing (TAP), which supplies peptides for loading and stable assembly of MHC class I molecules. Correctly folded MHC-peptide complexes are transported via the Golgi to the cell surface, where they are inspected by $CD8^+$ T lymphocytes. In case of a viral attack, additional peptides of viral origin are presented to $CD8^+$ T lymphocytes (CTL). This CTL response leads to lysis of infected cells.

In adaptation to the immune system, viruses have developed manifold strategies to evade the immune response. These mechanisms cause lifelong persistence in the host (reviewed by PLOEGH 1998). One of these mechanisms affects antigen presentation via MHC class I molecules by blocking TAP-mediated peptide translocation. The number of viruses known to interfere with TAP function is still growing. At least five members of the herpesvirus family including the porcine pseudorabies virus, bovine herpesvirus-1, the human cytomegalovirus, and the herpes simplex virus (type 1 and 2) are known to interfere with TAP function (AMBAGALA et al. 2000; HENGEL et al. 1996; HILL et al. 1994; HINKLEY et al. 1998; YORK et al. 1994). However, only two viral proteins have been characterized in detail: ICP47 encoded by herpes simplex virus type 1 and type 2 (HSV-1 and HSV-2) interacts with TAP on the cytosolic side and prevents peptide binding to TAP (AHN et al. 1996a; FRÜH et al. 1995; HILL et al. 1995; TOMAZIN et al. 1996). In contrast, the glycoprotein US6 of the human cytomegalovirus (HCMV) blocks peptide transport by binding to TAP in the ER lumen (AHN et al. 1997; HENGEL et al. 1997; LEHNER et al. 1997).

2 Pathway of Antigen Processing via MHC Class I

MHC class I molecules are composed of the heavy chain (HC), which forms the peptide-binding pocket, and the noncovalently associated β_2-microglobulin (β_2m) (reviewed by MADDEN 1995). The highly polymorphic gene products of the MHC class I locus are expressed in almost all nucleated mammalian cells. The maturation of MHC class I molecules is a multistep process (reviewed by CRESSWELL et al. 1999; GRANDEA and VAN KAER 2001). In the first step, calnexin binds to the newly synthesized glycosylated HC. Calnexin retains incompletely folded glycoproteins in the ER and, together with calreticulin, plays an important role in folding and formation of disulfide bonds mediated by ERp57 and PDI. After dissociation of calnexin and subsequent binding of β_2m to the prefolded HC, these MHC complexes, together with calreticulin, associate with TAP. The interaction between TAP and MHC molecules is mediated by tapasin. After peptide loading, MHC molecules dissociate from TAP and are exported out of the ER via Golgi to the cell surface, where their antigenic cargo is monitored by $CD8^+$ T lymphocytes.

The antigenic peptides are mainly generated by proteasomal degradation of endogenously synthesized proteins. These proteins are cleaved by the catalytic core particle (20S) of the proteasome complex (26S) (reviewed by BAUMEISTER et al. 1998; UEBEL and TAMPÉ 1999). The genes of two proteasomal subunits, LMP2 and LMP7 (low-molecular-weight peptides), and of TAP1 and TAP2 are located in the MHC II locus on human chromosome 6. Expression of these proteins is induced by interferon-γ (IFNγ), which causes a reorganization of de novo assembled proteasomes replacing the catalytically active subunits $\beta1$, $\beta5$, and $\beta2$ by LMP2, LMP7, and MECL1 (multicatalytic endopeptidase complex-like). Thereby, so-called immunoproteasomes are formed. This IFNγ-induced expression of TAP and immunoproteasomes results in an increase of the pool of antigenic peptides presented on the cell surface.

2.1 TAP – the Transporter Associated with Antigen Processing

The importance of TAP for the immune response was identified by studying various cell lines, which show a strongly reduced level of MHC class I molecules on the cell surface (LJUNGGREN et al. 1990; TOWNSEND et al. 1989). Surface expression was restored by transfection of *tap1* and/or *tap2*, suggesting that TAP is essential for antigen processing (CERUNDOLO et al. 1990; POWIS et al. 1991; SPIES and DEMARS 1991). TAP belongs to the superfamily of ATP-binding cassette (ABC) transporter (reviewed by DOIGE and AMES 1993; HIGGINS 1992). This family comprises polytopic integral membrane proteins transporting a very large spectrum of substrates in an ATP-dependent manner. ABC transporters possess two highly conserved cytoplasmic nucleotide-binding domains (NBDs) and two hydrophobic domains, which comprise 5–10 transmembrane helices (TMs) each, forming the putative translocation pore. In contrast to the NBDs, sequence homology of the hydrophobic domains is much lower. Human TAP consists of two subunits, TAP1 (71kDa) and TAP2 (75kDa), which form a stable heteromeric complex within the ER membrane (reviewed by ABELE and TAMPÉ 1999). Detailed analysis of the primary structure reveals that the transmembrane domains of TAP and MDR1 are similarly organized. The membrane topology of the eukaryotic ABC transporter MDR1 has been experimentally analyzed by many groups (reviewed by AMBUDKAR et al. 1999). MDR1 consists of two hydrophobic domains with six transmembrane helices each. The amino and carboxy termini face in the cytosol. Based on sequence alignments and membrane topology analysis a 2×6 transmembrane helix model similar to MDR1 can be derived for TAP (Fig. 1). Only the amino-terminal 175 and 140 residues of human TAP1 and TAP2, comprising four and three additional putative transmembrane helices, respectively, share no sequence homology with ABC transporters (LANKAT-BUTTGEREIT and TAMPÉ 1999; TAMPÉ et al. 1997). The peptide-binding region of TAP was mapped by trypsin and/or cyanogen bromide digestion of the cross-linked subunits following immunoprecipitation (NIJENHUIS and HÄMMERLING 1996). It comprises the cytosolic loop between TM4 and TM5 and 15 amino acids after the putative TM6 (Fig. 1).

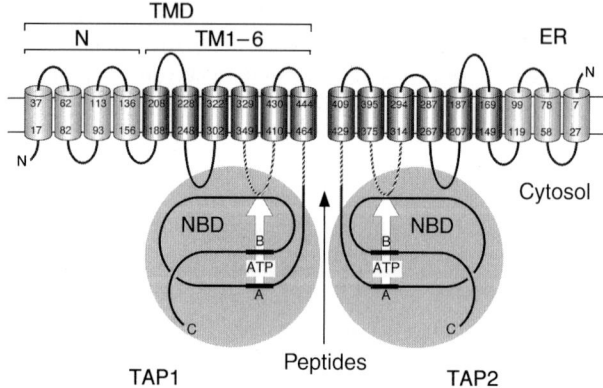

Fig. 1. Structural organization of the human TAP complex. The membrane helices are predicted from hydrophobicity analysis and sequence alignment with MDR1. (ABELE and TAMPÉ 1999; LANKAT-BUTT-GEREIT and TAMPÉ 1999). The N-terminal domain (*N*) comprises four and three transmembrane helices (*TM*), respectively. According to the 2 × 6 transmembrane helix model, the transmembrane helices are numbered TM1 to TM6. The nucleotide-binding domains (*NBDs*) consist of the highly conserved Walker A and B motifs (*A*, *B*). The *striped lines* illustrate the peptide-binding region mapped by photo-cross-linking studies. (NIJENHUIS and HÄMMERLING 1996)

Peptide transport by TAP is a multistep process consisting of an ATP-independent binding process and an ATP-dependent transport step (NEUMANN and TAMPÉ 1999; UEBEL et al. 1995; VAN ENDERT et al. 1994). Peptide binding requires a close arrangement of both TAP subunits (ANDROLEWICZ and CRESSWELL 1994; UEBEL et al. 1995; VAN ENDERT et al. 1994), whereas ATP and ADP can bind to both NBDs, even if they are expressed separately (MÜLLER et al. 1994; WANG et al. 1994). The pathway of peptide recognition and binding to TAP was followed in real time by fluorescence-labeled peptides (NEUMANN and TAMPÉ 1999). After a fast association step, a slow isomerization of the TAP complex was observed. Strikingly, peptide binding and ATP hydrolysis are tightly coupled, indicating that peptide binding induces a structural reorganization that subsequently triggers ATP hydrolysis and peptide translocation (GORBULEV et al. 2001).

By using semi-permeabilized cells or isolated microsomes, TAP selectivity was analyzed by peptide transport assays trapping peptides in the ER via N-glycosylation or by peptide binding assays (ANDROLEWICZ et al. 1993; NEEFJES et al. 1993; SHEPHERD et al. 1993; UEBEL et al. 1995; VAN ENDERT et al. 1994). Peptides with a length of 8–12 amino acids are most efficiently transported (KOOPMANN et al. 1996), whereas the optimum for peptide binding is between 8 and 16 amino acids (VAN ENDERT et al. 1994). Peptide epitopes are fixed in the peptide-binding pocket of TAP via the amino and carboxy termini (SCHUMACHER et al. 1994; UEBEL et al. 1997). The binding motif of human TAP was systematically analyzed by combinatorial peptide libraries and peptide analogues (UEBEL et al. 1997). Peptides with hydrophobic or basic amino acids (Phe, Leu, Arg, or Tyr) at the carboxy terminus are clearly preferred by human TAP, whereas Asp, Glu, Asn, and Ser are disfavored. At peptide positions 1 to 3, Lys, Asn, and Arg increase, whereas Asp and Glu destabilize TAP binding. Interestingly, the peptide binding motifs of TAP and MHC class I molecules are very similar. These results point to a coevolution of MHC class I and TAP, thereby optimizing the antigen processing machinery against attacks of pathogens (UEBEL and TAMPÉ 1999).

Despite the recent progress in resolving TAP function, many questions regarding the mechanism of peptide translocation stay open: How are peptide and ATP binding, ATP hydrolysis, and peptide translocation synchronized within the TAP complex? Do both NBDs hydrolyze ATP for peptide translocation? Do they cooperate in a parallel or sequential mode, in which one NBD supplies the energy for peptide translocation and the other supports restoration of the cytosolic peptide-binding pocket? In consequence, different mechanisms can be envisioned for evasion of the immune response by interfering with TAP function, such as inhibition of peptide binding, ATP binding and/or hydrolysis, trapping of a conformational state, blocking of the translocation channel, or modulating associated factors of the macromolecular transport and loading complex.

3 The Immune Evasion of HSV: Identification of ICP47

HSV causes lifelong persistent infections in the host. Different strains of the two subtypes HSV-1 and HSV-2 exist. For immune evasion HSV prevents CTL recognition by suppressing surface expression of MHC class I (KOELLE et al. 1993; POSAVAD and ROSENTHAL 1992). This decrease is caused by the retention of MHC class I molecules in the ER (HILL et al. 1994; YORK et al. 1994). In noninfected cells, MHC class I molecules are transported from the ER via the Golgi, where they become sialated by sialyl-transferases and convert from an endoglycosidaseH (*endo*H)-sensitive to an *endo*H-resistant form. In HSV-1- or HSV-2-infected human fibroblasts, MHC class I molecules still remained *endo*H-sensitive during a 90-min chase, in contrast to the situation found in non-infected cells (YORK et al. 1994). Inhibition of MHC class I maturation was observed within 2h after infection and was completed after 4h (YORK et al. 1994). Similar results were obtained by monitoring the sialation of HCs in HSV-2-infected cells (HILL et al. 1994). The maturation of MHC class I molecules was analyzed in various replication phases of the virus to identify factors affecting the peptide supply. The replication of HSV is divided into immediate-early, early, and late phase (HONESS and ROIZMAN 1974). In HSV-infected cells, phosphonoacetic acid (PAA), which inhibits viral DNA synthesis and therefore late gene expression, has no influence on MHC class I retention (HILL et al. 1994). In light of the observation that the maturation of MHC class I is inhibited early after infection, the responsible viral protein must be an immediate-early or early gene product. By analyzing various HSV mutants, it has been demonstrated that with respect to other viral proteins ICP47 is both necessary and sufficient for MHC I retention in the ER/*cis*-Golgi and for reduced lysis of infected human fibroblasts by $CD8^+$ T lymphocytes (YORK et al. 1994). The synthesis of

ICP47 is independent of the synthesis of other viral proteins. Therefore, ICP47 was identified as an immediate-early gene product. With the discovery of ICP47, the starting point was set to analyze the function and structure of this viral protein and to understand HSV persistence on the molecular level.

3.1 ICP47: From Gene to Function and Structure

ICP47 is encoded by the open reading frame (ORF) *US12* in the unique short *US* region of both subtypes HSV-1 and HSV-2, which have little differences in their genomes. In the genome of HSV-1 (strain 17), *US12* is located from base pair 145311 to 145577 (McGeoch et al. 1985), whereas ICP47-2 of HSV-2 (strain HG52) is encoded in the *Hin*dIII–I region from base pair 147775 to 148035 (McGeoch et al. 1987). ICP47-1 and ICP47-2 comprise 88 and 86 amino acids (10kDa), respectively. Both proteins share only 42% amino acid identity focused in the amino-terminal half. The low overall homology is mainly due to a frame shift mutation caused by a deletion of 13 base pairs starting from codon 59 of ICP47-1 compared with ICP47-2 (Fig. 2A).

Initial experiments elucidating the mechanism of TAP inhibition were performed with ICP47-1. Although ICP47 was originally localized in the cytosol and nucleus (York et al. 1994), low expression of ICP47 results in an immunofluorescence pattern indistinguishable from that of TAP (Früh et al. 1995). The

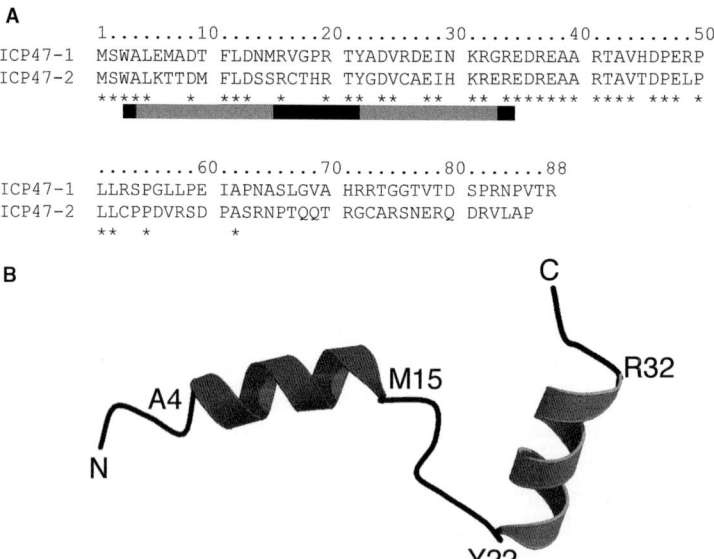

Fig. 2A,B. Structure of ICP47. **A** Sequence alignment of ICP47-1 and ICP47-2. Identical amino acids are indicated (*asterisk*). The active domain (*grey*) comprising two α-helices (*black*) is *underlined*. **B** The NMR structure of the active domain of ICP47. The structure consists of two helices extending from residues 4 to 15 and 22 to 32 linked by a flexible loop region. (Pfänder et al. 1999)

association of ICP47 and the TAP complex including HC and β_2m was demonstrated by immunoprecipitations of ICP47 from digitonin-lysed cells (FRÜH et al. 1995; HILL et al. 1995). Vice versa, the interaction of TAP with ICP47 was shown by anti-TAP1 immunoprecipitations obtained from Nonidet P40 (NP40) or digitonin lysates. Interestingly, the overexpression of TAP1 and TAP2 led to an increase of coprecipitated TAP and ICP47, indicating that the association of ICP47 with TAP seems not to be limited by a third factor (HILL et al. 1995). Most convincingly, ICP47 blocks TAP function in insect cells and yeast, which lack factors of the adaptive immune system (AHN et al. 1996b; TOMAZIN et al. 1996; URLINGER et al. 1997). No interaction of ICP47 with single TAP1 or TAP2 was observed, indicating that ICP47 binds to both TAP subunits (HILL et al. 1995).

There are several possibilities to encounter TAP from the cytosol. Beside the two ATP-binding sites, TAP possesses a single peptide-binding pocket formed by both subunits (ABELE and TAMPÉ 1999; NIJENHUIS and HÄMMERLING 1996). In the presence of ICP47 both TAP subunits can be crosslinked to 8-azido-(α-^{32}P)ATP, confirming that ICP47 does not affect ATP-binding to TAP (AHN et al. 1996a). By using peptide binding and photo-cross-linking assays it was demonstrated that ICP47 blocks peptide binding to TAP (AHN et al. 1996a; TOMAZIN et al. 1996). The binding affinity of ICP47-1 to TAP was determined to be 50nM (AHN et al. 1996a; TOMAZIN et al. 1996), which is orders of magnitude higher than the binding affinity of combinatorial peptide libraries for TAP (K_D=2.5µM) (UEBEL et al. 1997). The binding of ICP47 can be competitively inhibited by peptides (AHN et al. 1996a; TOMAZIN et al. 1996). Because the interaction sites of TAP and ICP47 have not been mapped, we cannot distinguish between a direct competition for the binding site and/or a conformational inhibition of TAP caused by ICP47. There are several arguments against the hypothesis that ICP47 acts like a larger peptide substitute. First, peptides with more than 16 residues have a very low affinity for TAP (VAN ENDERT et al. 1994). Second, ICP47 prevents chemical cross-linking of TAP1 and TAP2, whereas peptides promote it (LACAILLE and ANDROLEWICZ 1998). Third, binding of ICP47 seems to be asymmetric. Although the amino terminus of ICP47 interacts with both subunits of TAP, a ^{125}I-labeled Tyr21 polypeptide cross-links only TAP1 (GALOCHA et al. 1997).

ICP47 synthesized by solid phase or purified from *E. coli* inclusion bodies was found to be fully active in TAP inhibition. Therefore, functional and structural studies have been performed. By using overlapping fragments and systematic truncations at the amino and carboxy termini the active region of ICP47 was analyzed (GALOCHA et al. 1997; NEUMANN et al. 1997). A fragment of 32 residues, ICP47(3–34), was determined as minimal active region (NEUMANN et al. 1997). Interestingly, overlapping fragments of 25 residues, which cover the active region, neither block peptide binding to TAP nor affect a preformed ICP47-TAP inhibitory complex (NEUMANN et al. 1997). It should be stressed that the active domain does not include the region of highest homology between ICP47-1 and ICP47-2. To identify critical amino acid residues within the active domain, an alanine scan was performed (GALOCHA et al. 1997). Three regions, residues 8–12, 17–24, and 28–31, were classified as important for ICP47 function. Residues 17–24 appear to be most sensitive to alanine substitutions. With an alternative approach, a cluster of charged amino acids (residues 24, 31 and 32), which are conserved in ICP47-1 and ICP47-2, was found to be critical for ICP47 activity (NEUMANN et al. 1997).

After identification of the active domain and critical residues, structural aspects of the TAP inhibitor were analyzed by circular dichroism (CD), fluorescence, and nuclear magnetic resonance (NMR) spectroscopy. In aqueous solution ICP47 adopts a loosely folded, so-called random-coil structure (BEINERT et al. 1997; PFÄNDER et al. 1999). This observation is further supported by biochemical data demonstrating that ICP47 is still active after boiling or purification in organic solvents (NEUMANN et al. 1997). On the basis of the knowledge that N- or C-terminal signal sequences adopt an α-helical structure in a lipid environment, the secondary structure of ICP47 was analyzed in the presence of membrane mimetics or liposomes (BEINERT et al. 1997). Under these conditions, the CD spectra of ICP47 showed characteristic minima at 206 and 222nm, indicative of an α-helical structure. The α-helical content was proportional to the ratio of negatively charged lipids in the liposomes. Interaction of ICP47 with these membranes was followed by fluorescence spectroscopy and binding assays (BEINERT et al. 1997; PFÄNDER et al. 1999). The TAP inhibitor contains a single tryptophan that served as a sensitive probe to study membrane interaction (BEINERT et al. 1997). In aqueous solution, where ICP47 is highly dynamic and unstructured, the tryptophan is completely solvent-exposed and its fluorescence is quenched by collision with water. By binding to lipid membranes, ICP47 converts into an α-helical conformation. Thereby the tryptophan is shifted into a hydrophobic environment as deduced from an increase and blue shift of the fluorescence.

The structure of ICP47(2–34) was determined by NMR spectroscopy in a lipidlike environment (PFÄNDER et al. 1999). In micellar solution of deuterated sodium dodecyl sulfate, the viral TAP inhibitor adopts an ordered helix-linker-helix structure (Fig. 2B). The structure consists of two helices extending from residues 4 to 15 and from 22 to 32. A flexible loop region links both helices. Although the structure of ICP47 has not been determined in complex with TAP, there is good evidence that the helices are essential for ICP47 function. The perturbation of the amino-terminal α-helix by removal of the tryptophan and alanine residues at position 3 and 4, or by insertion of proline at position 11, leads to a loss of activity (NEUMANN et al. 1997). In summary, membrane association induces the helix-linker-helix conformation and an enrichment of ICP47 at the cytosolic face of the ER membrane, guiding the interaction with TAP.

A further interesting observation is the species specificity of ICP47. HSV fails to inhibit antigen presentation in murine fibroblasts (YORK et al. 1994). Moreover, expression of murine TAP in ICP47-positive cells restores class I surface expression (FRÜH et al. 1995). Further analysis demonstrated that ICP47 inhibits TAP of related animal species such as human, monkey, bovine, porcine, and canine. In contrast, TAP of mouse, rat, guinea pig, and rabbit are not affected by ICP47 (JUGOVIC et al. 1998). On the molecular level, the species specificity can be explained by a 100-fold higher affinity of ICP47 to human TAP (K_D=50nM) than to murine TAP (K_D=5μM) (AHN et al. 1996a; TOMAZIN et al. 1996). Possibly, the species specificity of ICP47 is linked to the peptide selectivity of human and mouse TAP. Clear evidence exists that human and murine TAP transport different sets of peptides (MOMBURG et al. 1994; SCHUMACHER et al. 1994). If ICP47 binds directly to the peptide-binding pocket, the peptide specificity of TAP may also affect the ICP47-TAP interaction. The species specificity of ICP47 may originate from coevolution of HSV with the human host. Thereby, the selection of the best-adapted TAP inhibitor was forced. Alternatively, the viral protein could be derived from the human host and, therefore, the protein is best adapted. However, no homologue of ICP47 has been found.

Although many functional and structural aspects of ICP47 have been clarified, some questions remain open. It is not known which regions of ICP47 and TAP interact with each other. Furthermore, there is evidence that the inhibition mechanism of ICP47 is very complex and not only based on competition with the peptide-binding pocket (Neumann and Tampé, unpublished results). In addition to competition with the peptide-binding pocket, ICP47 may form additional contacts with TAP by an induced fit, thereby trapping TAP in an inactive conformation.

4 The Immune Evasion of HCMV: Identification of US6

HCMV affects the cell surface expression of MHC class I molecules and therefore, reduces the antigen presentation for $CD8^+$ T lymphocytes (BARNES and GRUNDY 1992; BEERSMA et al. 1993; HENGEL et al. 1995; WARREN et al. 1994; YAMASHITA et al. 1993). HCMV-infected fibroblasts show a decrease in the peptide transport into the ER (HENGEL et al. 1996). A temperature-sensitive mutant of HCMV, ts9, which lacks the genes *US1–US15*, had lost the capacity to interfere with MHC class I assembly and to inhibit the peptide translocation function of TAP (HENGEL et al. 1996). By expression of each ORF of the *US* region *US2*, *US3*, *US6*, and *US11* were observed to reduce the MHC class I surface expression (AHN et al. 1997; HENGEL et al. 1997). However, US6 is the only protein that is responsible for blocking peptide translocation into the ER in HCMV-infected cells (AHN et al. 1997; HENGEL et al. 1997; LEHNER et al. 1997). In US6-expressing cells, nearly all MHC class I molecules are unstable at 37°C (AHN et al. 1997; HENGEL et al. 1997) and remain *endo*H-sensitive (AHN et al. 1997; HENGEL et al. 1997; LEHNER et al. 1997), which means they are retarded in the ER/*cis*-Golgi.

4.1 US6: From Gene to Function and Structure

The six ORFs of the *US6* gene family (*US6–US11*) are not essential for virus replication. US6 is encoded by the ORF from base pair 195949 to 195397 in the HCMV genome (strain Ad169) (CHEE et al. 1990). During viral infection, US6 expression starts in the early phase and reaches its maximum at 72h after infection in the late phase of the viral replication cycle (HENGEL et al. 1997). US6 consists of 183 amino acids. This type I membrane glycoprotein is glycosylated at Asn52 (21kDa) and is composed of a leader sequence (19aa), an ER-luminal domain (125aa), a membrane helix (21aa), and a short cytoplasmic tail (18aa) (Fig. 3).

The localization of US6 is similar to BiP and TAP, indicating that US6 is an ER-resident protein (HENGEL et al. 1997; LEHNER et al. 1997). Interestingly, US6 has no consensus motif for ER retention, and therefore the mechanism of ER-retention remains open. Because US6 blocks the intracellular peptide supply, the interaction of US6 and TAP was analyzed by immunoprecipitation. US6 binds to the TAP complex consisting of TAP, tapasin, and MHC class I molecules (AHN et al. 1997; HENGEL et al. 1997; LEHNER et al. 1997). By the analysis of mutant cell lines, it was demonstrated that US6 blocks TAP function even in the absence of MHC class I or tapasin (HENGEL et al. 1997). US6 was found to block TAP-dependent peptide transport in insect cells, indicating that US6 interacts with TAP directly and requires no additional factors (KYRITSIS et al. 2001). The effect of HCMV on MHC class I surface expression could be restored by treatment with IFNγ (HENGEL et al. 1995; LEHNER et al. 1997).

The molecular mechanism of TAP inhibition by US6 is poorly defined mostly because of the lack of isolated viral protein. By use of photoactive peptides it was shown that US6, in contrast to ICP47, does not prevent peptide binding to TAP (AHN et al. 1997; HENGEL et al. 1997). The analysis of chimeras with E3/19K or CD8, as well as of a carboxy-terminal truncation, led to the conclusion that the luminal domain, US6(20–139), is still functional (AHN et al. 1997).

During revision of this review, LEHNER and colleagues reported that US6 inhibits binding of ATP by TAP1 (HEWITT et al. 2001). US6 also stabilizes TAP at 37°C and prevents conformational rearrangements induced by peptide binding (HEWITT et al. 2001). In a parallel study, the molecular mechanism of TAP

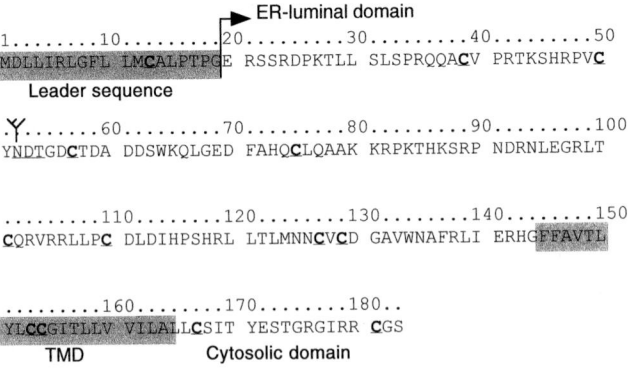

Fig. 3. Structural organization of US6. The type I membrane glycoprotein consists of 183aa including an N-terminal leader sequence (*grey*). The most likely cleavage site is between position 19 and 20. Cysteines and the N-glycosylation site are underlined. The putative transmembrane domain (*grey*) is followed by a short cytoplasmic sequence

inhibition by US6 was analyzed in vitro by using purified US6 and TAP coreconstituted in proteoliposomes (KYRITSIS et al. 2001). It was demonstrated that the recombinant ER-luminal domain is sufficient to block TAP-dependent peptide transport and glycosylation is not required for US6 function. Most importantly, US6 causes a specific arrest of the peptide-stimulated ATPase activity of the TAP complex by preventing binding of ATP but not of ADP. The affinity of the US6-TAP interaction was determined to be 1μM. The ER-luminal domain of US6 appears as a monomer in solution and consists of 19% α-helices, 25% β-sheets, and 27% β-turns. Notably, all eight cysteines form a stabilizing network of four intramolecular disulfide bonds (KYRITSIS et al. 2001). Therefore, US6 possesses structural properties distinct from US2 (GEWURZ et al. 2001) and ICP47 (PFÄNDER et al. 1999). Most likely, US6 binds to the short transmembrane loops of TAP1 and TAP2 facing the ER lumen, trapping TAP in a conformation that allows peptide binding but prevents ATP hydrolysis by blocking ATP-binding to TAP1.

5 Different Strategies, One Target

ICP47 and US6 are both expressed by members of the herpesvirus family. In contrast to the distinctive security system of HCMV consisting of four proteins (US2, US3, US6, and US11), HSV expresses only ICP47 to inhibit antigen processing and presentation by protein-protein interaction. Obviously, a single inhibitor is not sufficient to complete immune evasion of HCMV. Thus HCMV was forced to recruit proteins that target different steps in the antigen processing pathway. It should be stressed that the inhibition of TAP never leads to a complete depletion of MHC class I molecules on the cell surface, which may help to prevent lysis by natural killer cells. However, in the in vivo situation, it remains open whether a collection of CMV proteins has more advantages for immune evasion in comparison to the immediate-early expressed ICP47 of HSV.

Although these viral proteins have common interaction partners, they do not show any significant sequence homology with each other. Furthermore, no homologues among cellular proteins have been identified. Because intracellular peptide transport is a key step in antigen processing, it is quite logical that TAP is a major target for immune evasion. Surprisingly, however, ICP47 and US6 use completely different strategies to block TAP function. ICP47 inhibits TAP from the cytosolic side. It was proposed that the initial step of TAP inhibition is the binding of ICP47 to the ER membrane, converting a random-coil conformation into a helix-linker-helix structure. In the second step, ICP47 binds to TAP and blocks peptide binding, thereby preventing peptide-stimulated ATP-hydrolysis (GORBULEV et al. 2001). In contrast to ICP47, US6 is an ER-resident type I membrane glycoprotein attacking TAP from the ER-luminal side. US6 changes neither the peptide binding affinity nor the overall amount of bound peptides. However, both herpesviral proteins cause inhibition of the peptide-stimulated ATPase and transport activity of TAP (GORBULEV et al. 2001; KYRITSIS et al. 2001). In contrast to ICP47, the ER-luminal domain of US6 prevents ATP binding of TAP, whereas ICP47 acts indirectly on the ATPase activity by blocking peptide binding. The properties of both herpesviral TAP inhibitors are summarized in Fig. 4. However, many questions regarding the US6-TAP interaction as well as the structure of the luminal US6 domain wait to be answered.

6 Perspectives

Despite the improvement in our knowledge of peptide transport by TAP and its inhibition by ICP47 or US6, several key issues remain unsettled. A central question is how conformational changes are synchronized and transmitted between different regions within the macromolecular peptide transport and loading complex. It will be a major challenge to solve the question of which conformational state in the transport cycle of TAP is trapped by ICP47 and US6. Therefore, both viral proteins are helpful tools to investigate the transport mechanism of TAP. By analyzing the inhibition mechanism of ICP47 and US6 we will find new ways to understand and finally control TAP function. Detailed knowledge of the function and structure of these viral TAP inhibitors will help to design novel immune suppressors and vectors for gene transfer.

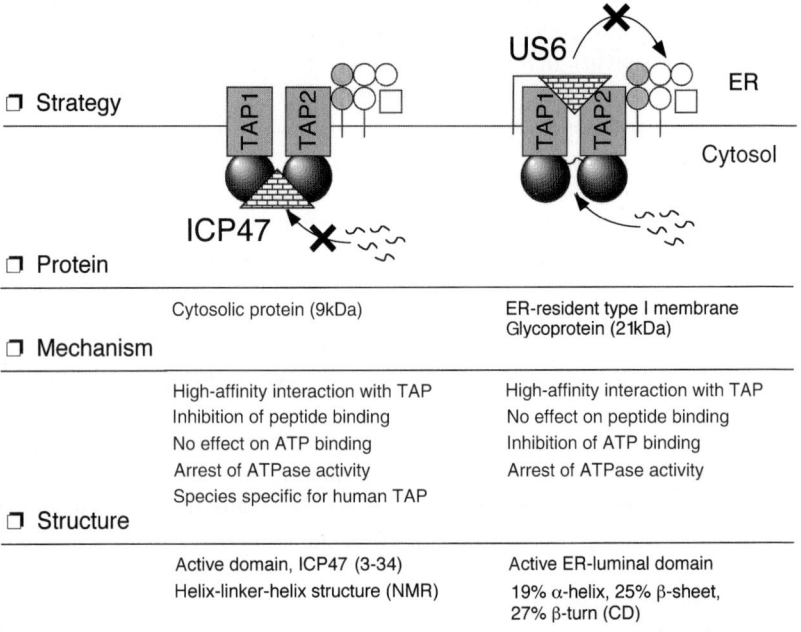

Fig. 4. Comparison of ICP47 and US6. Details and references are given in the text

Acknowledgements. We thank all lab members for helpful discussions. The work was supported by the Deutsche Forschungsgemeinschaft (DFG) and the Fonds der Chemischen Industrie (FCI).

References

Abele R, Tampé R (1999) Function of the transport complex TAP in cellular immune recognition. Biochim Biophys Acta 1461:405–419

Ahn K, Gruhler A, Galocha B, Jones TR, Wiertz EJHJ, Ploegh HL, Peterson PA, Yang Y, Früh K (1997) The ER-luminal domain of the HCMV glycoprotein US6 inhibits peptide translocation by TAP. Immunity 6:613–621

Ahn K, Meyer TH, Uebel S, Sempé P, Djaballah H, Yang Y, Peterson PA, Früh K, Tampé R (1996a) Molecular mechanism and species-specificity of TAP inhibition by herpes-simplex virus protein ICP47. EMBO J 15:3247–3255

Ahn KS, Angulo A, Ghazal P, Peterson PA, Yang Y, Früh K (1996b) Human cytomegalovirus inhibits antigen presentation by a sequential multistep process. Proc Natl Acad Sci USA 93:10990–10995

Ambagala AP, Hinkley S, Srikumaran S (2000) An early pseudorabies virus protein down-regulates porcine MHC class I expression by inhibition of transporter associated with antigen processing (TAP). J Immunol 164:93–99

Ambudkar SV, Dey S, Hrycyna CA, Ramachandra M, Pastan I, Gottesman MM (1999) Biochemical, cellular, and pharmacological aspects of the multidrug transporter. Annu Rev Pharmacol Toxicol 39:361–398

Androlewicz MJ, Anderson KS, Cresswell P (1993) Evidence that transporter associated with antigen processing translocate a major histocompatibility complex class I-binding peptide into the endoplasmic reticulum in an ATP-dependent manner. Proc Natl Acad Sci USA 90:9130–9134

Androlewicz MJ, Cresswell P (1994) Human transporters associated with antigen processing possess a promiscuous peptide binding site. Immunity 1:7–14

Barnes PD, Grundy JE (1992) Down-regulation of the class I HLA heterodimer and beta 2-microglobulin on the surface of cells infected with cytomegalovirus. J Gen Virol 73:2395–2403

Baumeister W, Walz J, Zühl F, Seemüller E (1998) The proteasome – paradigm of a self-compartmentalizing protease. Cell 92:367–380

Beersma MF, Bijlmakers MJ, Ploegh HL (1993) Human cytomegalovirus down-regulates HLA class I expression by reducing the stability of class I H chains. J Immunol 151:4455–4464

Beinert D, Neumann L, Uebel S, Tampé R (1997) Structure of the viral TAP-inhibitor ICP47 induced by membrane association. Biochemistry 36:4694–4700

Cerundolo V, Alexander J, Anderson K, Lamb C, Cresswell P, McMichael A, Gotch F, Townsend A (1990) Presentation of viral antigen controlled by a gene in the major histocompatibility complex. Nature 345:449–452

Chee MS, Bankier AT, Beck S, Bohni R, Brown CM, Cerny R, Horsnell T, Hutchison CAD, Kouzarides T, Martignetti JA, et al. (1990) Analysis of the protein-coding content of the sequence of human cytomegalovirus strain AD169. Curr Top Microbiol Immunol 154:125–169

Cresswell P, Bangia N, Dick T, Diedrich G (1999) The nature of the MHC class I peptide loading complex. Immunol Rev 172:21–28

Doige CA, Ames GF (1993) ATP-dependent transport systems in bacteria and humans: relevance to cystic fibrosis and multidrug resistance. Annu Rev Microbiol 47:291–319

Früh K, Ahn K, Djaballah H, Sempe P, van Endert PM, Tampé R, Peterson PA, Yang Y (1995) A viral inhibitor of peptide transporters for antigen presentation. Nature 375:415–418

Galocha B, Hill A, Barnett BC, Dolan A, Raimondi A, Cook RF, Brunner J, McGeoch DJ, Ploegh HL (1997) The active site of ICP47, a herpes simplex virus-encoded inhibitor of the major histocompatibility complex (MHC)-encoded peptide transporter associated with antigen processing (TAP), maps to the NH_2-terminal 35 residues. J Exp Med 185:1565–1572

Gewurz BE, Gaudet R, Tortorella D, Wang EW, Ploegh HL, Wiley DC (2001) Antigen presentation subverted: Structure of the human cytomegalovirus protein US2 bound to the class I molecule HLA-A2. Proc Natl Acad Sci USA 98:6794–6799

Gorbulev S, Abele R, Tampe R (2001) Allosteric crosstalk between peptide-binding, transport, and ATP hydrolysis of the ABC transporter TAP. Proc Natl Acad Sci USA 98:3732–3737

Grandea AG, 3rd, Van Kaer L (2001) Tapasin: an ER chaperone that controls MHC class I assembly with peptide. Trends Immunol 22:194–199

Hengel H, Esslinger C, Pool J, Goulmy E, Koszinowski UH (1995) Cytokines restore MHC class I complex formation and control antigen presentation in human cytomegalovirus-infected cells. J Gen Virol 76:2987–2997

Hengel H, Flohr T, Hämmerling GJ, Koszinowski UH, Momburg F (1996) Human cytomegalovirus inhibits peptide translocation into the endoplasmic reticulum for MHC class I assembly. J Gen Virol 77:2287–2296

Hengel H, Koopmann JO, Flohr T, Muranyi W, Goulmy E, Hämmerling GJ, Koszinowski UH, Momburg F (1997) A viral ER-resident glycoprotein inactivates the MHC-encoded peptide transporter. Immunity 6:623–632

Hewitt EW, Gupta SS, Lehner PJ (2001) The human cytomegalovirus gene product US6 inhibits ATP binding by TAP. EMBO J 20:387–396

Higgins CF (1992) ABC transporters: from microorganisms to man. Annu Rev Cell Biol 8:67–113

Hill A, Jugovic P, York I, Russ G, Bennink J, Yewdell J, Ploegh H, Johnson D (1995) Herpes simplex virus turns off the TAP to evade host immunity. Nature 375:411–415

Hill AB, Barnett BC, McMichael AJ, McGeoch DJ (1994) HLA class I molecules are not transported to the cell surface in cells infected with herpes simplex virus types 1 and 2. J Immunol 152:2736–2741

Hinkley S, Hill AB, Srikumaran S (1998) Bovine herpesvirus-1 infection affects the peptide transport activity in bovine cells. Virus Res 53:91–96

Honess RW, Roizman B (1974) Regulation of herpesvirus macromolecular synthesis. I. Cascade regulation of the synthesis of three groups of viral proteins. J Virol 14:8–19

Jugovic P, Hill AM, Tomazin R, Ploegh H, Johnson DC (1998) Inhibition of major histocompatibility complex class I antigen presentation in pig and primate cells by herpes simplex virus type 1 and 2 ICP47. J Virol 72:5076–5084

Koelle DM, Tigges MA, Burke RL, Symington FW, Riddell SR, Abbo H, Corey L (1993) Herpes simplex virus infection of human fibroblasts and keratinocytes inhibits recognition by cloned CD8+ cytotoxic T lymphocytes. J Clin Invest 91:961–968

Koopmann JO, Post M, Neefjes JJ, Hämmerling GJ, Momburg F (1996) Translocation of long peptides by transporters associated with antigen processing (TAP). Eur J Immunol 26:1720–1728

Kyritsis C, Gorbulev S, Hutschenreiter S, Pawlitschko K, Abele R, Tampé R (2001) Molecular mechanism and structural aspects of TAP inhibition by the cytomegalovirus protein US6. Submitted

Lacaille VG, Androlewicz MJ (1998) Herpes simplex virus inhibitor ICP47 destabilizes the transporter associated with antigen processing (TAP) heterodimer. J Biol Chem 273:17386–17390

Lankat-Buttgereit B, Tampé R (1999) The transporter associated with antigen processing TAP: structure and function. FEBS Lett 464:108–112

Lehner PJ, Karttunen JT, Wilkinson GW, Cresswell P (1997) The human cytomegalovirus US6 glycoprotein inhibits transporter associated with antigen processing-dependent peptide translocation. Proc Natl Acad Sci USA 94:6904–6909

Ljunggren HG, Stam NJ, Ohlen C, Neefjes JJ, Hoglund P, Heemels MT, Bastin J, Schumacher TN, Townsend A, Karre K, et al. (1990) Empty MHC class I molecules come out in the cold. Nature 346:476–480

Madden DR (1995) The three-dimensional structure of peptide-MHC complexes. Annu Rev Immunol 13:587–622

McGeoch DJ, Dolan A, Donald S, Rixon FJ (1985) Sequence determination and genetic content of the short unique region in the genome of herpes simplex virus type 1. J Mol Biol 181:1–13

McGeoch DJ, Moss HW, McNab D, Frame MC (1987) DNA sequence and genetic content of the HindIII l region in the short unique component of the herpes simplex virus type 2 genome: identification of the gene encoding glycoprotein G, and evolutionary comparisons. J Gen Virol 68:19–38

Momburg F, Roelse J, Howard JC, Butcher GW, Hämmerling GJ, Neefjes JJ (1994) Selectivity of MHC-encoded peptide transporters from human, mouse and rat. Nature 367:648–651

Müller KM, Ebensperger C, Tampé R (1994) Nucleotide binding to the hydrophilic C-terminal domain of the transporter associated with antigen processing (TAP). J Biol Chem 269:14032–14037

Neefjes JJ, Momburg F, Hämmerling GJ (1993) Selective and ATP-dependent translocation of peptides by the MHC-encoded transporter. Science 261:769–771

Neumann L, Kraas W, Uebel S, Jung G, Tampé R (1997) The active domain of the herpes simplex virus protein ICP47: a potent inhibitor of the transporter associated with antigen processing. J Mol Biol 272:484–492

Neumann L, Tampé R (1999) Kinetic analysis of peptide binding to the TAP transport complex: evidence for structural rearrangements induced by substrate binding. J Mol Biol 294:1203–1213

Nijenhuis M, Hämmerling GJ (1996) Multiple regions of the transporter associated with antigen processing (TAP) contribute to its peptide binding site. J Immunol 157:5467–5477

Pamer E, Cresswell P (1998) Mechanisms of MHC class I-restricted antigen processing. Annu Rev Immunol 16:323–358

Pfänder R, Neumann L, Zweckstetter M, Seger C, Holak TA, Tampé R (1999) Structure of the active domain of the herpes simplex virus protein ICP47 in water/sodium dodecyl sulfate solution determined by nuclear magnetic resonance spectroscopy. Biochemistry 38:13692–13698

Ploegh HL (1998) Viral strategies of immune evasion. Science 280:248–253

Posavad CM, Rosenthal KL (1992) Herpes simplex virus-infected human fibroblasts are resistant to and inhibit cytotoxic T-lymphocyte activity. J Virol 66:6264–6272

Powis SJ, Townsend AR, Deverson EV, Bastin J, Butcher GW, Howard JC (1991) Restoration of antigen presentation to the mutant cell line RMA-S by an MHC-linked transporter. Nature 354:528–531

Schumacher TN, Kantesaria DV, Serreze DV, Roopenian DC, Ploegh HL (1994) Transporters from H-2b, H-2d, H-2s, H-2k, and H-2g7 (NOD/Lt) haplotype translocate similar sets of peptides. Proc Natl Acad Sci USA 91:13004–13008

Shepherd JC, Schumacher TN, Ashton-Rickardt PG, Imaeda S, Ploegh HL, Janeway CA, Jr., Tonegawa S (1993) TAP1-dependent peptide translocation in vitro is ATP dependent and peptide selective [published erratum appears in Cell 1993 Nov 19;75(4):613]. Cell 74:577–584

Spies T, DeMars R (1991) Restored expression of major histocompatibility class I molecules by gene transfer of a putative peptide transporter. Nature 351:323–324

Tampé R, Urlinger S, Pawlitschko K, Uebel S. (1997). The transporters associated with antigen processing (TAP). In: Holland B (ed) Unusual secretory pathways: from bacteria to man. Springer, New York, pp 115–136

Tomazin R, Hill AB, Jugovic P, York I, van Endert P, Ploegh HL, Andrews DW, Johnson DC (1996) Stable binding of the herpes simplex virus ICP47 protein to the peptide binding site of TAP. EMBO J 15:3256–3266

Townsend A, Ohlen C, Foster L, Bastin J, Ljunggren HG, Karre K (1989) A mutant cell in which association of class I heavy and light chains is induced by viral peptides. Cold Spring Harbor Symp Quant Biol 54:299–308

Uebel S, Kraas W, Kienle S, Wiesmüller K-H, Jung G, Tampé R (1997) Recognition principle of the TAP-transporter disclosed by combinatorial peptide libraries. Proc Natl Acad Sci USA 94:8976–8981

Uebel S, Meyer TH, Kraas W, Kienle S, Jung G, Wiesmüller KH, Tampé R (1995) Requirements for peptide binding to the human transporter associated with antigen processing revealed by peptide scans and complex peptide libraries. J Biol Chem 270:18512–18516

Uebel S, Tampé R (1999) Specificity of the proteasome and the TAP transporter. Curr Opin Immunol 11:203–208

Urlinger S, Kuchler K, Meyer TH, Uebel S, Tampé R (1997) Intracellular location, complex-formation, and function of the transporter associated with antigen processing in yeast. Eur J Biochem 245:266–272

van Endert PM, Tampé R, Meyer TH, Tisch R, Bach JF, McDevitt HO (1994) A sequential model for peptide binding and transport by the transporters associated with antigen processing. Immunity 1:491–500

Wang K, Früh K, Peterson PA, Yang Y (1994) Nucleotide binding of the C-terminal domains of the major histocompatibility complex-encoded transporter expressed in *Drosophila melanogaster* cells. FEBS Lett 350:337–341

Warren AP, Ducroq DH, Lehner PJ, Borysiewicz LK (1994) Human cytomegalovirus-infected cells have unstable assembly of major histocompatibility complex class I complexes and are resistant to lysis by cytotoxic T lymphocytes. J Virol 68:2822–2829

Yamashita Y, Shimokata K, Mizuno S, Yamaguchi H, Nishiyama Y (1993) Down-regulation of the surface expression of class I MHC antigens by human cytomegalovirus. Virology 193:727–736

Yewdell J, Anton LC, Bacik I, Schubert U, Snyder HL, Bennink JR (1999) Generating MHC class I ligands from viral gene products. Immunol Rev 172:97–108

York IA, Roop C, Andrews DW, Riddell SR, Graham FL, Johnson DC (1994) A cytosolic herpes simplex virus protein inhibits antigen presentation to CD8+ T lymphocytes. Cell 77:525–535

Inhibition of the MHC Class II Antigen Presentation Pathway by Human Cytomegalovirus

D.C. Johnson and N.R. Hegde

Human cytomegalovirus (HCMV) causes serious disease in immunocompromised individuals. Normally, anti-HCMV immune response controls virus replication following reactivation from latency. However, HCMV, like other large herpesviruses, encodes immune evasion proteins that allow the virus to replicate, for a time or in specific tissues, and produce viral progeny in the face of robust host immunity. HCMV glycoproteins US2, US3, US6 and US11 all inhibit different stages of the MHC class I antigen presentation pathway and can reduce recognition by $CD8^+$ T lymphocytes. Here, we discuss two novel inhibitors of the MHC class II antigen presentation pathway, HCMV glycoproteins US2 and US3. Both US2 and US3 can inhibit presentation of exogenous protein antigens to $CD4^+$ T lymphocytes in in vitro assays. US2 causes degradation of MHC class II molecules: HLA-DR-α and HLA-DM-α, as well as class I heavy chain (HC), but does not affect DR-β or DM-β chains. Mutant forms of US2 have been constructed that can bind to DR-α and class I HC but do not cause their degradation, separating the binding step from other processes that precede degradation. We also found evidence that US2-induced degradation of class I and II proteins involves a cellular component, other than Sec61, that is limiting in quantity. Unlike US2, US3 binds newly synthesized class II α/β complexes, reducing the association with the invariant chain (Ii) and causing mislocalization of class II complexes in cells. US3 expression reduces accumulation of class II complexes in peptide-loading compartments and loading of peptides. Since US2 and US3 are expressed solely within HCMV-infected cells, it appears that these viral proteins have evolved to inhibit presentation of endogenous, intracellular viral antigens to anti-HCMV $CD4^+$ T cells. This is different from how the MHC class II pathway is normally viewed, as a pathway for presentation of exogenous, extracellular proteins. The existence of these proteins indicates the importance of class II-mediated presentation of endogenous antigens in signalling virus infection to $CD4^+$ T cells.

1 Introduction	102
2 The MHC Class II Antigen Presentation Pathway	103
3 Presentation of Endogenous Antigens by MHC Class II Proteins	104
4 Inhibition of the MHC Class II Antigen Presentation Pathway by HCMV	106
5 Further Characterization of HCMV US2	109
6 Effects of Other HCMV US-Encoded Glycoproteins on the MHC Class I and II Antigen Presentation Pathways	111
7 Concluding Remarks	112
References	113

Department of Molecular Microbiology and Immunology, Oregon Health & Science University, Portland, OR 97201, USA

1 Introduction

The genomes of herpesviruses range from large (~80 genes) to enormous (>250 genes). Only a fraction, in some cases only a small fraction, of these genes are required for replication in cultured cells. A substantial number of herpesvirus genes appear to be devoted to defending viruses from the myriad of host defenses and allowing viruses to interact with specific cell types so that there is a lifelong persistent or latent infection. This is especially true for the largest herpesviruses, e.g., human cytomegalovirus (HCMV), which replicates relatively slowly and in a variety of cell types and tissues. Without effective immune evasion or subterfuge, at local sites of infection and after reactivation in an immunocompetent host, one might expect effective control of these herpesviruses and reduced spread to other hosts. Moreover, in different tissues, the ability to block various facets of the immune response may have distinct consequences.

T lymphocytes play important roles in regulating and limiting the spread of herpesvirus infections (BORYCIEWICZ and SISSONS 1994; RIDDELL and GREENBERG 1995; SCHMID and ROUSE 1992). All of the herpesviruses that have been studied for this can block the major histocompatibility complex (MHC) class I antigen presentation pathway reducing recognition by $CD8^+$ T cells (ALCAMI and KOSZINOWSKI 2000; JOHNSON and MCFADDEN 2001; TORTORELLA et al. 2000). It is unlikely that this inhibition alters priming of $CD8^+$ T cells as antiviral T cells are observed in the host. This relates to observations that the effects of these viral proteins are not universally observed in all cells. Frequently, "professional" antigen presenting cells (APCs) are not infected or, in other cell types, components of the class I pathway are upregulated by inflammatory cytokines (FRUH et al. 1999; GOLDSMITH et al. 1998; HENGEL et al. 2000). Additionally, even if professional APCs are infected, these cells can express relatively high levels of MHC class I and the viral proteins may be unable to block the class I presentation pathway. Instead, these viral class I inhibitors probably influence virus replication at specific times (e.g., shortly after reactivation) or in specific cells or tissue types where inhibition of the class I pathway can be attained (reviewed in JOHNSON and MCFADDEN 2001). For example, with herpes simplex virus, there is evidence for a delay in the appearance of $CD8^+$ T cells in human lesions (CUNNINGHAM et al. 1985).

Shortly after we described the first of the herpesvirus inhibitors of the MHC class I pathway, herpes simplex virus cytoplasmic protein ICP47 (YORK et al. 1994), we wondered whether other herpesviruses might inhibit the MHC class II antigen presentation pathway. HCMV infects a number of cell types that can express class II proteins, including monocyte-macrophages, endothelial cells, and epithelial cells. There is extensive information consistent with the notion that HCMV glycoproteins and structural proteins are extensively introduced into the endocytic network of cells during virus egress (see below). Thus one would expect that considerable amounts of many HCMV proteins reach the MHC class II peptide loading compartments, also known as MIIC (NEEFJES 1999). Introduction of relatively large amounts of viral structural proteins into the endosomal network should lead to efficient presentation by the MHC class II pathway. This would signal the presence of viral antigens to $CD4^+$ T cells that can act cytolytically or to produce antiviral cytokines. Therefore, a viral protein capable of disrupting the class II pathway could influence the overall immune response to HCMV, by allowing certain virus-infected cells to escape detection by $CD4^+$ T cells.

2 The MHC Class II Antigen Presentation Pathway

The MHC class II proteins human leukocyte antigens (HLA)-DP, -DQ, and -DR are heterodimeric cell surface glycoproteins that bind antigenic peptides within the endocytic system and present them to $CD4^+$ T cells (reviewed in CRESSWELL 1994; CRESSWELL 1996; DENZIN and CRESSWELL 1995; PIETERS 2000). Class II-α and -β subunits associate in the endoplasmic reticulum (ER) with a third membrane glycoprotein, the invariant chain (Ii), that protects the peptide-binding groove of α/β (see Fig. 1). Three α/β/Ii complexes trimerize to form a nonameric complex that transits to the Golgi apparatus and is then sorted to endosomes by signals in the cytoplasmic domain of the Ii chain. Class II complexes move from endosomes into a lysosome-like compartment, denoted MIIC, where peptide loading occurs (NEEFJES 1999). The α/β complexes that reach the MIIC contain fragments of the Ii chain in

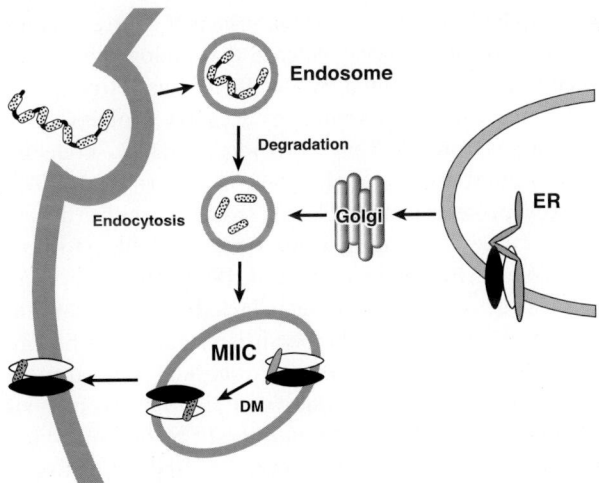

Fig. 1. The MHC class II antigen presentation pathway. MHC class II proteins (DR-α and DR-β) are inserted into the ER membrane and assemble with the invariant chain (Ii). Ii blocks the peptide binding groove of DR-α/β. These complexes are transported to the Golgi apparatus and then directed to TGN/endosomes by Ii. There is proteolysis of Ii, and DR α/β/CLIP complexes arrive in the MIIC. Antigenic proteins are taken up by endocytosis or phagocytosis into endosomes and degraded into peptides, which are also delivered to the MIIC. Peptides are loaded onto DR-α/β complexes by the action of DM, which binds to class II complexes and exchanges peptides for CLIP. Peptide-loaded class II complexes are transported to the cell surface for recognition by T cell receptors of $CD4^+$ T lymphocytes

the peptide-binding groove, among these one known as class II-associated invariant chain peptide (CLIP). The removal of CLIP and subsequent loading of antigenic peptides onto class II α/β complexes is facilitated by a nonpolymorphic class II-like molecule known as HLA-DM. DM is a heterodimeric protein composed of DM-α and DM-β, and it remains localized to the MIIC, acting as an enzyme to catalyze exchange of antigenic peptides onto class II. Without DM, peptide loading is highly inefficient and class II appears on the cell surface with CLIP.

The antigenic peptides bound to MHC class II α/β dimers in the MIIC are normally derived from proteins taken up by endocytosis or phagocytosis and delivered into the endosomal system. This is especially true for so-called "professional" APCs, such as B cells, monocyte-macrophages, and dendritic cells. The exogenous proteins are degraded into peptides by cathepsins, proteolytic enzymes that are constituents of endosomes, the *trans*-Golgi network (TGN), and lysosomes (VILLADANGOS et al. 1999). Antigenic peptides of 13–25 amino acids are exchanged onto MHC class II complexes in the MIIC, and these complexes move to the cell surface where presentation to $CD4^+$ T cells occurs.

3 Presentation of Endogenous Antigens by MHC Class II Proteins

Although exogenous or extracellular proteins can be the principal source of antigenic peptides in some cell types, there is ample evidence that endogenous viral, microbial, and cellular proteins can be presented by the MHC class II pathway. Endogenous, cytoplasmic proteins can be presented by class II proteins, in a transporter associated with antigen presentation (TAP)-independent, brefeldin A (BFA)-insensitive manner (JARAQUEMADA et al. 1990; MALNATI et al. 1992; NUCHTERN et al. 1990). Cytosolic proteins can be released from cells, either during cell lysis or by other mechanisms, and taken up by endocytosis or phagocytosis or as part of dead cells by professional APC (NUCHTERN et al. 1990). However, there is extensive in vitro evidence that cytosolic viral proteins including measles, influenza, and hepatitis C virus capsid proteins and herpesvirus proteins can be efficiently presented to $CD4^+$ T cells without entering an extracellular compartment (CHEN et al. 1998; JACOBSON et al. 1989; JARAQUEMADA et al. 1990; NUCHTERN et al. 1990; LOSS et al. 1993; MUNZ et al. 2000). Presentation of cytoplasmic viral proteins involves cytoplasmic proteases and is TAP-independent and insensitive to BFA (GUEGUEN and LONG 1996; MALNATI et al. 1992). One possibility involves transport from the cytoplasm into lysosomes/MIIC by unknown mechanisms, perhaps involving cytosolic heat shock proteins (CHAING et al. 1989). There is also evidence for efficient presentation of endogenous membrane proteins of both cellular and viral origin (reviewed in LICH et al. 2000). In this case, membrane proteins produced in the ER are transported to endosomes and the MIIC, either directly from the Golgi apparatus or after endocytosis from the cell surface.

Extraction of peptides from class II proteins has revealed that extracellular and membrane-bound antigens: proteins found in the extracellular compartment, as part of the plasma membrane or endosomal compartments, make up the majority of antigens presented by class II-expressing "professional" APC (CHICZ et al. 1993; RUDENSKY et al. 1991). For cells grown in culture, serum and host membrane proteins that intersect the endocytic pathway, e.g., MHC class I and II proteins, were the major source of class II-bound peptides. However, approximately 5%–10% of peptides bound to HLA-DR2, DR3, DR4, DR7, and DR8 are derived from cytoplasmic proteins, self-antigens such as fragments of ribosomal proteins, heat shock proteins, guanosine triphosphate-binding proteins, cytochromes, C-myc, and K-ras (CHICZ et al. 1993). Although the predominant peptides associated with class II are derived from host cell proteins, more relevant are peptides derived from viral and microbial polypeptides, antigens that are recognized by antiviral or -microbial $CD4^+$ T lymphocytes. Cytoplasmic viral proteins can be presented, although there is evidence that these proteins must be expressed at high levels in cells and be relatively long-lived (GUEGUEN and Long 1996; JARAQUEMADA et al. 1990). However, given the pattern of peptides extracted from class II molecules, presentation of viral membrane proteins is likely to be much more efficient than for cytoplasmic viral proteins.

Numerous studies, mostly involving model systems for vaccine development, have demonstrated that proteins targeted to the endocytic system are presented by MHC class II more efficiently and at lower antigen concentrations than proteins found in other membranes or in the cytoplasm. This makes sense given that processing and loading of class II antigens occurs in the endocytic system and MIIC. A human papilloma virus E7 protein fused to the cytoplasmic and transmembrane domain of lysosome-associated membrane protein-1 (LAMP-1) was presented more efficiently than E7 alone (WU et al. 1995). A vaccinia virus expressing the E7/LAMP-1 fusion protein produced markedly enhanced E7 $CD4^+$ T cell responses, enhanced antibodies, and cytotoxic T lymphocytes (CTLs) in mice. A secreted form of hen egg lysozyme (HEL), was presented more efficiently than an ER-retained form of HEL, $_{KDEL}$HEL (BONIFAZ et al. 1999). A HEL epitope, residues 52–61, fused to Ii efficiently stimulated T cell hybridomas, and this epitope outcompeted other epitopes when cells were incubated with exogenous HEL (SPONAAS et al. 1999). In some cell types, endogenous membrane proteins may be presented more efficiently than or by more distinct pathways than exogenous proteins. For example, when MHC class II proteins were induced in epithelial cells, the presentation of endogenous antigens, e.g., membrane-bound ovalbumin (OVA) was more efficient than that of exogenous OVA (HERSHBERG et al. 1997; MAILE et al. 2000).

Members of the herpesvirus family acquire their envelope in the final stages of assembly by wrapping of Golgi, TGN, or endosomal membranes around nucleocapsids (BRACK et al. 2000; MCMILLAN and JOHNSON 2001; SANCHEZ et al. 2000a,b). Thus, as virus particles bud into the TGN or endosomes, viral membrane glycoproteins, tegument, and capsid proteins are incorporated into the endosomal system. With HCMV the exact site of envelopment in the cytoplasm is less well defined than for the α-herpesviruses, but it is likely that envelopment occurs into the Golgi or TGN (FISH et al. 1998; SANCHEZ et al. 2000a,b; TOOZE et al. 1993). There is extensive intracellular traffic between the TGN/endosomes and lysosomes/ MIIC, and thus one might expect that HCMV proteins would be degraded to peptides and loaded onto MHC class II molecules. $CD4^+$ T cell recognition of HCMV peptides in MHC class II-expressing monocyte-macrophages or endothelial or epithelial cells would negatively impact virus replication, especially after reactivation in an immune host. In light of these observations, and widespread inhibition of MHC class I, we hypothesized that HCMV might express inhibitors of the MHC class II antigen presentation pathway.

4 Inhibition of the MHC Class II Antigen Presentation Pathway by HCMV

Initially, we performed experiments involving tetanus toxin-specific $CD4^+$ T cell clones that suggested inhibition of the MHC class II pathway by HCMV. These studies were based on unpublished experiments by Beth Walters and Stanley Riddell (Fred Hutchinson Cancer Research Center) suggesting that HCMV could reduce MHC class II antigen presentation. Our experiments, done in collaboration with Stanley Riddell and Jay Nelson (Oregon Health & Science University), involved infection of monocyte-macrophages prepared from human blood and restimulated either by concanavalin A or by alloreactivation as described previously (SODERBERG-NAUCLER et al. 1998). The overall picture from these experiments was that HCMV infection could reduce presentation of an exogenously added antigen (tetanus toxin) by monocyte-macrophages (results not shown). This effect was observed early after infection and before there was inhibition of MHC class II gene expression. However, the experiments produced variable results, the number of cells that were infected varied substantially, and the level of inhibition ranged from 0% to 70%, although in the majority of experiments there was some inhibition.

Efforts to ascertain whether HCMV inhibited the MHC class II pathway were limited by the fact that HCMV productively infects only a very limited number of cell types in culture. One cell that can be infected, at least to the extent that most cells express at least early genes, is U373 microglial cells. U373 cells express relatively low levels of MHC class II proteins. However, we could induce a higher level expression by treating cells with interferon-γ or by transfection with the CIITA transactivator gene, which normally acts to upregulate class II genes (TOMAZIN et al. 1999). When U373-expressing class II proteins were infected with HCMV, there was loss of expression of the class II complex when radiolabeled and immunoprecipitated with an anti-class II-α antibody (TOMAZIN et al. 1999). This experiment requires some clarification. To produce significant downregulation of class II proteins, it was necessary to add virus three times sequentially, each time centrifuging the virus onto the cell monolayer. We interpret these experiments to indicate that HCMV does not cause substantial expression of viral early genes in

U373 cells compared with fibroblasts unless relatively large amounts of virus are added to the cells. Indeed, there was substantially less HCMV gene expression in virus-infected U373 cells compared with virus-infected fibroblasts (R. Tomazin, unpublished data). However, it is important to note that these observations are not inconsistent with effects of HCMV on the MHC class II pathway in vivo. More accurately, the inefficient inhibition of the class II pathway we observe in these HCMV-infected cells is indicative of the poor replication of HCMV in these cultured cells. We would argue that in vivo, in the appropriate cells, effects on the class II pathway are more robust and likely vital.

Using HCMV mutant viruses produced by JONES et al. (1995), we mapped genes responsible for this reduction of class II to a region including the US2 and US3 genes (TOMAZIN et al. 1999). The effects on class II proteins were not related to reduced expression of class II genes through reduced transcription, as reported by Miller et al. (1998 and 1999). Inhibition of class II gene expression through effects on the

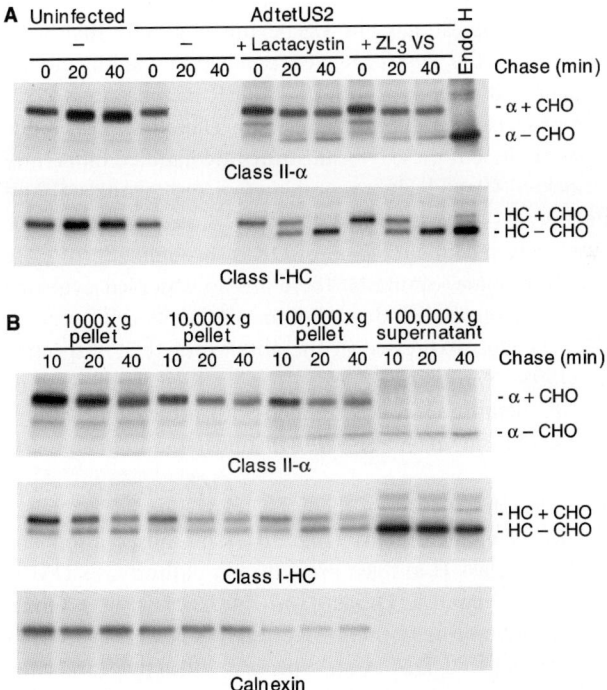

Fig. 2A,B. US2 induces degradation of class II-α chains by proteasomes without a cytoplasmic intermediate. **A** His16 cells were left uninfected or infected with Adtet-trans and AdtetUS2 using 20 and 100 PFU/cell, respectively. The cells were treated with proteasome inhibitors lactacystin or ZL$_3$VS and then radiolabeled for 1 min and the label was chased for 20 or 40 min. Class II-α or class I HC were immunoprecipitated after denaturing the samples. **B** His16 cells were infected, treated with ZL$_3$VS, and radiolabeled as in **A**. The cells were then subjected to subcellular fractionation, so that 1,000×g, 10,000×g, and 100,000×g pellets and 100,000×g supernatants were prepared. Membrane pellets or supernatant fractions were denatured and then MHC class II-α, MHC class I HC or calnexin immunoprecipitated. (Reprinted from TOMAZIN et al. 1999, with permission from *Nature Medicine*)

Jak/Stat pathway or effects on CIITA occurred later (24–48h after infection) than the effects on class II proteins observed here (6–12h after infection). Cell lines transfected with the US2 or US3 genes were produced, and US2 was found to reduce class II protein expression. Loss of expression of class II proteins was through rapid degradation of the α chain, and US2 binds directly to the α chain either in the absence or presence of β (Tomazin et al. 1999; N.R. Hegde and D.C. Johnson, unpublished data). The β chain was not degraded or directly bound by US2, although β turns over more rapidly in the absence of α (Tomazin et al. 1999). Degradation of α chains was through the action of cytosolic proteasomes, as α chains were stabilized in cells treated with inhibitors of the proteasome, as was US2 (Fig. 2A).

Previously, Wiertz et al. (1996a,b) demonstrated that HCMV US2 and US11 caused proteasome-mediated degradation of MHC class I heavy chains (HC). In US2-expressing cells, they demonstrated a transient interaction between class I HC and the Sec61 that forms a proteinaceous tunnel through which membrane proteins are translocated into the ER (Wiertz et al. 1996b). HC was retrotranslocated through Sec61 into the cytoplasm coincident with deglycosylation of the HC and proteolysis by the proteasome. Interestingly, in US2-expressing cells that were treated with proteasome inhibitors, the class I HC accumulated in a soluble form. By contrast, DR-α chains remained stably anchored in the ER membrane in US2-expressing cells treated with proteasome inhibitors (Fig. 2B). Therefore, US2 binds to both DR-α chains and class I HC but these two substrates are handled differently when the proteasome is blocked. Class I HC accumulates in the cytoplasm, and DR-α remains membrane bound. This suggested that the proteasome must be docked onto the ER membrane to mediate removal of DR-α through a Sec61 pore and into the lumen of the proteasome. Similarly, there are no observed cytosolic intermediates during proteasome-mediated degradation of the T cell receptor (TCR)-α chain and the DR-β chains expressed in the absence of their respective partners (Dusseljee et al. 1998; Yang et al. 1998). The TCR-α chain remains stably associated with the ER membrane in cells treated with proteasome inhibitors. These differences in mechanism may be useful in understanding how ER degradation occurs.

In addition to catalyzing the destruction of the MHC class I and II proteins, US2 caused proteolysis of the α chain of DM (Fig. 3B). This observation has important implications for MHC class II antigen presentation pathway, as DM is essential for loading of class II (reviewed in Denzin and Cresswell 1995). Given that DM acts much as an enzyme, and is normally expressed at lower levels than class II proteins, one might expect that DM is an especially attractive target for viruses attempting to thwart recognition by CD4[+] T cells. It was quite striking that US2 could cause degradation of DR-α, DM-α, and class I HC, proteins that share only limited (25%–30%) sequence identity (Fig. 3A). However, there is significant similarity in the overall folding and domain structure of these molecules. Because US2 binds to the α and not the β chains of class II-DR, as well as to MHC class I HC, we speculated that US2 might bind to the membrane distal α-1 domains of DR-α, DM-α, and class I HC (Tomazin et al. 1999). However, this suggestion requires confirmation by genetic and structural studies.

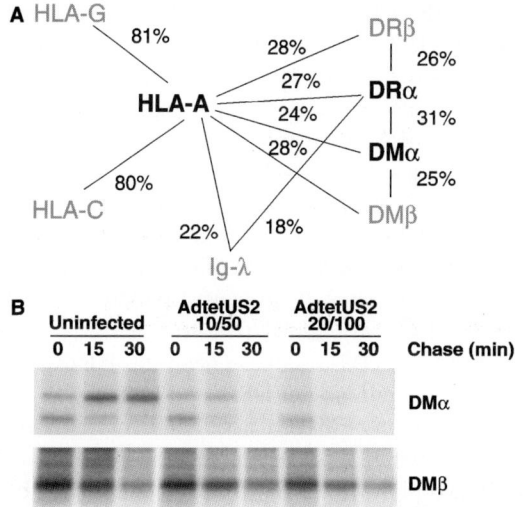

Fig. 3. A MHC Class I and II homology. Tree showing the amino acid homology (%) between classic MHC class I proteins HLA-A and HLA-C, nonclassic class I HLA-G, and MHC class II proteins DR-α, DR-β, DM-α, and DM-β. **B** Degradation of DM. His16 cells were left uninfected or infected with Adtet-trans and AdtetUS2, and the cells were radiolabeled as described in Fig. 2. Cell extracts were denatured and class II DM-α and DM-β immunoprecipitated with MAb 5C1 and MaP.DMB-C, respectively. (Reprinted from TOMAZIN et al. 1999, with permission from *Nature Medicine*)

It was reported that US2 does not cause degradation of HLA-G and HLA-C molecules, class I molecules normally expressed in trophoblast cells (SCHUST et al. 1998). This was a surprising result given that HLA-C and -G are much more closely related to HLA-A than DR-α and DM-α (Fig. 3A). We reevaluated the susceptibility of HLA-G and HLA-C in trophoblast cells, JEG-3 cells, that were infected with varying amounts of an adenovirus vector expressing US2. At higher concentrations of US2, levels that caused over 95% degradation of DR-α or class I HC, there was more than 50% degradation of HLA-C and no effect on HLA-G (N.R. Hegde, D.C. Johnson, unpublished data). Thus HLA-C can be degraded in US2-expressing cells, but less efficiently than HLA-A, whereas HLA-G is resistant to the effects of US2. It appears that US2 has evolved to destroy important components of the class I and II antigen presentation pathways, proteins that present viral antigens in important settings, but there was no pressure to destroy HLA-G.

5 Further Characterization of HCMV US2

It was of interest to determine whether US2 showed preferences for MHC class I HC, class II DR-α, or DM-α in cells that express all three of these proteins. U373 microglial cells stably transfected with the CIITA transactivator gene, denoted HIS16 cells, express both class I and class II proteins and efficiently present exogenous antigens to $CD4^+$ T cells (TOMAZIN et al. 1999). His16 cells express approximately equal amounts of DR and HC. Other transfected clones expressing lower levels of CIITA and class II proteins were isolated at the same time. In addition, HeLa cells were transfected with class II-encoding expression plasmids. US2 was introduced into each of these cells using Ad vectors so that different levels of US2 were delivered. The results indicated that US2 was slightly more active on class I HC than on DR-α (N.R. Hegde, D.C. Johnson, unpublished data). Therefore, it appears that US2 shows little preference for class I or class II proteins. In addition, US2 caused degradation of DR-α (Fig. 4) and DM-α (data not shown) in HeLa cells transfected with plasmids encoding DR-α or DM-α alone. Therefore, DR-β and DM-β are not required for US2-mediated degradation of DR-α and DM-α.

The details of how US2 causes degradation of MHC class I and II proteins are still poorly understood. This appears to involve a pathway by which ER resident proteins are retrotranslocated to the cytoplasm followed by degradation by the proteasome (WIERTZ et al. 1996b; JOHNSON and HAIGH 2000). By coprecipitation of class I and US2 proteins with cellular proteins, WIERTZ et al. (1996b) showed that Sec61 is involved in transport from the lumen of the ER into the cytoplasm, where degradation occurs. However, it is likely that other cellular proteins are also involved. US2 probably triggers some proofreading or quality-control machinery, perhaps involving chaperones, that normally functions to remove misfolded or aberrant proteins from the ER. The identities of other cellular proteins that participate in ER degradation is the subject of active investigation. Related to the hypothesis, we recently found that some cellular protein other than DR-α is limiting for the US2-mediated degradation process. Different concentrations of US2 could be expressed in His16 cells or transfected HeLa cells with various amounts of

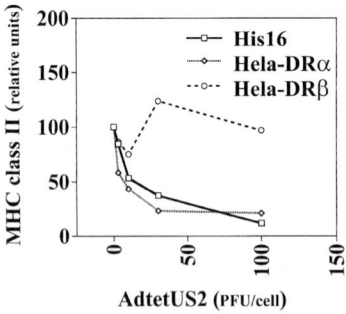

Fig. 4. Dose-dependence of US2-mediated degradation of HLA-DR in cells differing in DR expression. His16 or HeLa cells transiently transfected with plasmids encoding DR-α or DR-β were infected for 18h with AdtetUS2, and then cells were labeled for 3h with ^{35}S-methionine/cysteine and HLA-DR immunoprecipitated. Quantification by phosphoimager analyses indicated that His16 cells expressed 20- to 40-fold more DR than was expressed in transfected HeLa cells, and the levels of US2 expressed at each dose of AdtetUS2 were very similar. US2 at any given concentration was equally able to cause degradation of DR-α in transfected HeLa cells and in His16 cells. DR-α was degraded in the absence of DR-β

adenovirus (Ad) vectors. His16 cells express 20–40 times more DR-α than is expressed in transfected HeLa cells. Nonetheless, the rates of degradation at any given concentration of US2 are similar (Fig. 4). These results are consistent with the hypothesis that proteins other than DR-α play a role in the retrotranslocation process. It appears unlikely that Sec61 is limiting in these cells because there were no obvious differences in cotranslational insertion of class I or II proteins into the ER in the cells whether US2 was present or not.

A soluble form of US2 lacking the cytosolic tail has been shown to bind to class I HC but not cause its degradation (TORTORELLA et al. 1998). Similarly, we found that mutants lacking the cytoplasmic domain or cytoplasmic and transmembrane domains of US2 displayed reduced degradation of class I HC and DR-α, yet bound both proteins (M. Chevalier and D.C. Johnson, manuscript submitted). Moreover, studies involving N- and C-terminal truncated forms of US2 mapped the residues involved in binding both class I HC and DR-α to amino acids 40–130. Mutation of US2 residues 47–66 or 94–111 by substitution of a similar or homologous regions (hr1 and hr2) from US3 (AHN et al. 1996) abolished degradation but not binding of DR-α. Deletion of residues 132–135, a region shared by US3 and US11 (hr3), and including a cysteine residue (AHN et al. 1996), abrogated binding as well as degradation of DR-α. Therefore, the binding domain of US2 to DR-α involves a region located between amino acid residues 40 and 130 of US2, and a number of mutations in this central region of the extracellular domain affect binding and degradation of class I and class II proteins.

6 Effects of Other HCMV US-Encoded Glycoproteins on the MHC Class I and II Antigen Presentation Pathways

The US2–US11 region of HCMV potentially encodes 8–10 membrane glycoproteins: four of these, US2, US3, US6, and US11, are known to inhibit the MHC class I pathway (AHN et al. 1997; JONES et al. 1995), and one, US2, inhibits the class II pathway (TOMAZIN et al. 1999). To determine whether other glycoproteins in this region, either singly or in combination, might affect either the class I or class II pathways, recombinant Ad vectors were constructed expressing US2, US3, US7, US8, US9, US10, and US11. Human fibroblasts were infected with these Ad vectors and used as target cells in assays involving CD8$^+$ CTLs as described previously (YORK et al. 1994). US2, US3, and US11 all reduced CD8$^+$ T cell recognition and lysis, but none of the other glycoproteins had an effect (R. Tomazin, D.C. Johnson, unpublished data). Moreover, combinations of US7, US8, US9, and US10 glycoproteins did not obviously inhibit recognition by these CD8$^+$ T cell clones.

To determine whether other glycoproteins in the US2–US11 region affected the class II pathway, His16 cells were infected with Ad vectors expressing US2, US3, US7, US8, US9, US10, or US11 and tested in CD4$^+$ T cell assays. The CD4$^+$ T cell clones were specific for a tuberculosis antigen, Mtb39, and this protein was added exogenously as described previously (TOMAZIN et al. 1999). US2 effectively inhibited presentation of the Mtb39 antigen as reported. Moreover, US3 also reduced presentation of the Mtb39 antigen (R. Tomazin, C. Dunn, N.R. Hegde, D.C. Johnson, unpublished data). His16 cells infected with Ad vectors expressing the other HCMV glycoproteins or with control Ad vectors did not stimulate the CD4$^+$ T cells. Therefore, it appeared that HCMV US3 could also affect the MHC class II antigen presentation pathway. US3 was previously shown to cause retention of MHC class I proteins in the ER (AHN et al. 1996; JONES et al. 1996). In contrast to MHC class I, US3 and class II complexes reached the Golgi apparatus and attained endoglycosidase H resistance. However, the class II complexes were more extensively empty of peptides in US3-expressing cells. US3 was largely found in the Golgi apparatus and the TGN, with very little accumulation in the ER. Mechanistically, US3 reduced the association of Ii with class II α/β complexes, and this appeared to cause mislocalization of class II complexes so that they did not reach the loading compartment (N.R. Hegde, R. Tomazin, T. Wisner, C. Dunn, D.C. Johnson, manuscript submitted). As such, US3 has different effects on class II proteins than on class I proteins and is unique in reducing the association of Ii with class II complexes. Moreover, as with the class I antigen presentation pathway (AHN et al. 1996), US2 and US3 appear to collaborate to downregulate MHC class II proteins. In HCMV-infected cells, US3 may be expressed first, causing class II retention in the Golgi, and after US2 is expressed, DR-α and DM-α are degraded in the ER.

7 Concluding Remarks

Viral inhibitors of CD8$^+$ T cell recognition attack every step of the MHC class I antigen presentation pathway. There is limited information suggesting that this inhibition may be most important for herpesviruses during a narrow "window of opportunity", perhaps shortly after reactivation in an immune host, or in specific cell types. It also seems clear that priming of CD8$^+$ T lymphocyte responses does occur; relatively normal anti-viral CD8$^+$ T cells can be detected in HSV-infected hosts. However, it is a good bet that these inhibitors of the MHC class I pathway give herpesvirus a selective advantage, allowing early stages of disease to progress and progeny to be produced, especially after reactivation in an immune host. The inhibitors of the MHC class II pathway we have described may function in a similar manner. In class II-expressing cells, endogenous herpesvirus proteins, e.g., structural components of the virus particle that assembles in the TGN, will be degraded to peptides that can be loaded onto class II complexes. Recognition of HCMV-infected class II-expressing cells by CD4$^+$ T cells would normally lead to cell lysis, expression of antiviral cytokines, and increased production of antibodies. However, through the action of US2 and US3, HCMV can reduce MHC class II antigen presentation and resist these effects of CD4$^+$ T cells.

Note Added in Proof. After this manuscript was submitted, GEWURZ et al. (2001a) described the crystal structure of a soluble form of US2 produced in bacteria bound to MHC class I molecule HLA-A2. Our analyses of truncated forms of US2 expressed in mammalian cells have indicated a similar region involved in binding class I HC (M. Chevalier and D.C. Johnson, manuscript submitted). Additionally, mutant US2 (residues 40–130) bound equally well to MHC class II proteins (M. Chevalier, D. Tortorella, H.L. Ploegh, and D.C. Johnson, manuscript in preparation), whereas bacterially produced US2 does not bind to class II DR or DM in solution (GEWURZ et al. 2001b).

Acknowledgements. This work was supported by grants from the National Eye Institute (EY-11245). We are very grateful to David Lewinsohn, Elizabeth Mellins, John Trowsdale, Hidde Ploegh, Peter Cresswell, Tom Jones, and Stanley Riddell for reagents and advice.

References

Ahn K, Angulo A, Ghazal P, Peterson PA, Yang Y, Fruh K (1996) Human cytomegalovirus inhibits antigen presentation by a sequential multistep process. Proc Natl Acad Sci USA 93:10990–10995

Ahn K, Gruhler A, Galocha B, Jones TR, Wiertz EJ, Ploegh HL, Peterson PA, Yang Y, Fruh K (1997) The ER-luminal domain of the HCMV glycoprotein US6 inhibits peptide translocation by TAPl. Immunity 6:613–621

Alcami A, Koszinowski UH (2000) Viral mechanisms of immune evasion. Trends Microbiol 8:410–8

Bonifaz LC, Arzate S, Moreno J (1999) Endogenous and exogenous forms of the same antigen are processed from different pools to bind MHC class II molecules in endocytic compartments. Eur J Immunol 29:119–131

Borysiewicz LK, Sissons JG (1994) Cytotoxic T cells and human herpes virus infections. Curr Top Microbiol Immunol 189:123–150

Brack AR, Klupp BG, Granzow H, Tirabassi R, Enquist LW, Mettenleiter TC (2000) Role of the cytoplasmic tail of pseudorabies virus glycoprotein E in virion formation. J Virol 74:4004–4016

Chen MM, Shirai M, Liu Z, Arichi T, Takahashi H, Hishioka M (1998) Efficient class II major histocompatibility complex presentation of endogenously synthesized hepatitis C virus core protein by Epstein-Barr virus-transformed B-lymphoblastoid cell lines to CD4(+) T cells. J Virol 72: 8301–8308

Chiang HL, Terlecky SR, Plant CP (1989) A role for a 70-kilodalton heat shock protein in lysosomal degradation of intracellular proteins. Science 246:382–385

Chicz RM, Urban RG, Gorga JC, Vignali DA, Lane WS, Stominger JL (1993) Specificity and promiscuity among naturally processed peptides bound to HLA-DR alleles. J Exp Med 178:27–47

Cresswell P (1994) Assembly, transport, and function of MHC class II molecules. Annu Rev Immunol 2:259–293

Cresswell P (1996) Invariant chain structure and MHC class II function. Cell 84:505–507

Cunningham AL, Turner RR, Miller AC, Para MF, Merigan TC (1985) Evolution of recurrent herpes simplex virus lesions: an immunohistologic study. J Clin Invest 75:226–295

Denzin LK, Cresswell P (1995) HLA-DM induces CLIP dissociation from MHC class II α/β dimers and facilitates peptide loading. Cell 82:155–165

Dusseljee S, Wubbolts R, Verwoerd D, Tulp A, Janssen H, Calafat J, Neefjes J (1998) Removal and degradation of the free MHC class II β chain in the endoplasmic reticulum required proteasomes and is accelerated by BFA. J Cell Sci 111:2217–2226

Fish KN, Soderberg-Naucler C, Nelson JA (1998) Steady-state plasma membrane expression of human cytomegalovirus gB is determined by the phosphorylation of state of ser900. J Virol 72:6657–6664

Fruh K, Gruhler A, Krishna RM, Schoenhals GJ (1999) A comparison of viral immune escape strategies targeting the MHC class I assembly pathway. Immunol Rev 168:157–166

Goldsmith K, Chen W, Johnson DC, Hendricks RL (1998) Infected cell protein (ICP)47 enhances herpes simplex virus neurovirulence by blocking the CD8+ T cell response. J Exp Med 187:341–348

Gewurz BE, Gaudet R, Tortorella D, Wang EW, Ploegh HL, Wiley DC (2001a) Antigen presentation subverted: Structure of the human cytomegalovirus protein US2 bound to the class I molecule HLA-A2. Proc Natl Acad Sci USA 98:6794–6799

Gewurz BE, Wang EW, Tortorella D, Schust DJ, Ploegh HL (2001b) Human cytomegalovirus US2 endoplasmic reticulum-lumenal domain dictates association with major histocompatibility complex class I in a locus-specific manner. J Virol 75:5197–5204

Gueguen M, Long EO (1996) Presentation of a cytosolic antigen by major histocompatibility complex class II molecules requires a long-lived form of the antigen. Proc Natl Acad Sci USA 93:14692–14697

Hengel H, Reusch U, Geginat R, Holtappels R, Ruppert T, Hellebrand E, Koszinowski UH (2000) Macrophages escape inhibition of major histocompatibility complex class I-dependent antigen presentation by cytomegalovirus. J Virol 74:7861–7868

Hershberg RM, Framson PE, Cho DH, Lee LY, Kovats S, Beitz J, Blum J, Nepom GT (1997) Intestinal epithelial cells use two distinct pathways for HLA class II antigen processing. J Clin Invest 100:204–215

Jacobson S, Sekaly RP, Jacobson CL, McFarland HF, Long EO (1989) HLA class II-restricted presentation of cytoplasmic measles virus antigens to cytotoxic T cells. J Virol 63:1756–1762

Jaraquemada D, Marti M, Long EO (1990) An endogenous processing pathway in vaccinia virus-infected cells for presentation of cytoplasmic antigens to class II-restricted T cells. J Exp Med 172:947–954

Johnson AE, Haigh NG (2000) The ER translocon and retrotranslocation: is the shift into reverse manual or automatic? Cell 102:709–712

Johnson DC, McFadden G (2001) Viral immune evasion. In: Kaufmann SHE, Sher A, Ahmed R (eds) Immunology of infectious diseases. American Society for Microbiology (in press)

Jones TR, Hanson LK, Sun L, Slater JS, Stenberg RM Campbell AE (1995) Multiple independent loci within the human cytomegalovirus unique short region down regulate expression of major histocompatibility complex class I heavy chains. J Virol 69:4830–4841

Jones TR, Wiertz EJ, Sun L, Fish KN, Nelson JA, Ploegh HL (1996) Human cytomegalovirus US3 impairs transport and maturation of major histocompatibility complex class I heavy chains. Proc Natl Acad Sci USA 93:11327–11333

Lich JD, Elliott JF, Blum JS (2000) Cytoplasmic processing is a prerequisite for presentation of an endogenous antigen by major histocompatibility complex class II proteins. J Exp Med 191:1513–1524

Loss GE Jr, Elias CG, Fields PE, Ribaudo RK, McKisic M, Sant A (1993) Major histocompatibility complex class II-restricted presentation of an internally synthesized antigen displays cell-type variability and segregates from the exogenous class II and endogenous class I presentation pathways. J Exp Med 178:73–85

Maile R, Elsegood KA, Harding TC, Uney JB, Stewart G, Banting G, Dayan CM (2000) Effective formation of major histocompatibility complex class II- peptide complexes from endogenous antigen by thyroid epithelial cells. Immunology 99:367–374

Malnati MS, Marti M, LaVaute T, Jaraquemeda D, Biddison W, DeMars R, Long EO (1992) Processing pathways for presentation of cytosolic antigen to MHC class II restricted T cells. Nature 357:702–704

McMillan T, Johnson DC (2001) The cytoplasmic domain of herpes simplex virus gE causes accumulation in the trans-Golgi network, a site of virus envelopment and sorting of virions to cell junctions. J Virol 75:1928–1940

Miller DM, Rahill BM, Boss JM, Lairmore MD, Durbin JE, Waldman WJ, Sedman DD (1998) Human cytomegalovirus inhibits major histocompatibility complex class II expression by disruption of the Jak/Stat pathway. J Exp Med 187:675–683

Miller DM, Zhang Y, Rahill BM, Waldman WJ, Sedmak DD (1999) Human cytomegalovirus inhibits IFNα-stimulated antiviral and immunoregulatory responses by blocking multiple levels of IFN-α signal transduction. J Immunol 162:6107–6113

Munz C, Bickham KL, Subklewe M, Tsang ML, Chahroude A, Kurilla MG, Zhang D, O'Donnell M, Steinman RM (2000) Human CD4(+) T lymphocytes consistently respond to the latent Epstein-Barr virus nuclear antigen EBNA1. J Exp Med 191:1649–1660

Neefjes J (1999) CIIV, MIIC and other compartments for MHC class II loading. Eur J Immunol 29:1421–1425

Nuchtern JG, Biddison WE, Kalusner RD (1990) Class II MHC molecules can use the endogenous pathway of antigen presentation. Nature 343:74–76

Pieters J (2000) MHC class II-restricted antigen processing and presentation. Adv Immunol 75:159–208

Riddell SR, Greenberg PD (1995) Principles for adoptive T cell therapy of human viral diseases. Annu Rev Immunol 13:545–586

Rudensky A, Preston-Hurlbutt P, Hong SC, Barlow A, Janeway CA Jr (1991) Sequence analysis of peptides bound to MHC class II molecules. Nature 353:622–627

Sanchez V, Greis KD, Sztul E, Britt WJ (2000a) Accumulation of virion tegument and envelope proteins in a stable cytoplasmic compartment during human cytomegalovirus replication: characterization of a potential site of virus assembly. J Virol 74:975–986

Sanchez V, Sztul E, Britt WJ (2000b) Human cytomegalovirus pp28 (UL99) localizes to a cytoplasmic compartment which overlaps the endoplasmic reticulum-golgi-intermediate compartment. J Virol 74:3842–3851

Schmid DS, Rouse BT (1992) The role of T cell immunity in control of herpes simplex virus. Curr Top Microbiol Immunol 179:57–74

Schust D, Tortorella D, Seebach J, Phan C, Ploegh HL (1998) Trophoblast class I major histocompatibility complex (MHC) products are resistant to rapid degradation imposed by the human cytomegalovirus (HCMV) gene products US2 and US11. J Exp Med 188:497–503

Soderberg-Naucler C, Fish KN, Nelson JA (1998) Growth of human cytomegalovirus in primary macrophages. Methods Enzymol 16:126–138

Sponaas A, Carstens C, Koch N (1999) C-terminal extension of the MHC class II associated invariant chain by an antigenic sequence triggers activation of naive T cells. Gene Ther 6:1826–1834

Tomazin R, Boname J, Hegde NR, Lewinsohn DM, Altshuler Y, Jones TR, Cresswell P, Nelson JA, Riddell SR, Johnson DC (1999) Cytomegalovirus US2 destroys two components of the MHC class II pathway, preventing recognition by CD4+ T cells. Nat Med 5:1039–1043

Tooze J, Hollinshead M, Reis B, Radsak K, Kern H (1993) Progeny vaccinia and human cytomegalovirus particles utilize early endosomal cisternae for their envelopes. Eur J Cell Biol 60:163–178

Tortorella D, Gewurz BE, Furman MH, Schust DJ, Ploegh HL (2000) Viral subversion of the immune system. Annu Rev Immunol 18:861–926

Tortorella D, Story CM, Huppa JB, Wiertz EJ, Jones TR, Bacik I, Bennink JR, Yewdell JW, Ploegh HL (1998) Dislocation of type I membrane proteins from the ER to the cytosol is sensitive to changes in redox potential. J Cell Biol 142:365–376

Villadangos JA, Bryant RA, Deussing J, Driessen C, Lennon-Dumenil AM, Riese RJ, Roth W, Saftig P, Shi GP, Chapman HA, Peters C, Ploegh HL (1999) Proteases involved in MHC class II antigen presentation. Immunol Rev 172:109–120

Wiertz EJ, Jones TR, Sun L, Bogyo M, Geuze HJ, Ploegh HL (1996a) The human cytomegalovirus US11 gene product dislocates MHC class I heavy chains from the endoplasmic reticulum to the cytosol. Cell 84:769–779

Wiertz EJ, Tortorella D, Bogyo M, Yu J, Mothes W, Jones TR, Rapoport TA, Ploegh HL (1996b) Sec61-mediated transfer of a membrane protein from the endoplasmic reticulum to the proteasome for destruction. Nature 384:432–438

Wu TC, Guarnieri FG, Staveley-O'Carroll KF, Viscidi RP, Levitsky HI, Hedrick L, Cho KR, August R, Pardoll DM (1995) Engineering an intracellular pathway for major histocompatibility complex class II presentation of antigens. Proc Natl Acad Sci USA 92:11671–11675

Yang M, Omura S, Bonifacino JS, Weissman AM (1998) Novel aspects of degradation of T cell receptor subunits from the endoplasmic reticulum (ER) in T cells: importance of oligosaccharide processing, ubiquitination, and proteasome-dependent removal from ER membranes. J Exp Med 187:835–846

York I, Roop C, Andrews DW, Riddell SR, Graham FL, Johnson DC (1994) A cytosolic herpes simplex virus protein inhibits antigen presentation to CD8+ T lymphocytes. Cell 77:525–535

Viral Evasion of Natural Killer Cells During Human Cytomegalovirus Infection

V.M. Braud[1], P. Tomasec[2], and G.W.G. Wilkinson[2]

Cytotoxic T cells are major players in the immune defence against human cytomegalovirus (HCMV). The virus has, however, developed several mechanisms to escape from this control. In particular, it down-regulates cell surface expression of HLA class I molecules. Because natural killer (NK) cells recognize and eliminate cells that lack HLA class I molecules, HCMV-infected cells could be more susceptible to NK lysis. In this review, we discuss the role played by NK cells in immune defence against HCMV and we describe potential strategies the virus has developed to escape from NK cell-mediated lysis. We focus in particular on a newly described protein, HCMV gpUL40, that induces cell surface expression of HLA-E, a non-classical class I molecule known to regulate NK cell functions.

1 Introduction	117
2 Role of NK Cells in Defence Against HCMV	119
3 The Viral MHC Class I Homologue UL18	120
4 HCMV gpUL40 Induces HLA-E Cell Surface Expression	122
5 Concluding Remarks	124
References	125

1 Introduction

Human cytomegalovirus (HCMV) is a ubiquitous β-herpesvirus that will infect the majority of the population at some stage of life. As with other herpesviruses, primary infection is followed by lifelong persistence of the virus and disease can result from either primary infection or reactivation. The virus is well adapted to its host, with the vast majority of infections being sub-clinical. Nevertheless, HCMV is an important human pathogen, being the major viral cause of congenital abnormalities and a significant cause of infectious mononucleosis in normal adults, and has been implicated in vascular disease. Immunosuppressed or immunocompromised individuals, particularly transplant recipients and individuals with AIDS, are most at risk to HCMV disease, where the virus can be responsible for significant morbidity and mortality. HCMV is known to replicate in a wide variety of cells

---

[1] Centre National de la Recherche Scientifique, CNRS UMR 6097, Institut de Pharmacologie Moleculaire et Cellulaire, 660 Route des Lucioles, Sophia Antipolis, 06560 Valbonne, France
[2] Department of Medicine, University of Wales College of Medicine, Cardiff, CF14 4XN, UK

including fibroblasts, smooth muscle cells, endothelial cells, hepatocytes and tissue macrophages (PLACHTER et al. 1996; SINZGER et al. 1995). Although HCMV 'latency' is not well defined, the virus genome appears to persist in a sub-population of $CD34^+$ haematopoietic progenitors. Viral immediate-early gene expression is activated after differentiation to monocytes, and ultimately, virus production is associated with a macrophage population expressing dendritic cell markers (FISH et al. 1995; SODERBERG-NAUCLER et al. 1997). According to this model, HCMV reactivation may be a chronic event constantly stimulating and being controlled by the host immune response. The virus also has the potential to be disseminated by monocytes throughout the body. However, the selection of a 'professional antigen-presenting cell' as a site of virus reactivation and infection may be expected to exert a strong selective pressure on the virus to avoid priming host immunity.

The cellular immune response plays a crucial role in controlling HCMV disease. Immunocompromised patients lacking a detectable HCMV-specific $CD8^+$ cytotoxic T lymphocyte (CTL) response are susceptible to HCMV infection, whereas the presence of a specific CTL response correlates with resistance to infection (REUSSER et al. 1991). Adoptive transfer of in vitro expanded HCMV-specific CTL has proved effective in reconstituting a CTL response in bone marrow transplant recipients (RIDDELL et al. 1992; WALTER et al. 1995). In the murine model, CTL are also capable of controlling both primary infection and MCMV recurrence (BIRON et al. 1999; POLIC et al. 1998).

Although a strong CTL response is mounted in the host to control virus replication, HCMV has evolved to encode an impressive arsenal of evasion strategies. Some of these mechanisms are well characterised and result in downregulation of HLA class I molecules (see for review TORTORELLA et al. 2000). At least five HCMV proteins interfere with the presentation of viral antigens by MHC class I molecules. The pp65 tegument protein delivered by the infecting virus particle inhibits the processing of the major immediate early protein (IE1) (GILBERT et al. 1996). The US3 gene product is expressed with IE kinetics and acts by directly binding and retaining newly synthesised class I heavy chains in the endoplasmic reticulum (ER) (AHN et al. 1996; JONES et al. 1996). As the infection proceeds, gpUS6 acts after the generation of antigenic peptides by the proteasome, by blocking their transport into the lumen of the ER through the transporter associated with antigen processing (TAP) (AHN et al. 1997; HENGEL et al. 1997; LEHNER et al. 1997). Finally, gpUS2 and gpUS11 dislocate HLA class I heavy chains from the ER back to the cytosol through the translocon Sec61 complex, where they are rapidly degraded by the proteasome (WIERTZ et al. 1996).

Interestingly, not all HLA class I molecules are equally affected by gpUS2, gpUS3, gpUS6 and gpUS11. Whether the differential effects of HCMV genes on the expression of the various HLA alleles have functional consequences needs to be fully determined. The downregulation of HLA class I expression in HCMV-infected cells should prevent CTL recognition. Cells transfected with gpUS6 were resistant to killing by specific CTL (HENGEL et al. 1997; LEHNER et al. 1997). HCMV infection affected the recognition of an exogenously added influenza antigenic peptide by HLA-A2-restricted CTL (WARREN et al. 1994). The presentation of constitutively expressed minor or major H antigens was also affected by HCMV infection (HENGEL et al. 1995). However, the only evidence to date of a block of presentation of viral antigen was provided by GILBERT et al. (1993), who demonstrated weak CTL recognition of HCMV IE1-specific T cells during permissive HCMV infection. Given that absolute resistance to CTL activity could result in the death of the host and would thus be detrimental to the virus, it likely affects the extent and effectiveness of such escape mechanisms.

2 Role of NK Cells in Defence Against HCMV

As a result of HLA class I downregulation and because of the ability of NK cells to kill HLA class I-deficient cells, HCMV-infected cells should be more susceptible to NK-cell mediated lysis. NK cells are involved in both innate and adaptive immunity, generally playing a role in the early defence against pathogens. They have been shown to contribute to the protective response to a number of infections and cancers. During the first days after viral infections, a massive proliferation of NK cells is observed which is quickly followed by proliferation of $CD8^+$ T cells. During this early phase, NK cells exert their protective role via direct cellular cytotoxicity and via the production of cytokines, in particular IFN-γ. So far, very little direct evidence has been provided for a role of NK cells in protection against HCMV in humans. However, an important role for NK cells is suggested by the observation that a patient with a transitory NK cell deficiency developed abnormal sensitivity to herpesvirus infections including HCMV (BIRON et al. 1989). The best evidence for an antiviral role of NK cells is found in mice, where depletion of NK cells enhances MCMV replication (BUKOWSKI et al. 1984, 1985) and MCMV recurrence occurs after NK and $CD8^+$ T cell depletion (POLIC et al. 1998). Increased susceptibility to MCMV infection is also observed in NK cell-deficient beige mice (SHELLAM et al. 1981) and an antiviral activity of NK cells has been demonstrated in SCID mice (WELSH et al. 1991, 1994). Furthermore, some murine strains possess a gene (*Cmv-1*) that confers resistance to MCMV infection in the spleen (SCALZO et al. 1990). The effect of *Cmv-1* has been shown to be mediated by NK cells (SCALZO et al. 1992, 1995) and was recently identified as being Ly49H (BROWN et al. 2001; DANIELS et al. 2001; LEE et al. 2001).

According to the missing-self hypothesis proposed by Klas Karre (KARRE 1997), NK cells are able to spontaneously kill tumours or virus-infected cells that have lost HLA class I cell surface expression. NK cell cytotoxicity is controlled by a fine balance between activating and inhibitory signals, the later generally overriding the former. Both activating and inhibitory NK receptors can bind HLA class I molecules. One group of NK receptors belongs to the killer immunoglobulin (Ig)-like receptor family, and they interact with groups of HLA-C alleles, HLA-B alleles carrying the Bw4 specificity, or HLA-A3 and -A11 (LANIER 1998). A second group belongs to the C-type lectin family, and the receptors are formed of heterodimers of CD94 and an isoform of NKG2. CD94/NKG2A or B and CD94/

NKG2C interact with HLA-E (Braud and McMichael 1999). Within a third group belonging to the Ig superfamily and called Ig-like transcripts (ILTs) or leukocyte Ig-like receptors (LIRs), two receptors, ILT2/LIR1 and ILT4/LIR2, have been shown to bind to all HLA class I molecules (Borges et al. 1999; Colonna et al. 1999).

With such a diversity of receptors, it is important to assess the effect of downregulation of HLA class I molecules by HCMV on NK cell activity. Several studies have attempted to address this issue in vitro, but results are somewhat confusing. HLA-A and HLA-B alleles are downregulated efficiently by all the HCMV protein gpUS2, gpUS3, gpUS6 and gpUS11, but they are not the dominant ligands in NK cell recognition. HLA-C is not downregulated by gpUS2 and gpUS11 but is affected by gpUS3 and gpUS6 (Jun et al. 2000). HLA-C downregulation in HCMV-infected fibroblasts has been observed with the use of specific allosera (unpublished data) and of F4/326 antibody (Huard and Fruh 2000). In this later study, HLA-C downregulation was found to increase killing of HCMV-infected U373-MG cells by a KIR2DL1$^+$ NK clone, suggesting that downregulation of HLA-C renders infected cells more susceptible to NK lysis (Huard and Fruh 2000). However, this effect was found to be dependent on gpUS11 and not gpUS3, which is in contradiction with previous reports (Jun et al. 2000; Schust et al. 1998) but would agree with recent data (Lopez-Botet et al. 2001). Other groups failed to demonstrate a correlation between expression of MHC class I molecules and protection from NK killing. Reports indicate that cells infected with the laboratory strain AD169 are susceptible to NK lysis (Borysiewicz et al. 1985; Cerboni et al. 2000; Fletcher et al. 1998; Leong et al. 1998). The results are contradictory on cells infected with the HCMV strain Towne (Cerboni et al. 2000; Fletcher et al. 1998). Clinical strains seem to confer either protection or susceptibility (Cerboni et al. 2000; Fletcher et al. 1998). One study demonstrated that NK lysis of HCMV-infected cells was correlated with higher level of the adhesion molecule LFA3 (Fletcher et al. 1998), whereas another study correlated it with ICAM-1 expression (Leong et al. 1998). It is clear from these data that the in vitro studies lack consistency. This may be due to the lack of a suitable in vitro system. The isolate of HCMV, the timing, the type of cells infected, and the diversity of NK receptors expressed by NK cells are likely to modulate cytotoxic activity. Also, viral escape mechanisms may be in place in some of these studies and may explain the confusion. In the next two paragraphs, we comment on two possible strategies for evasion of NK cells.

3 The Viral MHC Class I Homologue UL18

During the sequencing of the HCMV strain AD169 genome Beck and Barrell identified an open reading frame, UL18, that exhibited low-level homology to

MHC class I heavy chain, approximately 24% in the α1 and α2 domains and 20% homology in the α3 domain (BECK and BARRELL 1988). Sequence comparisons suggest that its structure is similar to that of MHC class I molecules. Like class I molecules, UL18 heavy chain associates with β2 microglobulin (β2m) and binds endogenous peptides (BROWNE et al. 1990; FAHNESTOCK et al. 1995). However, its expression at the cell surface seems extremely low and may be regulated by intracellular levels of β2m (CHAPMAN and BJORKMAN 1998; LEONG et al. 1998; and unpublished data). In an attempt to identify receptors binding to gpUL18 and define its function, a soluble form of UL18 fused to the Fc portion of human IgG1 was generated. This UL18-Fc bound primarily to B cell lines and myelomonocytic cells and to a lesser extent to NK cells or T cells (COSMAN et al. 1997). The receptor was cloned and found to be a member of the previously mentioned LIR/ILT family. gpUL18 binds an inhibitory receptor ILT2/LIR1 that possesses immunoreceptor tyrosine-based inhibition motifs (ITIM) in its cytoplasmic domain. ILT2/LIR1 can interact both with gpUL18 and with all MHC class I molecules, the binding site being in a conserved region of the α3 domain (ALLAN et al. 1999; CHAPMAN et al. 1999; COLONNA et al. 1997; FANGER et al. 1998; NAVARRO et al. 1999). Interestingly, gpUL18 binds to ILT2/LIR1 with 1000-fold higher affinity than HLA class I molecules, suggesting that this viral product is likely to compete with host class I molecules (CHAPMAN et al. 1999).

Transfection of an HLA-A, -B, -C and –G-negative cell line (721.221cells) with UL18 was reported to elicit protection against killing by NK lines that could be reversed with an anti-CD94 blocking antibody (REYBURN et al. 1997). However, binding studies with the UL18-Fc fusion protein did not support a UL18 interaction with CD94/NKG2 receptors (COSMAN et al. 1997), and it is now known that HLA-E is the ligand for CD94/NKG2 receptors (BRAUD and MCMICHAEL 1999). HLA-E cell surface expression relies on the binding of conserved peptides from the leader sequence of HLA class I molecules (see Sect. 4). gpUL18 could potentially donate a peptide to enable HLA-E maturation to the cell surface. However, such a peptide has not been identified, and expression of UL18 did not protect target cells from lysis by CD94/NKG2A$^+$ NK clones (TOMASEC et al. 2000). An alternative explanation could be that the 721.221-UL18 transfectant was expressing HLA-E on the cell surface as an anti-β2m antibody was used to select clones that were expressing high level of gpUL18.

Although ILT2/LIR1 is now accepted as being the receptor for gpUL18, the role of UL18 during HCMV infection remains controversial. UL18 transcripts have been detected 72h after infection (HASSAN-WALKER et al. 1998). At that time, HLA class I molecules are downregulated. Although UL18 seems to be expressed at very low levels on the cell surface, its high affinity for ILT2/LIR1 may allow inhibitory signals to be delivered. In functional assays using NK clones, interaction of HLA class I molecules with ILT2/LIR1 protects target cells from lysis (COLONNA et al. 1997; VITALE et al. 1999). Paradoxically, UL18 expressing transfectants seem to be killed more efficiently by NK cells whereas HCMV-infected cells are more resistant to killing when UL18 is deleted (LEONG et al. 1998). Such contradictory results could reflect the lack of a suitable in vitro system to assess

HCMV escape from NK-cell lysis, as discussed above. Interestingly, MCMV also expresses a class I homologue, m114. The presence of m144 interferes with NK cell-mediated clearance of MCMV-infected cells in vivo and decreases NK cell rejection of lymphomas (CRETNEY et al. 1999; FARRELL et al. 1997). These data suggests that at least in mice, the class I homologue is involved in protection from NK attack.

ILT2/LIR1 is also expressed on B cells and monocytic cells such as monocytes, macrophages and dendritic cells (BORGES et al. 1999). Therefore, the role of UL18 may also involve the regulation of monocyte and dendritic cell functions during HCMV infection. In turn, inhibition of the secretion of cytokines such as IL-12, IL-15 and IFN-α by monocytes/macrophages may inhibit activation of NK cells. The role of UL18 in maintaining viral survival in the host remains to be clarified.

4 HCMV gpUL40 Induces HLA-E Cell Surface Expression

Another viral glycoprotein, gpUL40, was recently implicated in a viral escape mechanism from NK cells (TOMASEC et al. 2000; ULBRECHT et al. 2000). HCMV hijacks an important regulatory molecule, HLA-E, known to modulate NK cell function.

HLA-E is a non-classical class I molecule displaying very limited polymorphism (BRAUD and MCMICHAEL 1999). HLA-E associates with β2m but, in contrast to classical HLA class I molecules, binds a restricted set of peptides (BRAUD et al. 1997). Interestingly, these peptides are derived from the leader sequence of most classic class I molecules and the non-classical class I molecule HLA-G. HLA-E maturation follows the class I assembly pathway, where the leader peptide is transported by the TAP transporter into the lumen of the ER to be loaded onto HLA-E (BRAUD et al. 1998; LEE et al. 1998). Therefore, not only is HLA-E intrinsically linked to other HLA class I molecules by the source of peptides it binds to, it is also affected by blockage in the class I assembly pathway. HLA-E was shown to bind to the CD94/NKG2A and CD94/NKG2C/DAP12 NK cell receptors in both functional studies and studies using soluble tetrameric complexes of HLA-E (BORREGO et al. 1998; BRAUD et al. 1998; LEE et al. 1998). Like HLA class I molecules binding to KIR, interaction of HLA-E with CD94/NKG2A or CD94/NKG2C/DAP12 results respectively in inhibition or activation of NK cell-mediated cytotoxicity. NK cells are therefore able to assess HLA class I expression status using a single ligand/receptor interaction. Thus it is not particularly surprising that HCMV has found a way to evade this control.

A search in the database for peptides homologous to the MHC leader peptide binding to HLA-E revealed that the HCMV glycoprotein UL40 possess a nonamer peptide, VMAPRTLIL, found in HLA-C signal sequences (TOMASEC et al. 2000; ULBRECHT et al. 2000). The N-terminal sequencing of a recombinant UL40-Fc fusion protein confirmed the prediction that this peptide is located in a 37-amino acid-long signal sequence in gpUL40 (TOMASEC et al. 2000). We therefore hypothesised that gpUL40 may upregulate cell surface HLA-E and thus make HCMV-infected cells resistant to NK cell lysis. To analyse the function of gpUL40, replication-deficient adenovirus recombinants encoding UL40 and HLA-E were constructed and used to infect human fibroblasts. Expression of gpUL40 induced migration of HLA-E to the cell surface. Similarly, transient expression of gpUL40 in the HLA-A, -B, -C, -G-negative cell line 721.221 induced HLA-E cell surface expression. The upregulation of HLA-E by gpUL40 provided protection from killing by CD94/NKG2A$^+$ NK clones (TOMASEC et al. 2000; ULBRECHT et al. 2000). The inability of two-thirds of HLA-B alleles to increase expression of HLA-E at the cell surface has been correlated with a polymorphism at position 2 in the peptide (BRAUD et al. 1998). Whereas methionine is permissive, threonine dramatically reduces the peptide's binding affinity. When gpUL40 was mutated to insert a threonine at position 2 of the putative HLA-E-binding peptide sequence, the gpUL40 mutant failed to upregulate HLA-E (unpublished data). This demonstrated that the nonamer leader peptide is solely responsible for induction of HLA-E cell surface expression. Importantly, this upregulation of HLA-E by gpUL40 was found to be independent of the TAP transporter. HCMV protein gpUS6 blocks TAP transport, therefore preventing the loading of HLA-E with MHC leader peptides. This observation fits with a model in which gpUL40 provides the nonamer peptide to HLA-E directly in the ER, bypassing the requirement for TAP (Fig. 1). Recent work also supported this concept. We provided evidence that intramembrane proteolysis by a signal peptide peptidase (SPPase) is required for the generation of MHC class I leader peptides binding to HLA-E, whereas it is not needed for the generation of the UL40-derived leader peptide (LEMBERG et al. 2001; unpublished data).

To assess the role of gpUL40 during HCMV infection, we looked for UL40 expression in HCMV-infected cells in vitro. UL40 transcription was detected in

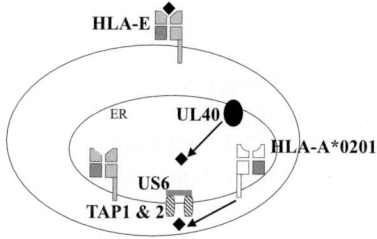

◆ HLA-E binding peptide

Fig. 1. HCMV gpUL40 induces HLA-E cell surface expression in a TAP-independent way. The leader peptide VMAPRTLVL from HLA-A*0201 signal sequence is generated and released in the cytosol. In HCMV-infected cells, US6 blocks the TAP transporter, preventing the HLA leader peptide from being transported into the lumen of the ER to be loaded onto HLA-E. However, gpUL40 signal sequence is processed and generates a peptide VMAPRTLIL directly liberated into the ER that binds to HLA-E, inducing its cell surface expression. Whereas classical HLA class I molecules are downregulated by HCMV, HLA-E is upregulated

the early and late phases of infection, and gpUL40 protein could be detected by immunoblotting. Interestingly, although HLA class I molecules were downregulated, HLA-E cell surface expression was increased (TOMASEC et al. 2000). Such induction of HLA-E by the virus was consistently seen with several HCMV strains. This indicates that gpUS2, gpUS3 and gpUS11, which are all implicated in the downregulation of classic MHC class I molecules, have limited effects on HLA-E. This is consistent with the recent observation that HLA-E is resistant to the actions of gpUS2 and gpUS11 (FURMAN et al. 2000; LOPEZ-BOTET et al. 2001). ZHU et al. (1997) also reported that the level of HLA-E mRNA was substantially increased during infection. This could significantly contribute to the upregulation of HLA-E.

The discovery of such viral mechanisms resulting in upregulation of HLA-E while classical HLA class I molecules are downregulated suggests that HCMV exploits gpUL40 as part of a strategy to evade NK cell-mediated cytotoxicity. Importantly, not only the laboratory strain AD169 but seven other HCMV strains and clinical strains possess a peptide binding to HLA-E in the signal sequence of their gpUL40. This conservation of the peptide in many strains, together with the delivery of the gpUL40 leader peptide directly in the ER, thus bypassing TAP, strongly implies that this mechanism has evolved in HCMV on a strong selective pressure. Significantly, productive HCMV infection can elicit efficient protection against lysis by CD94/NKG2A$^+$ NK cells and this protection is mediated by UL40-induced upregulation of HLA-E (manuscript in preparation).

5 Concluding Remarks

In this chapter, we have discussed the protective role of NK cells during HCMV infection and possible strategies the virus has developed to evade detection and destruction by NK cells. During clinical disease, conditions are made more complex by the direct and indirect effects of cytokines or chemokines on NK cell activation. Remarkable progress is being made in characterising the factors which are regulating NK cell responses during viral infection. IFN-α/β production has a direct antiviral effect and is also linked to activation of NK cell-mediated cytotoxicity. TNF-α, IL-6, IL-12 and IFN-γ also play an important role in inducing and maintaining NK cell responses. Targeting these cytokines would have a beneficial effect on virus survival, and it has been suggested that inhibition of IFN-α/β (MILLER et al. 1999) or IFN-γ signaling (MILLER et al. 1998) could be strategies that HCMV has developed to escape from such a control. Viral interference with apoptosis may also be another way to persist in the host. HCMV encodes a TNF receptor homologue UL144, but it is still unclear whether it can inhibit death (BENEDICT et al. 1999). The IE-1 and IE-2 genes have been implicated in the inhibition of TNF-induced apoptosis (ZHU et al. 1995). GpUL137 is an IE gene and an efficient inhibitor of apoptosis (GOLDMACHER et al. 1999).

Finally, a range of novel molecules are continually being identified and characterised that can activate or repress NK cells. The virus may manipulate such signals to efficiently control NK cell activity. The activating receptor NKG2D/ DAP10 has been recently shown to interact with stress-inducible MIC-A molecules and a new family of molecules, the gpUL16 binding proteins (ULBP) (BAUER et al. 1999; SUTHERLAND et al. 2001; WU et al. 1999). ULBP genes were cloned because of their interaction with the viral HCMV gpUL16 protein (COSMAN et al. 2001; KUBIN et al. 2001). gpUL16 expression in HCMV infected cells may therefore interfere with ULBP/MIC-NKG2D/DAP10 interaction and prevent NK cell activation. Further work should provide valuable information on the role of NK cells and viral escape mechanisms during HCMV infection.

References

Ahn K, Angulo A, Ghazal P, Peterson PA, Yang Y, Fruh K (1996) Human cytomegalovirus inhibits antigen presentation by a sequential multistep process. Proc Natl Acad Sci USA 93:10990–10995

Ahn K, Gruhler A, Galocha B, Jones TR, Wiertz EJ, Ploegh HL, Peterson PA, Yang Y, Fruh K (1997) The ER-luminal domain of the HCMV glycoprotein US6 inhibits peptide translocation by TAP. Immunity 6:613–621

Allan DS, Colonna M, Lanier LL, Churakova TD, Abrams JS, Ellis SA, McMichael AJ, Braud VM (1999) Tetrameric complexes of human histocompatibility leukocyte antigen (HLA)-G bind to peripheral blood myelomonocytic cells. J Exp Med 189:1149–1156

Bauer S, Groh V, Wu J, Steinle A, Phillips JH, Lanier LL, Spies T (1999) Activation of NK cells and T cells by NKG2D, a receptor for stress-inducible MICA. Science 285:727–729

Beck S, Barrell BG (1988) Human cytomegalovirus encodes a glycoprotein homologous to MHC class-I antigens. Nature 331:269–272

Benedict CA, Butrovich KD, Lurain NS, Corbeil J, Rooney I, Schneider P, Tschopp J, Ware CF (1999) Cutting edge: a novel viral TNF receptor superfamily member in virulent strains of human cytomegalovirus. J Immunol 162:6967–6970

Biron CA, Byron KS, Sullivan JL (1989) Severe herpesvirus infections in an adolescent without natural killer cells. N Engl J Med 320:1731–1735

Biron CA, Nguyen KB, Pien GC, Cousens LP, Salazar-Mather TP (1999) Natural killer cells in antiviral defense: function and regulation by innate cytokines. Annu Rev Immunol 17:189–220

Borges L, Fanger N, Cosman D (1999) Interactions of LIRs, a family of immunoreceptors expressed in myeloid and lymphoid cells, with viral and cellular MHC class I antigens. Curr Top Microbiol Immunol 244:123–136

Borrego F, Ulbrecht M, Weiss EH, Coligan JE, Brooks AG (1998) Recognition of human histocompatibility leukocyte antigen (HLA)-E complexed with HLA class I signal sequence-derived peptides by CD94/NKG2 confers protection from natural killer cell-mediated lysis. J Exp Med 187:813–818

Borysiewicz LK, Rodgers B, Morris S, Graham S, Sissons JG (1985) Lysis of human cytomegalovirus infected fibroblasts by natural killer cells: demonstration of an interferon-independent component requiring expression of early viral proteins and characterization of effector cells. J Immunol 134: 2695–2701

Braud V, Jones EY, McMichael A (1997) The human major histocompatibility complex class Ib molecule HLA-E binds signal sequence-derived peptides with primary anchor residues at positions 2 and 9. Eur J Immunol 27:1164–1169

Braud VM, Allan DSJ, O'Callaghan CA, Soderstrom K, D'Andrea A, Ogg GS, Lazetic S, Young NT, Bell JI, Phillips JH, Lanier LL, McMichael AJ (1998) HLA-E binds to natural killer cell receptors CD94/NKG2A, B and C. Nature 391:795–799

Braud VM, Allan DSJ, Wilson D, McMichael AJ (1998) TAP- and tapasin-dependent HLA-E surface expression correlates with the binding of an MHC class I leader peptide. Curr Biol 8:1–10

Braud VM, McMichael AJ (1999) Regulation of NK cell functions through interaction of the CD94/ NKG2 receptors with the nonclassical class I molecule HLA-E. Curr Top Microbiol Immunol 244:85–95

Brown MG, Dokun AO, Heusel JW, Smith HR, Beckman DL, Blattenberger EA, Dubbelde CE, Stone LR, Scalzo AA, Yokoyama WM (2001) Vital involvement of a natural killer cell activation receptor in resistance to viral infection. Science 292:934–937

Browne H, Smith G, Beck S, Minson T (1990) A complex between the MHC class I homologue encoded by human cytomegalovirus and beta 2 microglobulin. Nature 347:770–772

Bukowski JF, Warner JF, Dennert G, Welsh RM (1985) Adoptive transfer studies demonstrating the antiviral effect of natural killer cells in vivo. J Exp Med 161:40–52

Bukowski JF, Woda BA, Welsh RM (1984) Pathogenesis of murine cytomegalovirus infection in natural killer cell-depleted mice. J Virol 52:119–128

Cerboni C, Mousavi-Jazi M, Linde A, Soderstrom K, Brytting M, Wahren B, Karre K, Carbone E (2000) Human cytomegalovirus strain-dependent changes in NK cell recognition of infected fibroblasts. J Immunol 164:4775–4782

Chapman TL, Bjorkman PJ (1998) Characterization of a murine cytomegalovirus class I major histocompatibility complex (MHC) homolog: comparison to MHC molecules and to the human cytomegalovirus MHC homolog. J Virol 72:460–466

Chapman TL, Heikeman AP, Bjorkman PJ (1999) The inhibitory receptor LIR-1 uses a common binding interaction to recognize class I MHC molecules and the viral homolog UL18. Immunity 11:603–613

Colonna M, Navarro F, Bellon T, Llano M, Garcia P, Samaridis J, Angman L, Cella M, Lopez-Botet M (1997) A common inhibitory receptor for major histocompatibility complex class I molecules on human lymphoid and myelomonocytic cells. J Exp Med 186:1809–1818

Colonna M, Navarro F, Lopez-Botet M (1999) A novel family of inhibitory receptors for HLA class I molecules that modulate function of lymphoid and myeloid cells. Curr Top Microbiol Immunol 244:115–122

Cosman D, Fanger N, Borges L, Kubin M, Chin W, Peterson L, Hsu ML (1997) A novel immunoglobulin superfamily receptor for cellular and viral MHC class I molecules. Immunity 7:273–282

Cosman D, Mullberg J, Sutherland CL, Chin W, Armitage R, Fanslow W, Kubin M, Chalupny NJ (2001) ULBPs, novel MHC class I-related molecules, bind to CMV glycoprotein UL16 and stimulate NK cytotoxicity through the NKG2D receptor. Immunity 14:123–133

Cretney E, Degli-Espoti MA, Densley EH, Farell HE, Davis-Poynter NJ, Smyth MJ (1999) m144, a murine cytomegalovirus (MCMV)-encoded major histocompatibility complex class I homologue, confers tumor resistance to natural killer cell-mediated rejection. J Exp Med 190:435–443

Daniels KA, Devora G, Lai WC, O'Donnell CL, Bennett M, Welsh RM (2001) Murine cytomegalovirus is regulated by a discrete subset of natural killer cells reactive with monoclonal antibody to Ly49H. J Exp Med 194:29–44

Fahnestock ML, Johnson JL, Feldman RM, Neveu JM, Lane WS, Bjorkman PJ (1995) The MHC class I homolog encoded by human cytomegalovirus binds endogenous peptides. Immunity 3:583–590

Fanger NA, Cosman D, Peterson L, Braddy SC, Maliszewski CR, Borges L (1998) The MHC class I binding proteins LIR-1 and LIR-2 inhibit Fc receptor-mediated signaling in monocytes. Eur J Immunol 28:3423–3434

Farrell HE, Vally H, Lynch DM, Fleming P, Shellam GR, Scalzo AA, Davis-Poynter NJ (1997) Inhibition of natural killer cells by a cytomegalovirus MHC class I homologue in vivo. Nature 386:510–514

Fish KN, Stenglein SG, Ibanez C, Nelson JA (1995) Cytomegalovirus persistence in macrophages and endothelial cells. Scand J Infect Dis Suppl 99:34–40

Fletcher JM, Prentice HG, Grundy JE (1998) Natural killer cell lysis of cytomegalovirus (CMV)- infected cells correlates with virally induced changes in cell surface lymphocyte function-associated antigen-3 (LFA-3) expression and not with the CMV- induced down-regulation of cell surface class I HLA. J Immunol 161:2365–2374

Furman MH, Ploegh HL, Schust DJ (2000) Can viruses help us to understand and classify the MHC class I molecules at the maternal-fetal interface? Hum Immunol 61:1169–1176

Gilbert MJ, Riddell SR, Li CR, Greenberg PD (1993) Selective interference with class I major histocompatibility complex presentation of the major immediate-early protein following infection with human cytomegalovirus. J Virol 67:3461–3469

Gilbert MJ, Riddell SR, Plachter B, Greenberg PD (1996) Cytomegalovirus selectively blocks antigen processing and presentation of its immediate-early gene product. Nature 383:720–722

Goldmacher VS, Bartle LM, Skaletskaya A, Dionne CA, Kedersha NL, Vater CA, Han J, Lutz RJ, Watanabe S, McFarland ED, Kieff ED, Mocarski ES, Chittenden T (1999) A cytomegalovirus-encoded mitochondria-localized inhibitor of apoptosis structurally unrelated to Bcl-2. Proc Natl Acad Sci USA 96:12536–12541

Hassan-Walker AF, Cope AV, Griffiths PD, Emery VC (1998) Transcription of the human cytomegalovirus natural killer decoy gene, UL18, in vitro and in vivo. J Gen Virol 79:2113–2116

Hengel H, Esslinger C, Pool J, Goulmy E, Koszinowski UH (1995) Cytokines restore MHC class I complex formation and control antigen presentation in human cytomegalovirus-infected cells. J Gen Virol 76, Iss 12:2987–2997

Hengel H, Koopmann JO, Flohr T, Muranyi W, Goulmy E, Hammerling GJ, Koszinowski UH, Momburg F (1997) A viral ER-resident glycoprotein inactivates the MHC-encoded peptide transporter. Immunity 6:623–632

Huard B, Fruh K (2000) A role for MHC class I down-regulation in NK cell lysis of herpes virus-infected cells. Eur J Immunol 30:509–515

Jones TR, Wiertz EJ, Sun L, Fish KN, Nelson JA, Ploegh HL (1996) Human cytomegalovirus US3 impairs transport and maturation of major histocompatibility complex class I heavy chains. Proc Natl Acad Sci USA 93:11327–11333

Jun Y, Kim E, Jin M, Sung HC, Han H, Geraghty DE, Ahn K (2000) Human cytomegalovirus gene products US3 and US6 down-regulate trophoblast class I MHC molecules. J Immunol 164:805–811

Karre K (1997) How to recognize a foreign submarine. Immunol Rev 155:5–9

Kubin M, Cassiano L, Chalupny J, Chin W, Cosman D, Fanslow W, Mullberg J, Rousseau AM, Ulrich D, Armitage R (2001) ULBP1, 2, 3: novel MHC class I-related molecules that bind to human cytomegalovirus glycoprotein UL16, activate NK cells. Eur J Immunol 31:1428–1437

Lanier LL (1998) NK cell receptors. Annu Rev Immunol 16:359–393

Lee N, Goodlett DR, Ishitani A, Marquardt H, Geraghty DE (1998) HLA-E surface expression depends on binding of TAP-dependent peptides derived from certain HLA class I signal sequences. J Immunol 160:4951–4960

Lee N, Llano M, Carretero M, Ishitani A, Navarro F, Lopez-Botet M, Geraghty DE (1998) HLA- E is a major ligand for the natural killer inhibitory receptor CD94/NKG2A. Proc Natl Acad Sci USA 95:5199–5204

Lee SH, Girard S, Macina D, Busa M, Zafer A, Belouchi A, Gros P, Vidal SM (2001) Susceptibility to mouse cytomegalovirus is associated with deletion of an activating natural killer cell receptor of the C-type lectin superfamily. Nat Genet 28:42–45

Lehner PJ, Karttunen JT, Wilkinson GW, Cresswell P (1997) The human cytomegalovirus US6 glycoprotein inhibits transporter associated with antigen processing-dependent peptide translocation. Proc Natl Acad Sci USA 94:6904–6909

Lemberg MK, Bland FA, Weihofen A, Braud VM, Martoglio B (2001) Intramembrane proteolysis of signal peptides: an essential step in the generation of HLA-E epitopes. J Immunol (in press)

Leong CC, Chapman TL, Bjorkman PJ, Formankova D, Mocarski ES, Phillips JH, Lanier LL (1998) Modulation of natural killer cell cytotoxicity in human cytomegalovirus infection: the role of endogenous class I major histocompatibility complex and a viral class I homolog [published erratum appears in J Exp Med 1998 Aug 3; 188(3): following 614]. J Exp Med 187:1681–1687

Lopez-Botet M, Llano M, Ortega M (2001) Human cytomegalovirus and natural killer-mediated surveillance of HLA class I expression: a paradigm of host-pathogen adaptation. Immunol Rev 181: 193–202

Miller DM, Rahill BM, Boss JM, Lairmore MD, Durbin JE, Waldman JW, Sedmak DD (1998) Human cytomegalovirus inhibits major histocompatibility complex class II expression by disruption of the Jak/Stat pathway. J Exp Med 187:675–683

Miller DM, Zhang Y, Rahill BM, Waldman WJ, Sedmak DD (1999) Human cytomegalovirus inhibits IFN-alpha-stimulated antiviral and immunoregulatory responses by blocking multiple levels of IFN-alpha signal transduction. J Immunol 162:6107–6113

Navarro F, Llano M, Bellon T, Colonna M, Geraghty DE, Lopez-Botet M (1999) The ILT2(LIR1) and CD94/NKG2 A NK cell receptors respectively recognize HLA-G1 and HLA-E molecules co-expressed on target cells. Eur J Immunol 29:277–283

Plachter B, Sinzger C, Jahn G (1996) Cell types involved in replication and distribution of human cytomegalovirus. Adv Virus Res 46:195–261

Polic B, Hengel H, Krmpotic A, Trgovcich J, Pavic I, Luccaronin P, Jonjic S, Koszinowski UH (1998) Hierarchical and redundant lymphocyte subset control precludes cytomegalovirus replication during latent infection. J Exp Med 188:1047–1054

Function of CMV-Encoded MHC Class I Homologues

H.E. Farrell[1], N.J. Davis-Poynter[1], D.M. Andrews[2], and M.A. Degli-Esposti[2]

Homologues of MHC class I proteins have been identified in the genomes of human, murine and rat cytomegaloviruses (CMVs). Given the pivotal role of the MHC class I protein in cellular immunity, it has been postulated that the viral homologues subvert the normal antiviral immune response of the host, thus promoting virus replication and dissemination in an otherwise hostile environment. This review focuses on recent studies of the CMV MHC class I homologues at the molecular, cellular and whole animal level and presents current hypotheses for their roles in the CMV life cycle.

1 Introduction 131
2 Structural and Biochemical Characteristics
  of the CMV-Encoded MHC Class I Homologues 133
3 Cellular Receptors for MHC Class I Proteins 136
4 Consequences of UL-18-LIR-1 Interactions on the HCMV Life Cycle 140
5 Possible Effects of the Viral MHC Class I Homologues on the Functions
  of Monocytes and DCs 140
6 Role of CMV-Encoded MHC Class I Homologues as Surrogate Inhibitory Ligands
  for NK Cells 142
7 Use of Rodent Models of CMV Infection to Determine the Function
  of CMV MHC Class I Homologues In Vivo 143
8 Concluding Remarks 147
References 148

1 Introduction

It is now over 25 years since the discovery of the central role of the cellular MHC class I molecule in host T cell immunity (Doherty and Zinkernagel 1975). In particular, studies of MHC class I-CD8$^+$ CTL interactions have provided the

---

[1] The Animal Health Trust, Virology Section, Kentford, Newmarket, Suffolk CB8 7UU, UK
[2] The University of Western Australia, Department of Microbiology, Queen Elizabeth II Medical Centre, Nedlands 6907, Western Australia, Australia

cornerstone for our understanding of how a wide range of virus infections of humans and animals are successfully eliminated. Elegant experimentation has elucidated many of the intracellular processing events leading to the presentation of viral peptides in association with MHC class I, as well as MHC class I-TCR interactions that lead to CTL activation (LEHNER and CRESSWELL 1996). In addition, much has been learned concerning the cytolytic mechanisms of CTL that are responsible for target cell death and the cascade of antiviral and pro-inflammatory mediators released from CTL and accessory cells that further limit the spread of virus infection (GARCIA 1999; YEWDELL et al. 1999).

In recent years it has become apparent that the MHC class I molecule also plays a dominant role in shaping our first line of defence – our innate immunity – to either virus-infected or transformed cells. Specifically, the discovery that cell surface expression of MHC class I molecules can modulate the function of natural killer (NK) cells (KARRË 1995) has provided new insights into the immunoregulatory role of these molecules and has rekindled interest in the importance of NK cells in antiviral defense mechanisms. Indeed, the identification of large superfamilies of receptors for MHC class I on a wide range of lymphoid and myeloid cells suggests that MHC class I has a greater global effect on immune regulation than was first appreciated (BORGES et al. 1997; CELLA et al. 1997; COLONNA et al. 1997, 1998). The complexity of MHC class I-mediated regulation of NK cells is highlighted by observations that receptors for MHC class I can be either inhibitory or stimulatory for NK cells (LANIER 1998). Such pairs of related stimulatory and inhibitory molecules have been identified in most of the rodent and human NK receptor families (reviewed by LANIER 1998; LONG 1999; MORETTA et al. 2000; TOMASELLO et al. 2000; COLONNA et al. 2000). Furthermore, an individual NK cell may co-express multiple MHC class I receptors, with stimulatory and inhibitory receptors sharing related ligands. Dissecting the nature of specific MHC class I-NK cell interactions that contribute to the maintenance of homeostasis, or early defence against virus infection or malignancy, is the subject of intense investigation. Although the 'experimental' outcome of such interactions is likely to result from the balance of inhibitory and stimulatory signals that control NK cell function, their relative importance in the physiological setting is also likely to be affected by a number of other factors, including (a) the cytokine milieu that modulates NK cell effector function and trafficking, (b) the in vivo distribution of NK cell subsets that possess different MHC class I receptors and (c) host age and genotype.

Given the central role of MHC class I in antiviral immunity, it is perhaps not surprising that the ability to interfere with MHC class I antigen expression is a property of a number of different virus groups, particularly slowly replicating or persistent viruses (reviewed by FARRELL and DAVIS-POYNTER 1998; FRÜH et al. 1999). Indeed, in the case of the alpha-, beta- and gammaherpesviruses, interference with MHC class I processing and antigen presentation is a common feature of infected cells, with direct effects identified on either T cell or NK cell responses in vitro and, in some cases, in vivo. Remarkably, this common goal in immune evasion is achieved by different mechanisms between each of the virus subfamilies and indeed, members of the same virus sub-family may utilise alternative processes. Details of the mechanisms of action of such genes are described elsewhere in this issue and are also the subject of a number of recent reviews (DAVIS-POYNTER and FARRELL 1997; FARRELL and DAVIS-POYNTER 1998; HENGEL et al. 1998, 1999).

Concomitant with investigations aimed at determining the importance of herpesvirus genes that modulate MHC class I expression on immune evasion, the role of herpesvirus-encoded homologues of MHC class I proteins has been the subject of considerable study. Whereas functions which disrupt cellular MHC class I expression have been identified in members of each of the three herpesvirus subfamilies, the identification of viral MHC class I homologues has been restricted to members of the betaherpesvirus family. Specifically, each of the three cytomegaloviruses (CMVs) that have been fully sequenced, namely human, mouse and rat CMV, has been shown to encode MHC class I homologues (reviewed by FARRELL et al. 2000). Notably, the other members of the betaherpesvirus group, human herpesvirus 6 and 7, do not possess such homologues, suggesting that these proteins perform functions during the virus life cycle that are characteristic of the CMVs. An MHC class I homologue has also been identified for the poxvirus Molluscum Contagiosum virus (SENKEVICH et al. 1996; SENKEVICH and MOSS 1998) but will not be discussed further in this review.

The aim of this review is to describe the current understanding and questions that have arisen concerning the role of the cytomegalovirus MHC class I homologues. Particular emphasis will be placed on addressing the possible relevance of the MHC class I homologues expressed in cell types that are known to feature in the life cycle of CMV in the natural setting.

2 Structural and Biochemical Characteristics of the CMV-Encoded MHC Class I Homologues

The CMV MHC class I homologues of HCMV, MCMV and RCMV (encoded by UL18, m144 and r144, respectively) are located proximal to glycoprotein gene family blocks present within each virus genome (BECK and BARRELL 1988; RAWLINSON et al. 1996; VINK et al. 2000). Although the amino acid sequences of the MHC class I homologues encoded by rodent CMV are well conserved, neither show significant sequence similarity with the HCMV counterpart. In addition, the HCMV MHC class I homologue is located at the opposite end of the genome from the rodent CMVs. It is therefore possible that the rodent and human CMV MHC class I homologues were acquired by independent events subsequent to the divergence of these viruses within primate and rodent hosts. However, the similar location of each of the MHC class I homologues relative to blocks of tandemly repeated glycoprotein genes (albeit that for HCMV the MHC class I homologue is located at the opposite end of the genome from the rodent CMVs) suggests that they may have been derived from a common ancestral CMV gene. Although the glycoprotein gene families of HCMV lack obvious sequence homology with those of murine or rat CMV, the identification of homologous functions being encoded by members of these families for HCMV and MCMV (reviewed by HENGEL et al. 1999) suggests that the glycoprotein gene families themselves may have diverged significantly from a common gene(s) present in an ancestral CMV. Moreover, the presence of one member (m17) of the right-end glycoprotein gene family adjacent to the left-end gene block in MCMV suggests an earlier recombination event between the left and right ends of the genome. Such an event might explain the presence of the HCMV and MCMV class I homologues at the opposite ends of their respective genomes. Evidence in support of or opposition to this argument will come from analysis of CMV genomes from other species.

The MHC class I homologues encoded by rodent and human CMVs exhibit conserved structural features which are characteristic of cellular MHC class I proteins (CHAPMAN and BJORKMAN 1998; BJORKMAN and PARHAM 1990; BEISSER et al. 2000). First, the viral MHC class I homologues are predicted to form three globular domains, with domains 1 and 2 comprising β-sheet and α-helical structures that are similarly present in cellular MHC class I molecules. Second, there is a bias of amino acids conserved with their cellular counterparts in the predicted α-3 domain of the viral class I homologues. This is the region involved in binding to β2-microglobulin (β2m); hence it is predicted that the viral homologues, like their cellular counterparts, will associate with β2m. Third, the location of cysteine residues which, in cellular MHC class I, are critical for tertiary structure through the formation of disulphide bonds is conserved in the MHC class I homologues. Finally, an N-linked glycosylation site located between the α-1 and α-2 domains of all three viral polypeptides is present at a position similar to a glycosylation site conserved in all cellular MHC class I proteins.

There are, however, a number of significant differences between the CMV-encoded MHC class I homologues and cellular MHC class I molecules. First, gpUL18 is unusual in that it is highly glycosylated; it possesses 13 predicted N-linked glycosylation sites, compared with the 1–3 sites that are found in cellular MHC class I molecules and the 4 and 3 glycosylation sites identified in gpm144 and gpr144, respectively (BECK and BARRELL 1988; FARRELL et al. 1997; BEISSER et al. 2000). This level of gpUL18 glycosylation has been confirmed by protein analysis of gpUL18 expressed in eukaryotic cells or in HCMV-infected cell lysates (BROWNE et al. 1990; CHAPMAN and BJORKMAN 1998; LEONG et al. 1998). Second, gpm144 and gpr144 differ from cellular MHC class I and gpUL18 in that they both contain a substantial deletion within their predicted α-2 domains (FARRELL et al. 1999; BEISSER et al. 2000). In classic MHC class I molecules, peptides bind in a groove located between the two α-helices of the α-1 and α-2 domains that span an eight-stranded β pleated sheet (BJORKMAN and PARHAM 1990). The N-and C-termini of the peptides bind to pockets of conserved residues at the left and right ends of this groove, respectively, thereby restricting the size of peptides able to be accommodated to octamers or nonamers. The impact of a truncated α-2 domain in the predicted structure of gpm144 has been analysed in detail. The truncation in gpm144 was predicted to affect several of the β-strands located at the base of the groove, suggesting that it would form a distinctive structure unable to bind peptide (FARRELL et al. 1997). Although gpUL18 does not possess this major α-2 deletion, it nevertheless possesses small insertions and deletions which were predicted to affect the right end of the groove, suggesting that the binding of peptides at their C-termini may be different to that of MHC class I proteins.

Biochemical characterisation of both gpUL18 and gpm144 has confirmed the above predictions with respect to their β2m and peptide binding properties (BROWNE et al. 1990; FAHNESTOCK et al. 1995; CHAPMAN and BJORKMAN 1998). Using antisera specific to either human/mouse β2m or to the viral MHC class I homologues, immunoprecipitation experiments demonstrated that both gpUL18 and gpm144 associate with β2m. Comparisons of acid eluates derived from soluble gpm144/β2m expressed in CHO cells with eluates derived from soluble counterparts of cellular MHC class I, gpUL18, and the rat neonatal Fc receptor (FcRn) (the latter being a non-classic MHC class I protein that does not bind peptide) demonstrated that gpUL18, but not gpm144, was able to bind peptide (CHAPMAN and BJORKMAN, 1998). Although endogenous peptides with N-terminal characteristics similar to those eluted from cellular MHC class I proteins were eluted from gpUL18, they were of a varied length, demonstrating that the C-termini of bound peptides were not constrained by residues located at the right end of the peptide groove. As predicted, endogenous peptides were not isolated from either gpm144 or FcRn. Analysis of the melting behaviour of gpm144/β2m demonstrated that this complex was thermally stable in the absence of peptide, a situation quite unlike most classic cellular MHC class I molecules and gpUL18, both of which require bound peptide for stability. Although the β2m- and peptide-binding properties of gpr144 await characterisation, its similarity with the primary sequence of gpm144 suggests that it will share features with this molecule.

How might these biochemical and structural features of the viral MHC class I homologues relate to their functional roles, and what is the significance in the structural divergence between the human and rodent CMV counterparts? CMV mutants lacking their respective MHC class I homologue have been shown to replicate to wild-type levels in fibroblast cultures, demonstrating that the homologues are not essential for virus replication in vitro (BROWNE et al. 1992; FARRELL et al. 1997; BEISSER et al. 2000). For gpUL18, its ability to bind both peptide and β2m suggested that it might play a role in disrupting the CTL recognition of HCMV-infected cells. Indeed, it was initially proposed that gpUL18 might sequester β2m in HCMV-infected cells away from cellular MHC class I molecules, thereby impairing normal MHC class I-mediated antigen presentation (BROWNE et al. 1990). However, studies of a UL18-disrupted HCMV mutant (ΔUL18) have demonstrated that cellular MHC class I downregulation occurs normally after infection in the absence of gpUL18 (BROWNE et al. 1992).

Coinciding with the discovery of multiple genes of HCMV that participate in the downregulation of cellular MHC class I molecules, the inhibitory effect of MHC class I on NK cell function was realised (KARRE 1995). As a consequence, FAHNESTOCK et al. (1995) proposed the following hypotheses. The first hypothesis was that the cell-surface downregulation of MHC class I in HCMV-infected cells might be sufficient to render them susceptible to NK cell-mediated cytolysis. Secondly, given the ability of gpUL18 to bind both β2m and peptide, these authors reasoned that gpUL18 may be sufficiently similar to MHC class I to act as a molecular 'decoy' for NK cells, with the ability to inhibit NK cell function. These hypotheses have also been raised subsequently for gpm144 (FARRELL et al. 1997), and more recently, for gpr144 (BEISSER et al. 2000). A number of reports, discussed below, have provided evidence for or against these hypotheses over recent years. Furthermore, the recent identification of families of receptors specific for cellular MHC class I proteins (BORGES et al. 1997; CELLA et al. 1997; COLONNA et al. 1997) has provoked new hypotheses concerning the immunomodulatory role of cellular MHC class I proteins and, consequently, their viral counterparts (COSMAN et al. 1999; FARRELL et al. 1999, 2000). Notably, the divergence between the human and rodent CMV MHC class I homologues suggests that either they serve a different function during the virus life cycle in vivo or, alternatively, they may possess a conserved function that is, however, mediated via a different mechanism in their respective host species.

3 Cellular Receptors for MHC Class I Proteins

The T cell receptor, responsible for defining the specificity of cytotoxic T cells for MHC class I/peptide ligands, has been well characterised. More recently, both human and mouse NK cells have been shown to possess multiple receptor families that bind MHC class I proteins, certain of which are also found on other immune effector cells (reviewed by YOKOYAMA 1998; LANIER, 1998; LÓPEZ-BOTET and BELLÒN 1999); examples of four such receptor-MHC class I interactions are displayed schematically in Fig. 1. Of these, one family appears to be conserved in both species: the CD94/NKG2 heterodimers. CD94 is believed to be the major determinant of binding specificity, whereas the disulphide-linked NKG2 molecule is critical in transducing intracellular signals via the cytoplasmic domain; thus, whereas the CD94/NKG2A receptor is inhibitory, the CD94/NKG2C receptor is stimulatory. The nature of the intracellular motifs responsible for transducing either inhibitory or stimulatory signals is conserved between different NK cell receptor families and is described in more detail below with reference to the 'killer immunoglobulin-like receptors'. The CD94/NKG2A heterodimer, expressed selectively on NK cells and some subsets of cytotoxic T cells, has been shown to bind non-classic class I molecules, HLA-E (human; BRAUD et al. 1998; LEE et al. 1998) and Qa-1^b (mouse; VANCE et al. 1998). Both HLA-E and Qa-1^b molecules possess a number of similar features; notably, each protein exhibits a restricted peptide repertoire which mainly comprises leader sequences from other MHC class I molecules, and like other MHC class I proteins, peptide binding is required for cell surface expression and stability. Genes for CD94/NKG2 have also been identified in the rat (DISSEN et al. 1997; BERG et al. 1998), although their ability to

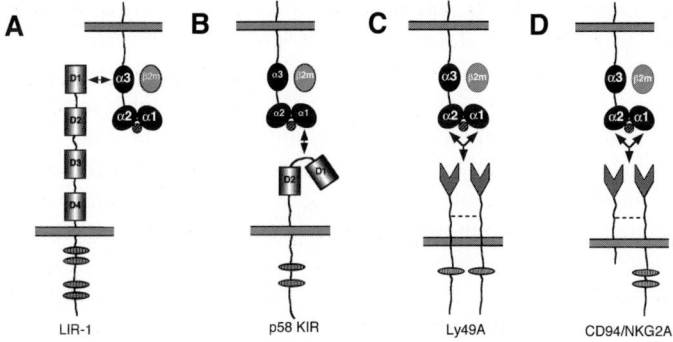

Fig. 1A–D. Interactions between MHC class I and inhibitory receptors. Four examples of interactions between inhibitory receptors and MHC class I are shown. In each case, the MHC class I-β2m-peptide complex is shown, with the α1, α2 and α3 extracellular domains of the class I heavy chain labelled (*solid ovals*), β2m shown as a *grey oval* and peptide depicted as a *small circle*. Panels **A** and **B** depict interaction with immunoglobulin superfamily receptors LIR-1 and p58 KIR respectively. The extracellular immunoglobulin-like domains (*rectangles*) are labelled (D1, D2, etc.). Panels **C** and **D** depict interaction with C-type lectin superfamily receptors, the Ly49A homodimer and CD94/NKG2A heterodimer, respectively. The extracellular carbohydrate recognition domains are depicted as *grey echelons*, and disulphide bonds are indicated by *dashed lines*. In all cases, the ITIM motifs present in the cytoplasmic domains of the receptors are depicted as *vertically hatched ovals*. Regions of interaction between the receptors and the MHC class I-β2m-peptide complex are indicated by *arrows*

form a complex with similar specificity remains to be determined. Given the conservation of these NK receptors and ligands between species, it has been postulated that MHC class I recognition by NK cells is a primitive function, occurring before the divergence of human and murine ancestors.

The predominant receptor family for MHC class I proteins on murine NK cells consists of homodimeric type II lectin-like cell surface molecules. Sixteen related Ly49 receptors (Ly49A–P), encoded by genes within the NK gene complex on mouse chromosome 6, have been identified to date (HELD et al. 1999; MAKRIGIANNIS et al. 1999). The closest functional relatives of the Ly49 gene family in human are the CD94/NKG2 receptors, with no structural orthologues identified at present except for the human Ly49L pseudogene (WESTGAARD et al. 1998). Ly49 receptors have been shown to bind MHC class I proteins in an allele-specific manner. The prototype Ly49 receptor, Ly49A, binds D^d or D^k and has inhibitory functions. The MHC specificities of Ly49 molecules have been determined for at least 5 of the 16 described receptors, and some overlap in specificity has been noted (HANKE et al. 1999). Analysis of the crystal structure of Ly49A bound to H-$2D^d$ has shown that Ly49 binding to MHC requires recognition of the α1 and α2 domains and that the presence of peptide in the peptide-binding groove is essential (TORMO et al. 1999).

In contrast to the lectin-like receptors on mouse NK cells, a large immunoglobulin superfamily (now termed 'killer immunoglobulin-like receptors', or KIRs) has been identified and its members characterised for their ability to regulate human NK cell functions. Comprising either two or three extracellular immunoglobulin domains, the KIRs appear to recognise allele-specific classic MHC class I molecules (reviewed by LANIER 1998; YOKOYAMA 1998). Notably, a particular MHC class I allotype may be recognised by more than one KIR. Although MHC class I specificity is determined by the extracellular immunoglobulin region, the ability to either activate or inhibit NK cell function is dependent on the presence of particular protein motifs within their cytoplasmic domains. KIRs with inhibitory function possess long cytoplasmic domains which invariably exhibit cytoplasmic immunoreceptor tyrosine-based inhibitory motifs (ITIMs; V/IxYxxL/V, with x denoting any amino acid), which, on tyrosine phosphorylation, recruit tyrosine phophatases (SHP-1/SHP-2) which initiate the cascade of intracellular events leading to NK cell inhibition (reviewed by LEIBSON 1997; LÓPEZ-BOTET and BELLÓN 1999). Activating KIR isoforms possess shorter intracytoplasmic domains and contain a charged amino acid residue in the transmembrane region. The activating isoforms are multimeric complexes which recruit a small disulphide-bonded homodimer (DAP-12) that contains an immunoreceptor tyrosine-based activation motif (ITAM) in its cytoplasmic domain (OLCESE et al. 1997; LANIER et al. 1998). The crosslinking of phosphorylated DAP-12 with the activatory KIRs leads to the recruitment of tyrosine kinases (ZAP70, syk) that provide the intracellular activation signals. Although triggering of inhibitory receptors appears to provide a dominant inhibitory signal to NK cells, it has been postulated that the triggering receptors are required to stimulate NK cytotoxicity and cytokine release when the threshold of engagement of inhibitory receptors falls below a critical level. Given that the affinity of inhibitory receptors for MHC class I is greater than for their activatory counterparts, the ability of triggering receptors to override the inhibitory signal would imply that the availability of inhibitory ligand on the target cell and/or the level of expression of inhibitory and activatory receptors can be modulated. Further studies in this area are required to resolve these questions. Murine gp49 receptors, distant homologues of human KIRs (and LIRs, described below), encode both inhibitory (gp49B) and activatory (gp49A) isoforms (WANG et al. 1997). These proteins are expressed on NK cells as well as mast cells. The ligands for the gp49 molecules are unknown, as is their potential interaction with viral MHC class I homologues.

A second family of human immunoglobulin receptors which modulate NK cell function has been identified. Members of the leucocyte immunoglobulin-like receptors (LIR) family (LIR1–8; also known as immunoglobulin-like transcripts or ILTs) possess two or four extracellular immunoglobulin domains that are related to the KIR family (BORGES et al. 1997; COSMAN et al. 1999). In addition, they also contain intracytoplasmic ITIM motifs which inhibit activatory signalling. However, unlike KIRs, which recognise allele-specific HLA molecules, a number of LIRs bind a broad range of classic and non-classic MHC class I proteins (VITALE et al. 1999). Remarkably, this family of receptors is not restricted to NK cells but may be found on a range of leucocytes, including B and T cells, myeloid and dendritic cells (DC) (BORGES et al. 1997; COLONNA et al. 1997, 1998; FANGER et al. 1998). The LIR/ILT have been shown to downregulate NK and T cell cytotoxicity and to inhibit Ca^{2+} mobilisation in B cells and macrophages (CELLA et al. 1997; COLONNA et al. 1997, 1998; FANGER et al. 1998). Thus they have the potential for global modulation of immune responses.

Studies by COSMAN et al. (1997) have established that LIR-1 binds both MHC class I molecules and HCMV gpUL18. To date, LIR-1 has the broadest cellular distribution of this subfamily, being identified on dendritic cells, monocytes and B cells, with a lower distribution on T cells and on only a subset of NK cells. LIR-1 is an inhibitory receptor; its interaction with MHC class I on NK cells and T cells protects target cells from NK- and T cell-mediated cytolysis and inhibits intracellular Ca^{2+} mobilisation in other cell types (CELLA et al. 1997; COLONNA et al. 1997, 1998; FANGER et al. 1998). No other LIR (or KIR) has been shown to bind gpUL18, suggesting that this molecule has been specifically targeted by HCMV. Indeed, recent binding studies have shown that gpUL18 binds LIR-1 with > 1,000-fold affinity compared with MHC class I proteins (CHAPMAN et al. 1999). However, it should be noted that the affinity of LIR-1 for cellular MHC class I is of the order of that observed for the KIRs and class I, demonstrating a remarkable feature of the gpUL18/LIR-1 interaction. The binding affinity of gpUL18 for LIR-1 was not significantly affected by altering the level of carbohydrate on gpUL18, suggesting that LIR-1 interacts with a protein rather than a carbohydrate epitope. Furthermore, recognition of gpUL18 was shown to be independent of bound peptide (CHAPMAN et al. 1999).

Immunoglobulin domain-swapping experiments between gpUL18 and HFE (a cellular immunoglobulin-β2m-class I-like heterodimer unable to bind LIR-1) showed that the α-3 region of gpUL18, the domain with the highest level of conservation with MHC class I molecules, is required for the interaction with LIR-1 (CHAPMAN et al. 1999). Similarly, LIR-1 interacts with the α-3 domain of cellular MHC class I proteins, albeit with lower affinity. The α-3 domain is the least polymorphic of the three domains of cellular MHC class I proteins, which may account for the broad MHC class I specificity of LIR-1. LIR-1 possesses four extracellular immunoglobulin domains, of which the N-terminal domain alone has been shown to comprise the primary binding sites for both UL18 and MHC class I. In contrast to LIR-1-MHC class I interactions, two KIR immunoglobulin domains are required for binding to MHC class I proteins. Furthermore, the KIRs bind to the polymorphic α-1/α-2 domains of class I proteins (FAN et al. 1997; LANIER et al. 1997), which may therefore account for their allelic specificity.

In the mouse, the paired immunoglobulin-like receptors (PIRs) represent possibly the closest homologues of the human LIRs (KUBAGAWA et al. 1997). Like the LIRs, these receptors are distributed on myeloid and B cell lineages (KUBAGAWA et al. 1997, 1999). Unlike the LIRs, however, PIRs are not expressed on NK cells (BLERY et al. 1998; KUBAGAWA et al. 1999). Functionally, PIR-A isotypes have been shown to mediate activation, whereas PIR-B has inhibitory functions. The cellular ligands for PIRs have not been identified, but interaction with the MCMV MHC class I homologue, m144, has been excluded (T. Chapman and P. Bjorkman, personal communication).

4 Consequences of UL-18-LIR-1 Interactions on the HCMV Life Cycle

The finding that UL18 binds LIR-1 with much greater affinity than cellular MHC class I molecules provides strong evidence that UL18/LIR-1 interactions play a significant immunomodulatory role. Indeed, this finding reconciles observations that UL18 transcript and protein expression in HCMV-infected cells is low (HASSAN-WALKER et al. 1998; LEONG et al. 1998). The high affinity of the interaction may be sufficient to signal to cells via LIR-1, particularly as cellular MHC class I proteins are downregulated during HCMV infection. Nevertheless, despite a proven association between UL18 and LIR-1, the mode of action of UL18 on host immunity and/or HCMV replication has yet to be established.

5 Possible Effects of the Viral MHC Class I Homologues on the Functions of Monocytes and DCs

It has been demonstrated that monocytes and dendritic cells are major reservoirs in the natural life cycle of HCMV (TAYLOR-WEIDEMAN et al. 1994; HAHN et al. 1998). Reactivation of HCMV in long-term cultures of allogeneically stimulated PBL from latently infected individuals has shown that infectious virus can be recovered from myeloid cells expressing both macrophage and dendritic cell markers (SÖDERBERG-NAUCLÉR et al. 1997). Lytic-phase HCMV transcripts are expressed only when such cultures are stimulated to differentiate with growth or pro-inflammatory cytokines. These findings are consistent with the view that HCMV becomes latent in a haemopoietic cell precursor from which it reactivates after cellular differentiation. Indeed, HCMV has been detected in myeloid $CD33^+$ progenitor cells, including populations expressing dendritic cell markers, CD1a and CD10 (HAHN et al. 1998). After differentiation by various stimuli, infectious virus was recovered, lending further support to the notion that cellular differentiation is an important factor in HCMV reactivation.

Among the lytic-phase transcripts detected in peripheral blood mononuclear cells is UL18, although levels were reported to be low (HASSAN-WALKER et al. 1998). Although the effect of UL18-LIR-1 interactions in monocytes and dendritic cells has not been reported, it has been speculated (COSMAN et al. 1999) that UL18-LIR-1 interactions might modulate cellular differentiation and hence either drive, or prevent, the establishment of HCMV latency. Clearly, studies of the intracellular signalling events that arise after UL18-LIR-1 interactions in cells relevant to clinical infection will be required to address these issues.

In vitro, it has been shown that infection of monocyte-derived DC (mDC) with HCMV results in the production of infectious virus (RIEGLER et al. 2000a). Efficient infection of mDC is highly dependent on the virus having been propagated on endothelium-derived cells, with virus having been propagated on fibroblasts showing an 80- to 90-fold reduction in efficiency of infection (RIEGLER et al. 2000b). Indirect immunoperoxidase techniques revealed that the temporal cascade of HCMV protein expression (i.e. at immediate-early, early and late times post-infection) was slightly delayed compared with infected fibroblast cultures. Notably, HCMV infection was shown to downregulate the expression of CD80, CD86 and MHC class I, but no alteration was observed in the levels of CD1a, CD83 and MHC class II (RIEGLER et al. 2000b). Functional impairment of HCMV-infected DC involves an inhibition of their ability to induce the proliferative response of PBLs in a mixed lymphocyte reaction and an inhibited capacity of infected DC to mature in response to TNF-α (RIEGLER et al. 2000b).

Recent studies have shown MCMV infection of immature DCs at titres exceeding those observed after infection of embryonic fibroblasts (ANDREWS et al. 2001). Furthermore, like HCMV-infected DCs, the kinetics of MCMV replication in murine DCs were delayed when compared with standard fibroblast infections. This 'slowing' of replication is further accentuated when mature DCs are infected. As for HCMV, MCMV infection appears to downregulate the expression of CD80 and MHC class I, although unlike HCMV, MCMV also downregulates the expression of MHC class II and has no effect on CD86 expression. The contribution of gpm144 to these events has yet to be determined. Notably, recent data have identified gpm144 expression (using antisera directed against the gpm144-β2m complex) in the follicular compartments of MCMV-infected spleens (D. Andrews and M. Degli-Esposti, unpublished observations). The proximity of m144-positive cells with cells derived from the myelomonocytic lineage would provide an opportunity for interactions in vivo and is the subject of current investigation.

The effect of murine and human CMV infection on a variety of immunoregulatory DC antigens is interesting considering the important roles these cells play in early immune defence against virus infections. Present in the skin, mucosa and numerous tissues, DCs are amongst the earliest members of the immune system to come into contact with virus and are known to play a critical role in the immune response through their ability to process antigen, migrate to lymphoid organs and, assisted by the release of a cascade of cytokines such as IFN-γ and IL-12, initiate antiviral T cell responses (BANCHEREAU and STEINMAN 1998; KLAGGE and SCHNEIDER-SCHAULIES 1999). More recently, it has been demonstrated that murine DCs are capable of stimulating NK cell cytolytic activity and IFN-γ production via direct cell–cell contact (FERNANDEZ et al. 1999). Given the pivotal role of DCs in shaping the immune response to virus infection, they may provide a fertile site for HCMV and MCMV to deploy their armory of immune subversive proteins, thus allowing efficient virus dissemination within the natural host. As LIR-1, the cellular receptor for gpUL18, is found at high levels on DCs, it is possible that DCs are major targets for the action of gpUL18. The potential of HCMV and MCMV to modulate DC-mediated regulation of CTL and NK cell responses and the contribution of gpUL18 and gpm144 in modulating the functions of monocytes and/or DCs are important aspects of future studies.

6 Role of CMV-Encoded MHC Class I Homologues as Surrogate Inhibitory Ligands for NK Cells

Given the comprehensive strategies employed by HCMV to downregulate cellular MHC class I proteins, it has also been suggested that UL18 might act as a surrogate MHC class I molecule able to inhibit NK cell-mediated killing of infected cells. Studies testing this hypothesis have provided conflicting results.

Initial studies using human MHC class I-negative cells (721.221) transfected with UL18 showed that gpUL18 blocked NK cell activation by triggering CD94/NKG2Al (REYBURN et al. 1997). Notably, however, HLA-E transcripts are detected in 721.221 cells, and HLA-E/β2m complexes can be stabilised and expressed at the cell surface after the binding of signal sequence peptides derived from cellular MHC class I molecules. Although the UL18 signal sequence does not conform to the consensus of cellular MHC class I, it is possible that the resistance of UL18-transfected cells to NK cell killing was mediated by the stabilization of HLA-E rather than UL18 expression. Indeed, HLA-E-mediated inhibition of NK cells has been postulated as an alternative NK cell evasion mechanism for HCMV. An involvement of CD94/NKG2a-mediated inhibition of target cell lysis has been recently demonstrated for another HCMV gene product, gpUL40, which possesses a signal sequence conserved with cellular MHC class I signal sequence-derived peptides known to be ligands of HLA-E (TOMASEC et al. 2000; ULBRECHT et al. 2000). Notably, although classic MHC class I molecules are downregulated during HCMV infection, HLA-E is upregulated. Expression of gpUL40 in HLA-E-positive target cells conferred resistance to NK cell lysis via engagement of the CD94/NKG2a receptor. In addition, although binding of HLA-E to cellular MHC class I derived peptides is TAP-dependent, the stabilization of HLA-E by gpUL40 was shown to be TAP-independent, suggesting that HLA-E expression may still occur in HCMV-infected cells in which gpUS6 interferes with TAP1/2 function. Accordingly, it has been proposed that gpUL40-mediated stabilisation of HLA-E at the cell surface may be a mechanism by which HCMV-infected cells acquire resistance to NK cell-mediated lysis.

Subsequent to studies implicating a role for UL18 in protecting infected cells from NK cell-mediated lysis, it was reported that fibroblasts infected with a UL18 knockout virus did not gain susceptibility to NK cell killing (LEONG et al. 1998). Indeed, NK cell recognition of HCMV-infected fibroblasts has been found to correlate with upregulation of molecules important for NK cell adhesion and/or activation, thus questioning the importance of MHC class I shutdown as a factor contributing to the triggering of NK cells (LEONG et al. 1998; FLETCHER et al. 1998). Recent studies have demonstrated that fibroblasts may lack the positive signals required for NK cell triggering, and thus NK cells might fail to respond to the downregulation of cellular MHC class I molecules in these target cells. Subsequent studies employing U373 malignant glioma cells that possess such positive signals (i.e. these cells could be lysed by KIR2D2$^+$ NK cell clones by the addition of anti-MHC class I or anti-KIR antibodies) have demonstrated increased NK cell susceptibility after MHC class I shutdown during HCMV infection (HUARD and FRÜH 2000). Furthermore, HCMV gpUS11, which diverts newly synthesized MHC class I from the ER lumen to the cytosol, is sufficient to trigger NK cell cytotoxicity against US11-transfected HeLa cells. Whether gpUL18 has an effect on NK cell killing in this system remains to be determined. Nevertheless, these experiments demonstrate that the outcome of interactions between HCMV-infected cells and NK cells is dependent on both virus-induced upregulation of adhesion/activatory molecules and downregulation of MHC class I molecules and the presence of appropriate NK cell ligands on the targets.

Other mechanisms of HCMV-induced modulation of infected cells to NK cell killing, not related to MHC class I shutdown or upregulation of NK cell adhesion molecules, may also be important in infections with clinical HCMV strains (CERBONI et al. 2000). Notably, these studies showed that in fibroblasts, resistance to NK cell-mediated lysis is not dependent on engagement of LIR-1. Furthermore, recent reports have stated that, like infected fibroblasts, HCMV-infected endothelial cells and macrophages are resistant to NK cell lysis and that this resistance is independent of virally induced downregulation of either cellular MHC class I or adhesion molecules, as well as UL18 expression (ODEBERG et al. 2000).

The finding that the resistance of HCMV-infected cells to NK cell-mediated lysis is independent of UL18 expression suggests that it may not participate as a MHC class I 'decoy'. Indeed, the fact that the receptor for UL18, LIR-1, is present on only a small proportion of NK cells is consistent with the notion that UL18 does not interact directly with NK cells unless a LIR$^+$ NK subpopulation is induced during HCMV infection and is critical for virus clearance. Although transcripts of UL18 have been identified in mononuclear PBL during HCMV viraemia (HASSAN-WALKER et al. 1998), the location and nature of cells expressing the gpUL18 glycoprotein in vivo have yet to be demonstrated. Further studies that address gpUL18 and LIR-1$^+$ antigen expression in cells from HCMV-infected tissues will shed some light on the function of gpUL18 in vivo.

7 Use of Rodent Models of CMV Infection to Determine the Function of CMV MHC Class I Homologues In Vivo

Both rodent CMVs provide useful systems to analyse virus-host interactions and hence to discern the role of immune-evasive proteins encoded by their genomes. In the case of MCMV, a wealth of knowledge concerning the importance of host genetic factors and dissection of various compartments of the immune response in limiting virus dissemination and disease in vivo has been generated over the past two decades.

Although a functional role for gpUL18 has yet to be clearly demonstrated, in vivo studies with an MCMV mutant deleted of gpm144 (Δm144) have demonstrated that it is cleared more rapidly than wild-type MCMV during the acute stages of infection (FARRELL et al. 1997). This increased clearance was not explained by a replication defect for the mutant virus. In fact, despite relatively low levels of replication during the first 5 days after infection, the gpm144 null virus was able to disseminate to salivary glands (10 days after infection), replicating at this site to titres similar to those of wild-type virus. Virulence of Δm144 was restored by in vivo depletion of host NK cells but was largely unaffected by depletion of T cells, indicating that gpm144 plays a role in inhibition of NK cell-mediated viral clearance.

The importance of NK cells in controlling MCMV replication has been well established. NK cells are induced shortly after MCMV infection, although the contribution of NK cells towards MCMV clearance varies according to the genetic constitution of the host (BANCROFT et al. 1981). Mouse strains with low or absent endogenous NK cell levels are highly susceptible to MCMV infection (SHELLAM et al. 1981; BUKOWSKI et al. 1984), and adoptive transfer studies have demonstrated a protective role for NK cells against MCMV (BUKOWSKI et al. 1985). Genetic analysis of MCMV-susceptible and -resistant mouse strains has lead to the identification of the *Cmv1* locus, which restricts the early replication of MCMV in the spleen through the cytotoxic action of NK cells (SCALZO et al. 1992). Notably, the *Cmv1* locus maps to the NK gene complex (NKC) on mouse chromosome 6, proximal to an array of genes encoding regulatory molecules that are expressed predominantly on NK cells (SCALZO et al. 1995). Interestingly, loci that affect (a) resistance to infection with the mouse poxvirus ectromelia (*Rmp1*) (DELANO and BROWNSTEIN 1995), (b) magnitude of acute and latent herpes simplex virus infections (*Rhs1*) (SIMMONS et al. 2000), and (c) possibly resistance to *Leishmania* (BEEBE et al. 1997; ROBERTS et al. 1999), are also NKC-linked, suggesting that this genetic region may be critical to innate resistance against a broad spectrum of pathogens. A syntenic NKC region is located on human chromosome 12, and it is possible that a functional homologue of *Cmv1* may contribute to NK cell-mediated responses in humans.

Some investigators have speculated that the MCMV-encoded MHC class I homologue may bind one of the Ly49 inhibitory receptors, thus exerting its inhibitory effect on NK cells (LONG 1999). This assumption is based on the fact that *Cmv1* maps to the NKC where the Ly49 genes are encoded. To date, however, there is no evidence that the *Cmv1* protein is a receptor for gpm144, and indeed, given that the gpm144 effects are also observed in BALB/c mice that lack the *Cmv1* resistance allele (FARRELL et al. 1997), it is unlikely that the *Cmv1* locus encodes the gpm144 ligand. Binding of gpm144 to a subset of Ly49 receptors has been tested, and the findings from those studies are discussed later.

In mice that lack the *Cmv1* resistance allele ($Cmv1^s$), $CD8^+$ T cells are critical for MCMV clearance from visceral organs during a primary acute infection. In contrast, mouse strains possessing the *Cmv1* resistance allele (*Cmv1^r*) dramatically reduce the virus load in visceral organs during acute infection through the action of NK cells. Furthermore, evidence exists that the mechanisms of NK cell control during acute infection of the spleen and liver are quite different. In the liver, NK cell-activatory cytokines, such as IFN-γ and IL-12, are critical for NK cell-mediated virus clearance, whereas in the spleen, virus clearance by NK cells is mediated by cytotoxic mechanisms (TAY and WELSH 1997). Nevertheless, despite the action of NK cells, ultimate clearance of infectious virus in *Cmv1^r* mice is also dependent on CD8$^+$ T cells.

In *Cmv1^r* mice depleted of NK1.1$^+$ cells, the Δm144 mutant was shown to replicate to levels approaching those observed in untreated animals infected with wild-type MCMV, demonstrating an important role of NK cells in mediating early clearance of MCMV in the absence of gpm144 (FARRELL et al. 1997). In addition, treatment of *Cmv1^r* mice with neutralising anti-IL-12 antibody in vivo was found to increase Δm144 titres to levels approaching those of wild-type MCMV in the liver, suggesting that gpm144 might interfere with IL-12-mediated activation of lymphocytes, including NK cells (FARRELL et al. 1999). The importance of studying cell types relevant to NK cell activation/inhibition in vivo during MCMV infection has been demonstrated recently by findings showing that myelomonocytic cell subsets represent a major portion of the cells recruited to the sites of MCMV infection (D. Andrews, H. Farrell and M. Degli-Esposti, unpublished observations). The relevance of these cells in modulating NK cell function is currently under investigation.

Studies of the in vivo function of the RCMV gpr144 have also been performed (BEISSER et al. 2000). In contrast to the results obtained with MCMV gpm144, no reduction in replication efficacy was observed after infection with an r144 null (Δr144) recombinant compared with wild-type virus. However, as noted by the authors, the RCMV infection system used requires immunosuppression (via whole body irradiation) of the rats before infection. This pre-treatment will clearly have a major effect on cellular immunity, including NK cell function, and hence any potential attenuation of Δr144 via increased sensitivity to cellular immunity may be obscured by ablation of the relevant cellular effectors. In accordance with the MCMV studies, Δr144 showed no alteration with respect to virus dissemination to the salivary glands and virus persistence.

As for HCMV studies, there is currently a lack of suitable in vitro models to test a direct effect of m144 (or r144) in inhibiting NK cell-mediated cytotoxicity in virus-infected cells. Furthermore, as with gpUL18, cell surface expression of m144 in fibroblasts in vitro is very low, suggesting that (a) m144 exerts its effect because of high-affinity interactions with its receptor; (b) m144 is expressed at higher levels in other cell types which are involved in inhibiting NK cell function in vivo or (c) m144 initiates a cascade of immunomodulatory events which interfere with NK cell function by an indirect mechanism. Because m144 is expressed in specific myelomonocytic cells within the marginal zones of splenic follicles in vivo (D. Andrews, H. Farrell and M. Degli-Esposti, unpublished observations), the relevance of these cell types in modulating NK cell function either directly or indirectly is the subject of current investigation.

Unlike gpUL18, whose ectopic expression has proved very difficult to maintain, gpm144 can be stably expressed. Transient or stable expression of gpm144 in a variety of transfected cells has demonstrated a partial inhibitory effect of gpm144 on NK cell cytotoxicity, suggesting that gpm144 may indeed exert a direct effect on NK cell function (CRETNEY et al. 1999; KUBOTA et al. 1999). Like gpUL18, it is possible that gpm144 shares a receptor with a cellular MHC class I molecule. Unfortunately, the identification of such a receptor has thus far remained elusive. Notably, antibodies directed against known murine NK cell inhibitory receptors, Ly49A, C, G and I, did not interfere with the inhibitory effect of gpm144, suggesting that gpm144 engages a receptor other than the above Ly49s (KUBOTA et al. 1999). These studies were confirmed and extended by studies using soluble gpm144 to test for binding to a number of the Ly 49 receptors. In these studies, gpm144 was shown not to bind to Ly49A, B, D, E and G, indicating that none of the tested Ly49 molecules is the receptor for gpm144 (D. Andrews and M. Degli-Esposti, unpublished observations). As discussed above, a potential interaction between gpm144 and the PIRs was also tested with a soluble gpm144-β2m complex. Binding of both PIR-A and PIR-B proved negative, re-enforcing the possibility that gpm144 binds a novel receptor (T. Chapman and P. Bjorkman, personal communication).

Interestingly, studies of the effects of gpm144 on tumour clearance have suggested that this molecule plays a crucial role in reducing the recruitment and activation of NK cells. Using a tumour model in which gpm144 was expressed on the surface of MHC class I-deficient RMA-S cells, we have shown that the presence of gpm144 affects the recruitment and activation of NK cell effectors in the peritoneum – the site of tumour challenge (CRETNEY et al. 1999). Thus challenge with RMA-S-m144 tumour cells resulted in a twofold decrease in the number of NK cells which accumulate in the peritoneum and in a threefold decrease in their lytic potential. Complementary effects have been noted in viral infection, where challenge with the recombinant virus lacking gpm144 was found to lead to an increase in the number and cytolytic activity of NK cell effectors in the spleen, compared with challenge with wild-type, gpm144-expressing virus (E. Densley, H. Farrell and M. Degli-Esposti, unpublished observations).

The results of the above in vitro and in vivo studies suggest that the m144 effects on NK cells may be operating at a level other than direct interaction between NK cells and virus-infected target cells. Potential models for either direct or indirect inhibition of NK cell activity against MCMV-infected cells are displayed in Fig. 2. By analogy with HCMV, a possible scenario involves interaction with monocytes and/or DC, which in humans are known to express the gpUL18 ligand LIR-1. The sequence divergence between gpUL18 and gpm144 suggests that they may bind distinct ligands, although it should be noted that for both molecules the major region of homology to cellular MHC class I lies in the α-3 domain, which, in addition to associating with β2m, is known to be essential for gpUL18 binding to LIR-1. Identification and characterisation of the gpm144 ligand will clarify these speculations and shed further light on the role of viral MHC class I homologues in evasion of host immune responses.

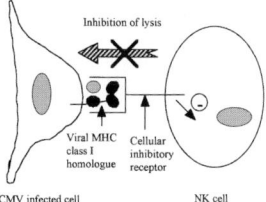

Fig. 2A,B. Possible functions of the CMV-encoded MHC class I homologues: models for inhibition of NK cell activity against CMV-infected cells. Three alternative models for the function of a CMV-encoded MHC class I homologue are depicted, involving either direct (**A**) or indirect (**B**) inhibition of NK cell activity. In all cases, it is proposed that the viral MHC class I homologue engages a cellular inhibitory receptor, which transduces inhibitory signals via the intracellular domain. In panel **A**, the inhibitory receptor is expressed on NK cells, resulting in direct inhibition of NK cell-mediated cytolysis of CMV-infected cells. In panel **B**, the inhibitory receptor is expressed on monocytes or dendritic cells (*DC*), resulting in indirect inhibition of NK cell activity by either (i) inhibition of production of cytokines which are required for optimal NK cell recruitment/activity or (ii) inhibition of direct stimulation of NK cells via cell–cell contact

8 Concluding Remarks

It is now over a decade since an MHC class I homologue encoded by HCMV was identified. Initial theories concerning the function of such a molecule focussed on the possibility that it may interfere with processing of endogenous cellular MHC class I, thereby interfering with CTL recognition of CMV-infected cells via the T cell receptor. These initial proposals have not been supported by subsequent experimentation, however, and it is now recognised that both HCMV and MCMV possess a variety of distinct mechanisms for downregulation of cell surface presentation of peptides by MHC class I. After the discovery of diverse receptors for MHC class I expressed by a variety of cell types, in particular the inhibitory receptors of NK cells, attention has now switched to the possibility that the CMV-encoded MHC class I homologues may inhibit NK cell activity against virus-infected cells, either directly or via an intermediary cell type. For MCMV, in vivo studies have indeed demonstrated a role for m144 in interfering with NK cell-mediated clearance. There is still considerable debate concerning the functions of the MHC class I homologues, however, and direct demonstration of protective effects of these molecules against NK cell-mediated lysis of CMV-infected cells in tissue culture has proven elusive. It is anticipated that clarification of these functions will await future improvements in our knowledge of cellular receptors for MHC class I and of the interaction between the CMVs and cells expressing such receptors.

References

Andrews DM, Andoniou CE, Granucci F, Ricciardi-Castagnoli P, Degli-Esposti MA (2001) Infection of dendritic cells by murine cytomegalovirus induces functional paralysis. Nat Immunol 11:1077–1084
Bancherau J, Steinman RM (1998) Dendritic cells and the control of immunity. Nature 392:245–252
Bancroft GJ, Shellam GR, Chalmers, JE (1981) Genetic influences on the augmentation of natural killer (NK) cells during murine cytomegalovirus infection: correlation with patterns of resistance. J Immunol 126:988–994
Beck S, Barrell BG (1988) Human cytomegalovirus encodes a glycoprotein homologous to MHC class-I antigens. Nature 331:269–272
Beebe AM, Mauze S, Schork NJ, Coffman RL (1997) Serial backcross mapping of multiple loci associated with resistance to *Leishmania major* in mice. Immunity 6:551–557
Beisser PS, Kloover JS, Grauls GELM, Blok MJ, Bruggeman CA, Vink C (2000) The r144 major histocompatibility complex class I-like gene of rat cytomegalovirus is dispensable for both acute and long-term infection in the immunocompromised host. J Virol 74:1045–1050
Berg SF, Dissen E, Westgaard IH, Fossum S (1998) Two genes in the rat homologous to human NKg2. Eur J Immunol 28:444–450
Bjorkman PJ, Parham P (1990) Structure, function and divergence of class I major histocompatibility complex molecules. Annu Rev Biochem 90:253–288
Blery M, Kubagawa H, Chen CC, Vely F, Cooper MD, Vivier E (1998) The paired Ig-like receptor PIR-B is an inhibitory receptor that recruits the protein-tyrosine phosphatase SHP-1. Proc Natl Acad Sci USA 95:2446–2451
Borges L, Hsu ML, Fanger N, Kubin M, Cosman D (1997) A family of human lymphoid and myeloid Ig-like receptors, some of which bind to MHC class I molecules. J Immunol 159:5192–5196
Braud VM, Allen DSJ, O'Callaghan CA, Söderström K, D'Andrea A, Ogg GS, Lazetic S, Young NT, Bell JI, Phillips JH, McMichael AJ (1998) HLA-E binds to natural killer-cell receptors CD94/NKG2A. Nature 391: 795–799
Browne H, Churcher M, Minson T (1992) Construction and characterization of a human cytomegalovirus mutant with the UL18 (class I homolog) gene deleted. J Virol 66:6784–6787
Browne H, Smith G, Beck S, Minson T (1990) A complex between the MHC class I homologue encoded by human cytomegalovirus and β2 microglobulin. Nature 347:770–772
Bukowski JF, Warner JF, Dennert G, Welsh RM (1985) Adoptive transfer studies demonstrating the antiviral effects of natural killer cells in vivo. J Exp Med 161:40–52
Bukowski JF, Woda BA, Welsh RM (1984) Pathogenesis of murine cytomegalovirus infection in natural killer cell-depleted mice. J Virol 52:119–128
Cella M, Dohring C, Samaridis J, Dessing M, Brockhaus M, Lanzavecchia A, Colonna M (1997) A novel inhibitory receptor (ILT3) expressed on monocytes, macrophages, and dendritic cells involved in antigen processing. J Exp Med 185: 1743–1751
Cerboni C, Mousavi-Jazi M, Linde A, Söderström K, Brytting M, Wahren B, Kärre K, Carbone E. (2000) Human cytomegalovirus strain-dependent changes in NK cell recognition of infected fibroblasts. J Immunol 164:4775–4782

Chapman TL, Heikema AP, Bjorkman PJ (1999). The inhibitory receptor LIR-1 uses a common binding interaction to recognize class I MHC molecules and the viral homolog UL18. Immunity 11: 603–613

Chapman TL, Bjorkman PJ (1998) Characterization of a murine cytomegalovirus class I MHC homolog: comparison to MHC molecules and the human cytomegalovirus MHC homolog. J Virol 72: 460–466

Colonna M, Nakajima H, Cella M (2000) A family of inhibitory and activating Ig-like receptors that modulate function of lymphoid and myeloid cells. Semin Immunol 12:121–127

Colonna M, Navarro G, Bellón T, Llano M, Garcia P, Samaridis J, Angman L, Cella M, López-Botet M (1997) A common inhibitory receptor for major histocompatibility complex class I molecules on human lymphoid and myelomonocytic cells . J Exp Med 186:1809–1818

Colonna M, Samaridis J, Cella M, Angman L, Allen RL, O'Callaghan CA, Dunbar R, Ogg GS, Cerundolo V, Rolink (1998) Human myelomonocytic cells express an inhibitory receptor for classical and nonclassical MHC class I molecules. J Immunol 160:3096–3100

Cosman D, Fanger N, Borges L, Kubin M, Chin W, Peterson L, Hsu ML (1997) A novel immunoglobulin superfamily receptor for cellular and viral MHC class I molecules. Immunity 7: 273–282

Cosman D, Fanger N, Borges L (1999) Human cytomegalovirus, MHC class I and inhibitory signalling receptors: more questions than answers. Immunol Rev 168:177–185

Cretney E, Degli-Esposti MA, Densley EH, Farrell HE, Davis-Poynter NJ, Smyth MJ (1999) m144, a murine cytomegalovirus (MCMV)-encoded major histocompatibility complex class I homologue, confers tumor resistance to natural killer cell-mediated rejection in the peritoneum. J Exp Med 190:435–444

Davis-Poynter NJ, Farrell HE (1997) Human and murine cytomegalovirus evasion of cytotoxic T lymphocyte and Natural Killer cell-mediated immune responses. Semin Virol 8:369–376

Delano ML, Brownstein DG (1995) Innate resistance to lethal mousepox is genetically linked to the NK gene complex on chromosome 6 and correlates with early restriction of virus replication by cells with an NK phenotype. J Virol 69:5875–5877

Dissen E, Berg SF, Westgaard IH, Fossum S (1997) Molecular characterization of a gene in the rat homologous to human CD94. Eur J Immunol 27: 2080–2086

Doherty PC, Zinkernagel RM (1975) H-2 compatibility is required for T-cell-mediated lysis of target cells infected with lymphocytic choriomeningitis virus. J Exp Med 141:502–507

Fahnestock ML, Johnson JL, Feldman RMR, Neveu JM, Lane WS, Bjorkman PJ (1995) The MHC class I homolog encoded by human cytomegalovirus binds endogenous peptides. Immunity 3:583–590

Fan QR, Mosyak L, Winter CC, Wagtmann N, Long EO, Wiley DC (1997) Structure of the inhibitory receptor for human natural killer cells resembles haematopoietic receptors. Nature 389:96–100

Fanger NA, Cosman D, Peterson L, Braddy SC, Maliszewski CR, Borges L (1998) The MHC class I binding proteins LIR-1 and LIR-2 inhibit Fc receptor-mediated signaling in monocytes. Eur J Immunol 28:3423–3434

Farrell H, Degli-Esposti M, Densley E, Cretney E, Smyth M, Davis–Poynter N (2000) Cytomegalovirus MHC class I homologues and natural killer cells: an overview. Microbes Inf 2:521–532

Farrell HE, Degli-Esposti MA, Davis-Poynter NJ (1999) Cytomegalovirus evasion of natural killer cell responses. Immunol Rev 168: 187–197

Farrell HE, Davis-Poynter NJ (1998) From sabotage to camouflage: viral evasion of cytotoxic T lymphocyte and natural killer cell-mediated immunity. Sem Cell Develop Biol 9: 369–378

Farrell HE, Vally H, Lynch DM, Fleming P, Shellam GR, Scalzo AA, Davis-Poynter NJ (1997) Inhibition of natural killer cells by a cytomegalovirus MHC class I homologue in vivo. Nature 386: 510–514

Fernandez NC, Lozier A, Flament C, Ricciardi-Castagnoli P, Bellet D, Suter M, Perricaudet M, Tursz T, Maraskovsky E, Zitvogel L (1999) Dendritic cells directly trigger NK cell functions: cross-talk relevant in innate anti-tumor immune responses in vivo. Nature Med 5:405–411

Fletcher JM, Prentice HG, Grundy JE (1998) Natural killer cell lysis of cytomegalovirus (CMV)-infected cells correlates with virally induced changes in cell surface lymphocyte function-associated antigen-3 (LFA–) expression and not with the CMV-induced down-regulation of cell surface class I HLA. J Immunol 161:2365–2374

Früh K, Gruhler A, Krishna RM, Schoenhals GJ (1999) A comparison of viral immune escape strategies targeting the MHC class I assembly pathway. Immunol Rev 168: 167–176

Garcia KC (1999) Molecular interactions between extracellular components of the T-cell receptor signalling complex. Immunol Rev 172:73–85

Hahn G, Jores R, Mocarski ES (1998) Cytomegalovirus remains latent in a common precursor of dendritic and myeloid cells. Proc Natl Acad Sci USA 95:3937–3942

Hanke T, Takizawa H, McMahon CW, Busch DH, Pamer EG, Miller JD, Altman JD, Liu Y, Cado D, Lemonnier FA, Bjorkman PJ, Raulet DH (1999) Direct assessment of MHC class I binding by seven Ly49 inhibitory NK cell receptors. Immunity 11:67–77

Hassan-Walker AF, Cope AV, Griffiths PD, Emery VC (1998) Transcription of the human cytomegalovirus natural killer decoy gene, UL18 in vitro and in vivo. J Gen Virol 79:2113–2116

Held W, Kunz B, Ioannidis V, Lowin-Kropf B (1999) Mono-allelic Ly49 NK cell receptor expression. Sem Immunol 11:349–355

Hengel H, Brune W, Koszinowski UH (1998) Immune evasion by cytomegalovirus – survival strategies of a highly adapted opportunist. Trends Microbiol 6:190–197

Hengel H, Reusch U, Gutermann A, Ziegler H, Jonjic S, Lucin P, Koszinowski UH (1999) Cytomegaloviral control of MHC class I function in the mouse. Immunol Rev 168: 167–176

Huard B, Früh K (2000) A role for MHC class I down-regulation in NK cell lysis of herpes virus-infected cells. Eur J Immunol 30:509–515

Karrë K (1995) Express yourself or die: peptides, MHC molecules, and NK cells. Science 267:978–979

Klagge IM, Schneider-Schaulies S (1999) Virus interactions with dendritic cells. J Gen Virol 80:823–833

Kubagawa H, Burrows PD, Cooper MD (1997) A novel pair of immunoglobulin-like receptors expressed by B cells and myeloid cells. Proc Natl Acad Sci USA 94:5261–5266

Kubagawa H, Chen CC, Ho LH, Shimada TS, Gartland L, Mashburn C, Uehara T, Ravetch JV, Cooper MD (1999) Biochemical nature and cellular distribution of the paired immunoglobulin-like receptors, PIR-A and PIR-B. J Exp Med 189:309–318

Kubota A, Kubota S, Farrell HE, Davis-Poynter N, Takei F (1999) Inhibition of NK cells by murine CMV-encoded class I MHC homologue, m144. Cell Immunol 191: 145–151

Lanier LL (1998) NK cell receptors. Annu Rev Immunol 16:359–393

Lanier LL, Corliss B, Phillips JH (1997) Arousal and inhibition of human NK cells. Immunol Rev 155:145–154

Lanier LL, Corliss BC, Wu J, Leong C, Phillips JH (1998) Immunoreceptor DAP12 bearing tyrosine-based activation motif is involved in activating NK cells. Nature 391:703–707

Lee N, Llano M, Carretero M, Ishitani A, Navarro F, López-Botet M, Geraghty DE (1998) HLA-E is a major ligand for the NK inhibitory receptor CD94/NKG2A. Proc Natl Acad Sci USA 5199–5204

Lehner PJ, Cresswell P (1996) Processing and delivery of peptides by MHC class I molecules. Curr Opin Immunol 8:59–67

Leibson PJ (1997) Signal transduction during natural killer cell activation: inside the mind of a killer. Immunity 6:655–661

Leong CC, Chapman TL, Bjorkman PJ, Formankova D, Mocarski ES, Phillips JH, Lanier LL (1998) Modulation of natural killer cell cytotoxicity in human cytomegalovirus infection: the role of endogenous class I major histocompatibility complex and a viral class I homolog. J Exp Med 10:1681–1687

Long E (1999) Regulation of immune responses through inhibitory receptors. Ann Rev Immunol 17: 875–904

López-Botet M, and Bellòn T. (1999) Natural killer cell activation and inhibition by receptors for MHC class I. Curr Opin Immunol 11: 301–307

Makrigiannis AP, Gosselin P, Mason LH, Taylor LS, McVicar DW, Ortaldo JR, Anderson SK (1999) Cloning and characterisation of a novel activating Ly49 closely related to Ly49A. J Immunol 163:4931–4938

Moretta A, Biassoni R, Bottino C, Moretta L. (2000) Surface receptors delivering opposite signals regulate the function of human NK cells. Semin Immunol 12:129–138

Odeberg J, Cerboni C, Browne H, Carboni E, Soderberg-Naucler C. (2000) Resistance to NK lysis of human cytomegalovirus infected endothelial cells and macrophages is not dependent on downregulation of HLA class I molecules or UL18 expression. Abstract, 25th International Herpesvirus Workshop, Portland

Olcese L, Cambiaggi A, Semenzato G, Bottino C, Moretta A, Vivier E (1997) Human killer cell activatory receptors for MHC class I molecules are included in a multimeric complex expressed by natural killer cells. J Immunol 158:5083–5086

Rawlinson WD, Farrell HE, Barrell BG (1996) Analysis of the complete DNA sequence of murine cytomegalovirus. J Virol 70:8833–8849

Riegler S, Hebart H, Einsele H, Brossart P, Jahn G, and Sinzger C. (2000a) Monocyte-derived dendritic cells are permissive to the complete replicative cycle of human cytomegalovirus. J Gen Virol 81: 393–399

Riegler S, Hebart H, Einsele H, Grigoleit U, Jahn G, Sinzger C. (2000b) Dendritic cells infected by endothelial cell propagated strains of HCMV are functionally altered. Abstract, 25th International Herpesvirus Workshop, Portland

Reyburn HT, Mandelboim O, Vales-Gomez M, Davis DM, Pazmany L, Strominger JL (1997) The class I MHC homologue of human cytomegalovirus inhibits attack by natural killer cells. Nature 386:514–517

Roberts LJ, Baldwin TM, Speed TP, Handman E, Foote S. (1999) Chromosomes X, 9 and the H2 locus interact epistatically to control *Leishmania major* infection. Eur J Immunol 29:3047–3050

Scalzo AA, Fitzgerald NA, Wallace CR, Gibbons AE, Smart YC, Burton RC, Shellam GR (1992) The effect of the Cmv-1 resistance gene, which is linked to the natural killer cell gene complex, is mediated by natural killer cells. J Immunol 149:581–589

Scalzo AA, Lyons PA, Fitzgerald NA, Forbes CA, Yokoyama WM, Shellam GR (1995) Genetic mapping of *Cmv1* in the region of mouse chromosome 6 encoding the NK gene complex-associated loci *Ly49* and *musNKR-P1*. Genomics 27:435–441

Senkevich TG, Bugert JJ, Sisler JR, Koonin EV, Darai G, Moss B (1996) Genome sequence of human tumorigenic poxvirus: prediction of specific host response-evasion genes. Science 273:813–816

Senkevich TG, Moss B (1998) Domain structure, intracellular trafficking, and β2-microglobulin binding of a major histocompatibility complex class I homolog encoded by molluscum contagiosum virus. Virology 250:397–407

Shellam GR, Allan JE, Papadimitriou JM, Bancroft GJ (1981) Increased susceptibility to cytomegalovirus infection in beige mutant mice. Proc Natl Acad Sci USA 78:5104–5108

Simmons A, Scalzo A, Pereira R (2000) *Rhs1*, a novel genetic locus proximal to *Ly55* and *Cmv1*, influences the magnitude of acute and latent herpes simplex infections of the nervous system. Abstract, 25th International Herpesvirus Workshop, Portland

Söderberg-Nauclér C, Fish KN, Nelson JA (1997) Reactivation of latent human cytomegalovirus by allogeneic stimulation of blood cells from healthy donors. Cell 91:119–126

Tay CH, Welsh RM (1997). Distinct organ-dependent mechanisms for the control of murine cytomegalovirus infection by natural killer cells. J Virol 71: 267–275

Taylor-Weideman J, Sissons P, Sinclair J (1994) Induction of endogenous human cytomegalovirus gene expression after differentiation of monocytes from healthy carriers. J Virol 68:1597–1604

Tomasec P, Braud VM, Rickards C, Powell MB, McSharry BP, Gadola S, Cerundolo V, Borysiewicz LK, McMichael AJ, Wilkinson GWG (2000). Surface expression of HLA-E, an inhibitor of natural killer cells, enhanced by human cytomegalovirus gpUL40. Science 287:1031–1033

Tomasello E, Blery M, Vely E, Vivier E (2000) Signalling pathways engaged by NK cell receptors: double concerto for activating receptors, inhibitory receptors and NK cells. Semin Immunol 12:139–147

Tormo J, Natarajan K, Margulies DH, Mariuzza RA (1999) Crystal structure of a lectin-like natural killer cell receptor bound to its MHC class I ligand. Nature 402:623–631

Ulbrecht M, Martinozzi S, Grzeschik M, Hengel H, Ellwart JW, Pla M, and Weiss EH (2000) The human cytomegalovirus UL40 gene product contains a ligand for HLA-E and prevents NK cell-mediated lysis. J Immunol 164:5019 5022

Vance RE, Kraft JR, Altman JD, Jensen PE, Raulet DH (1998) Mouse CD94/NKG2A is a natural killer cell receptor for the nonclassical major histocompatibility complex (MHC) class I molecule Qa-1b. J Exp Med 188:1841–1848

Vink C, Beuken E, Bruggeman CA (2000) Complete DNA sequence of the rat cytomegalovirus genome. J Virol 74:7656–7665

Vitale M, Castriconi R, Parolini S, Pende D, Hsu ML, Moretta L, Cosman D, Moretta A (1999) The leukocyte Ig-like receptor (LIR)-1 for the cytomegalovirus UL18 protein displays a broad specificity for different HLA class I alleles: analysis of LIR-1 + NK cell clones. Int Immunol 11:29–35

Wang LL, Mehta IK, LeBlanc PA, Yokoyama WM (1997). Mouse natural killer cells express gp49B1, a structural homologue of human killer inhibitory receptors. J Immunol 158:13–17

Westgaard IH, Berg SF, Orstavik S, Fossum S, Dissen E (1998) Identification of a human member of the Ly49 multigene family. Eur J Immunol 28:1839–1846

Yewdell JW, Norbury CC, Bennink JR (1999) Mechanisms of exogenous antigen presentation by MHC class I molecules in vitro and in vivo: Implications for generating CD8(+) T cell responses to infectious agents, tumors, transplants, and vaccines. Adv Immunol 73:1–77

Yokoyama WM (1998) Natural killer cell receptors. Curr Opin Immunol 10:298–305

Human Cytomegalovirus Inhibition of Major Histocompatibility Complex Transcription and Interferon Signal Transduction

D.M. MILLER, C.M. CEBULLA, and D.D. SEDMAK

Pathogens have evolved diverse mechanisms for escaping host innate and adaptive immunity. Viruses that maintain a persistent infection are particularly effective at disabling key arms of the host immune response. For example, the herpesviruses establish a persistent infection in human and animal hosts, in part through critical immunoevasive strategies. Cytomegalovirus, a beta-herpesvirus, impairs major histocompatibility complex (MHC) class I and class II antigen presentation by decreasing MHC expression on the surface of the infected cell, thus enabling infected cells to escape $CD8^+$ and $CD4^+$ T lymphocyte immunosurveillance. Moreover, cytomegalovirus blocks the interferon signal transduction pathway, thereby limiting the direct and indirect antiviral effects of the interferons. In this review, we focus on an emerging paradigm in which the effectiveness of viruses, particularly human cytomegalovirus, to escape antiviral immune responses is significantly enhanced by their ability to inhibit MHC transcription and interferon (IFN)-stimulated (JAK/STAT) signal transduction.

1	Introduction	154
2	Cytomegalovirus Inhibition of MHC Class II Transcription	154
2.1	The Role of MHC Class II in Antiviral Immune Responses	154
2.2	Induction of MHC Class II Transcription	155
2.3	Cytomegalovirus Inhibition of IFN-Stimulated MHC Class II Transcription	155
3	Cytomegalovirus Inhibition of MHC Class I Transcription	157
3.1	The Role of MHC Class I in Antiviral Immune Responses	157
3.2	Induction of MHC Class I Transcription	158
3.3	Cytomegalovirus Inhibition of IFN-Stimulated MHC Class I Transcription	159
3.3.1	IFN-γ	159
3.3.2	IFN-α	160
4	Cytomegalovirus Inhibition of IFN Signal Transduction	160
4.1	The Role of IFNs in Antiviral Immunity and IFN-Stimulated Gene Expression	160
4.2	IFN Signal Transduction	161
4.3	Cytomegalovirus Inhibition of IFN-Stimulated Gene Expression and Antiviral Responses	162
4.3.1	IFN-γ	162
4.3.2	IFN-α	164
5	Summary	165
References		166

The Department of Pathology, The Ohio State University College of Medicine, 1645 Neil Avenue, Room 129, Columbus, OH 43210, USA

1 Introduction

Cytomegalovirus (CMV) is a ubiquitous beta-herpesvirus that has the distinguishing characteristic of all herpesviruses, the ability to persist in the host. Persistent CMV causes serious infections such as interstitial pneumonia, gastrointestinal mucosal ulceration, retinitis, and hepatitis in immunosuppressed patients. Elucidating mechanisms of CMV persistence is critical to a complete model of pathogenesis as CMV infection, morbidity, and mortality often are the result of dissemination of virus acquired before the onset of immunosuppression.

The response to viral infection involves cellular and cytokine components of the host innate and adaptive immune system. However, viruses have developed protean means of subverting the host immune response. This has been an explosive area of research that has revealed striking examples of the diverse strategies CMV has evolved to counteract host immunity. The MHC is the ultimate interface between virus and adaptive immunity. $CD8^+$ and $CD4^+$ T lymphocytes recognize peptides derived from viral proteins presented in the context of MHC class I and class II molecules, respectively (DOHERTY 1995). This recognition triggers the cell-mediated immune system armamentarium whereby virus-specific T lymphocyte clones proliferate and mediate cytolysis of virally infected cells and/or release cytokines that inhibit viral replication or recruit T cells and monocytes to amplify the antiviral response.

The interferons (IFNs) are potent stimulators of MHC expression and antigen presentation. Type I IFNs (IFN-α/β) induce MHC class I expression, whereas type II IFNs (IFN-γ) induce MHC class I and class II expression as well as components of the antigen processing machinery (Boss 1997; GLIMCHER and KARA 1992; TING and BALDWIN 1993). Moreover, IFNs mediate direct antiviral effects through several distinct effector molecules that block viral infection or replication (KIMURA et al. 1994, 1996; SAMUEL 1991).

It is increasingly evident that viruses have evolved mechanisms to inhibit MHC expression and IFN-mediated antiviral effects. Here we review CMV-mediated disruption of MHC transcription and the IFN signal transduction pathway.

2 Cytomegalovirus Inhibition of MHC Class II Transcription

2.1 The Role of MHC Class II in Antiviral Immune Responses

The primacy of $CD4^+$ T lymphocytes in controlling and clearing viral infections is evident in diverse viral infections. Lymphocytic choriomeningitis virus (LCMV)-specific $CD4^+$ T lymphocytes are generated in LCMV infection, and more than 10% of this population secrete IFN-γ in response to class II-restricted LCMV peptides (MULLER et al. 1992; VARGA and WELSH 1998). Control of hepatitis C virus (HCV) correlates with a vigorous anti-HCV CD4$^+$ T cell response (CHEN et al. 1998a). Mice depleted of CD8$^+$ T lymphocytes halt disseminated CMV disease with kinetics similar to those of nondepleted controls, and the clearance of CMV from select organs is dependent on the CD4$^+$ T cell population (JONJIC et al. 1989, 1990). Finally, cytolysis and antiviral cytokine production by CMV-specific class II-restricted CD4$^+$ T cells play a prominent role in controlling CMV infection (DAVIGNON et al. 1996; HENGEL et al. 1994; LUCIN et al. 1992).

2.2 Induction of MHC Class II Transcription

MHC class II gene expression is regulated in a tissue-specific and cytokine-inducible manner. Three human MHC class II isotypes, HLA-DR, HLA-DQ, and HLA-DP, are coordinately controlled at the transcriptional level through shared, conserved upstream regulatory elements (KARA and GLIMCHER 1991). Professional antigen-presenting cells such as dendritic cells, macrophages, B cells, and thymic epithelial cells constitutively express MHC class II (Boss 1997). Cytokines, particularly IFN-γ, induce MHC class II expression in a variety of class II-negative cells such as endothelial cells, fibroblasts, and epithelial cells (Boss 1997).

IFN-γ is the most potent inducer of MHC class II expression in class II-negative cells. IFN-γ stimulates MHC class II transcription through three distinct steps: IFN-γ signal transduction and transcription factor formation, class II transactivator (CIITA) expression, and transcriptional activation at MHC class II promoters (Fig. 1). IFN-γ binding at its receptor activates the JAK/STAT signal transduction pathway (discussed in detail below). These signals result in the formation of signal transducer and activator of transcription 1 (STAT-1) homodimers. STAT-1 homodimers translocate to the nucleus, where they induce the expression of interferon regulatory factor-1 (IRF-1), a transcription factor important for MHC class I and class II expression. IRF-1, STAT-1 homodimers, and the ubiquitously expressed upstream factor-1 (USF-1) cooperatively activate the transcription and expression of CIITA (MUHLETHALER-MOTTET et al. 1998). CIITA is the master regulator of constitutive and inducible MHC class II expression and is found only in class II-positive cells, and its expression pattern correlates with MHC class II gene expression qualitatively and quantitatively (OTTEN et al. 1998; SILACCI et al. 1994; STEIMLE et al. 1993, 1994). CIITA activates MHC class II transcription through protein-protein interactions with the ubiquitously expressed MHC class II promoter binding proteins RFX, CREB, and NF-Y (Boss 1997; RILEY et al. 1995; ZHOU et al. 1997; ZHOU and GLIMCHER 1995).

2.3 Cytomegalovirus Inhibition of IFN-Stimulated MHC Class II Transcription

Flow cytometry analyses demonstrate that IFN-γ-inducible MHC class II expression is blocked in HCMV-infected endothelial cells (ECs) of arterial, venous,

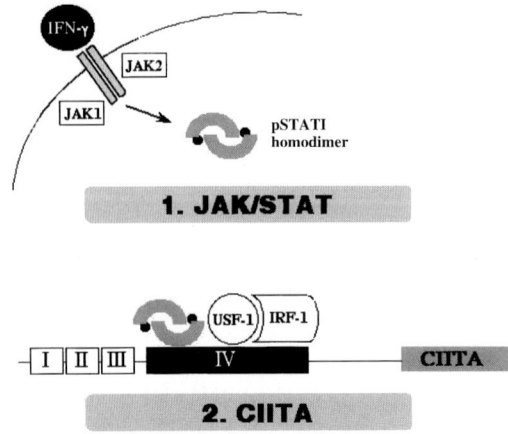
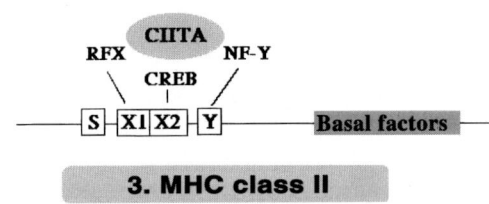

Fig. 1. Mechanisms of IFN-γ-stimulated MHC class II transcription. *Step 1*: IFN-γ binds to its receptor and triggers the phosphorylation of the receptor-associated kinases JAK1 and JAK2. The JAKs phosphorylate latent cytoplasmic STAT-1, which forms a homodimer that translocates to the nucleus to activate IFN-γ-stimulated genes. *Step 2*: IFN-γ-stimulated STAT-1 homodimers and IRF-1 cooperatively bind with the ubiquitously expressed USF-1 at CIITA promoter IV to activate CIITA expression. *Step 3*: CIITA activates MHC class II transcription through protein-protein interactions with ubiquitously expressed class II-promoter-binding transcription factors RFX, CREB, and NF-Y

microvascular, adult, and fetal origin (KNIGHT et al. 1997; SCHOLZ et al. 1992; SEDMAK et al. 1994). Moreover, IFN-γ-stimulated MHC class II RNA levels are significantly decreased in HCMV-infected ECs compared with noninfected controls, suggesting an HCMV-mediated mechanism for the block in IFN-γ-stimulated class II RNA expression (SEDMAK et al. 1994).

Viral infection is a potent stimulus for production of the type I interferons (IFN-α/β), which independently downregulate MHC class II expression at the transcriptional level by a mechanism downstream of CIITA expression (LU et al. 1995). Experiments with IFN-α/β-blocking antibodies show that the disruption of IFN-γ-stimulated MHC class II expression in HCMV-infected cells is not due to a IFN-α/β-mediated effect on class II transcription (SEDMAK et al. 1995). However, IFN-α/β, secreted in response to HCMV infection, decreases inducible and constitutive MHC class II expression in surrounding noninfected cells in vitro and in vivo (POLLOCK et al. 1997; SEDMAK et al. 1995).

Further investigations focused on evaluating the three steps of IFN-γ-stimulated class II expression: signal transduction and transcription factor activation, CIITA expression, and class II transcription in CMV-infected cells. Ribonuclease protection assays (RPA) and RT-PCR experiments reveal that neither MHC class II nor CIITA RNA is upregulated in response to IFN-γ treatment in HCMV-infected ECs at 72h after infection (MILLER et al. 1998). Moreover, electrophoretic mobility shift assays (EMSA) demonstrate that STAT-1 homodimers are not induced by IFN-γ treatment in HCMV-infected ECs (MILLER et al. 1998). Therefore, the defect in IFN-γ induction of class II transcription occurs proximal to the MHC class II promoter, CIITA expression, and transcription factor activation, suggesting an HCMV-mediated lesion in the IFN-γ signal transduction pathway (MILLER et al. 1998).

Additionally, there is evidence that HCMV represses CIITA mRNA expression. At 6h after HCMV infection, IFN-γ signal transduction and STAT-1 nuclear translocation are intact, whereas IFN-γ-stimulated MHC class II expression is blocked (LEROY et al. 1999). RPA demonstrates that CIITA mRNA expression is downregulated in HCMV infected cells at early times after infection. Moreover, the decrease in MHC class II expression affects the efficiency of presentation of HCMV-derived peptides to HCMV-specific CD4+ T lymphocytes, whereas antigen presentation can be restored by transfection of CIITA (LEROY et al. 1999). These data suggest that a HCMV protein or HCMV-induced signal inhibits IFN-γ stimulated CIITA expression at the level of the CIITA promoter IV, the IFN-γ-inducible component of the CIITA promoter (LEROY et al. 1999). A similar effect on MHC class II transcription is found with MCMV infection of differentiated bone marrow macrophages. At early times after infection, MCMV infection inhibits IFN-γ-stimulated MHC class II expression downstream of the IFN-γ signal transduction pathway, suggesting that MCMV represses CIITA expression or function and thus inhibits MHC class II transcription (HEISE et al. 1998).

In summary, HCMV blocks IFN-γ-stimulated MHC class II transcription through two means with distinct mechanisms and kinetics. At relatively early times after infection, there is an HCMV-mediated repression of CIITA expression or function (HEISE et al. 1998; LEROY et al. 1999). At later times after infection, HCMV blocks IFN-γ-stimulated transcription factor activation, CIITA expression, and class II transcription (MILLER et al. 1998). The HCMV gene products mediating these effects remain to be identified; however, experiments have shown that immediate-early and/or early gene products are involved (HEISE et al. 1998; LEROY et al. 1999; MILLER et al. 1998). This represents a critical HCMV immunoevasive strategy as the virus has evolved two mechanisms, operative over distinct stages of the viral replication cycle, for blocking the IFN-γ-inducible MHC class II transcription system.

3 Cytomegalovirus Inhibition of MHC Class I Transcription

3.1 The Role of MHC Class I in Antiviral Immune Responses

CD8+ T lymphocytes recognize peptides presented in the context of MHC class I molecules. This triggers virus-specific cytotoxic T lymphocyte-mediated cytolysis of infected cells as well as secretion of proinflammatory cytokines such as the IFNs. Class I molecules are heterodimeric complexes, consisting of a transmembrane glycoprotein heavy chain and β_2-microglobulin (β_2m). Proteasome-generated peptides are transported into the endoplasmic reticulum (ER) by the transporters associated with antigen processing (TAP). After their synthesis in the ER, class I heterodimers are loaded with peptides 8–11 amino acids in length, forming a stable trimolecular complex that trafficks to the cell surface to display the peptide to $CD8^+$ T lymphocytes (HEEMELS and PLOEGH 1995).

3.2 Induction of MHC Class I Transcription

MHC class I transcription is coordinately regulated by conserved promoter elements and transcription factors. MHC class I promoters may be divided into upstream and downstream regulatory groups (Fig. 2). The downstream group consists of an S-X-Y motif containing *cis*-elements shared by class I and class II promoters (GIRDLESTONE 1996; SHIRAYOSHI et al. 1988; VAN DEN ELSEN et al. 1998; VOGEL et al. 1986). This region mediates constitutive class I expression as well as IFN-γ-stimulated MHC class I expression through the interaction of CIITA with the S-X-Y module-binding proteins, including RFX, CREB/ATF-like factors, and NF-Y, which are involved in class II transcription (GIRDLESTONE 1996; VAN DEN ELSEN et al. 1998; VOGEL et al. 1986).

The upstream group of the class I promoter controls constitutive and IFN-α-induced expression and consists of the enhancer A element and the IFN-stimulated response element (ISRE). The enhancer A region contains binding sites for transcription factors NF-κB and Sp1. The ISRE element contains binding sites for IFN-α-stimulated transcription factors such as IRF-1 and IFN-stimulated gene factor-3 (ISGF3), an IFN-α-stimulated transcription factor that is a key activator of IFN-α-stimulated genes (GOBIN et al. 1997; GOBIN and VAN DEN ELSEN 1999). IFN-α

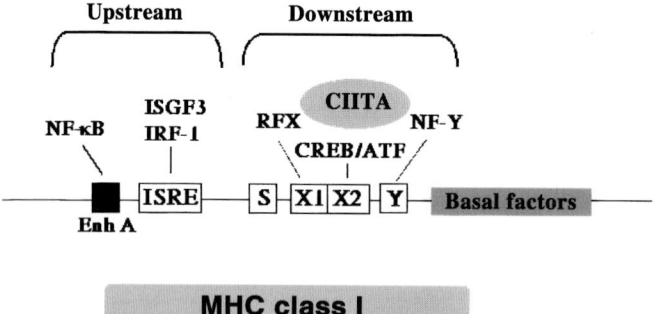

Fig. 2. Mechanism of IFN-stimulated MHC class I transcription. IFN-γ upregulates CIITA expression, which transactivates MHC class I transcription through protein-protein interactions with the S-X-Y motif-bound transcription factor complexes. In addition, IFN-γ upregulates IRF-1, which transactivates class I transcription through the ISRE element within the upstream promoter region. IFN-α stimulates formation of ISGF3, which binds to the ISRE element in MHC class I promoters to increase class I transcription. In addition, IFN-α upregulates expression of IRF-1, which also binds to ISRE sites to increase class I expression

upregulates MHC class I transcription through the upstream promoter elements. IFN-α stimulates formation of ISGF3, which binds to the ISRE element in MHC class I promoters and increases class I transcription (TING and BALDWIN 1993). In addition, IFN-α upregulates IRF-1 expression that also binds to ISRE sites and increases class I expression (LEHTONEN et al. 1997; TING and BALDWIN 1993).

IFN-γ upregulates MHC class I transcription by the induction of transcription factors that act at upstream and downstream *cis*-elements of MHC class I promoters. In vivo studies have shown that IRF-1 and CIITA mediate the IFN-γ upregulation of MHC class I expression (HOBART et al. 1997; MARTIN et al. 1997; RAMASSAR et al. 1996). IFN-γ stimulates the upregulation of CIITA expression that transactivates MHC class I transcription through protein-protein interactions with the S-X-Y motif-bound transcription factor complexes (MARTIN et al. 1997). In addition, IFN-γ upregulates IRF-1 that transactivates class I transcription through the ISRE element within the upstream promoter region (MARTIN et al. 1997).

3.3 Cytomegalovirus Inhibition of IFN-Stimulated MHC Class I Transcription

3.3.1 IFN-γ

Given the overlap in transcriptional control elements between MHC class I and class II expression, it is possible that the defect in IFN-γ-stimulated MHC class II expression may extend to IFN-γ-stimulated MHC class I expression. MHC class I and class II promoters share analogous S-X-Y motifs (TING and BALDWIN 1993; VAN DEN ELSEN et al. 1998). Moreover, CIITA and IRF-1 are IFN-γ-induced transcription factors that are critical for class I and class II transcription (BOSS 1997; MUHLETHALER-MOTTET et al. 1998).

Previous experiments demonstrated that IFN-γ-stimulated CIITA and IRF-1 expression is blocked in HCMV-infected ECs, and this is associated with disruption of IFN-γ-stimulated MHC class I RNA upregulation in HCMV-infected fibroblasts and ECs (MILLER et al. 1998, 2000). In addition, IFN-γ-stimulated β_2m expression is blocked in infected cells (MILLER et al. 2000).

IFN-γ upregulates MHC class I antigen presentation not only through its direct effects on class I heavy chains and β_2m but also by upregulating regulatory elements of the proteasome, LMP2 and LMP7, and the TAP complex (JOHNSEN et al. 1998; KOOTSTRA et al. 1997). Northern blot experiments show that IFN-γ-stimulated TAP1, TAP2, LMP2, and LMP7 expression is blocked in HCMV-infected fibroblasts and ECs (MILLER et al. 2000). Therefore, HCMV infection blocks the upregulation of class I heavy chains, β_2m, regulatory proteasome subunits, and TAP.

In summary, a defect in IFN-γ-stimulated responses in HCMV-infected cells shuts down IFN-γ-stimulated MHC class I and MHC class II RNA upregulation, β_2m expression, regulatory proteasome subunits, and the TAP complex. Therefore, a defect that occurs proximal to activation of transcription factors affects these critical antigen presentation processes in infected cells.

3.3.2 IFN-α

IFN-α stimulates MHC class I expression by upregulating transcription through the ISRE site of the MHC class I promoter (TING and BALDWIN 1993; VAN DEN ELSEN et al. 1998). Northern blot analyses demonstrate that HCMV infection blocks IFN-γ-stimulated MHC class I expression in infected fibroblasts and ECs (MILLER et al. 1999). Moreover, gene expression and EMSA analyses reveal that IFN-α-stimulated IRF-1 expression and transcription factor activation are blocked in HCMV-infected cells, a phenotype similar to the block in IFN-γ-stimulated responses in HCMV-infected cells (MILLER et al. 1999).

4 Cytomegalovirus Inhibition of IFN Signal Transduction

4.1 The Role of IFNs in Antiviral Immunity and IFN-Stimulated Gene Expression

In the early 1950s, Isaacs and Lindenmann discovered a family of proteins capable of interfering with viral replication, the interferons (IFNs) (ISAACS and LINDENMANN 1957). IFNs mediate three major classes of biological activity: antiproliferative, antiviral, and immunoregulatory effects (DARNELL et al. 1994; DAVID 1995; KOROMILAS et al. 1992; MEURS et al. 1992; SAMUEL 1991; SCHINDLER and DARNELL 1995). IFNs rapidly induce the transcription of previously quiescent genes without the necessity of protein synthesis, i.e., immediate-early gene expression (DARNELL et al. 1994).

Type I (IFN-α/β) and type II (IFN-γ) IFNs are major lines of defense against viral infection. IFNs mediate direct antiviral effector mechanisms that inhibit multiple steps of viral replication (SAMUEL 1991; VILCEK and SEN 1996). For example, 2′,5′-oligoadenylate synthetase (2′,5′-OAS) activates ribonuclease L, which degrades mRNA and limits the accumulation of viral transcripts. Protein kinase R blocks translation of viral transcripts by phosphorylating translation initiation factor eIF-2. Mx proteins block influenza, vesicular stomatitis virus, and herpes simplex virus replication by an unknown mechanism.

IFNs are critical regulators of the antiviral immune response. IFNs are potent stimulators of antigen presentation, thereby augmenting cell-mediated antiviral immune responses. Type I IFNs upregulate MHC class I protein expression, and type II IFNs upregulate MHC class I and II (Boss 1997; DARNELL et al. 1994). Moreover, IFNs upregulate components of the antigen processing machinery, such as proteasome expression and function (MARTIN et al. 1997).

There are in vivo and in vitro data supporting the critical role of the IFNs in anti-CMV immunity. Intramuscular IFN-α reduces the replication of MCMV in the spleen and liver of mice, and IFN-α receptor knockout mice are 800-fold more susceptible to MCMV infection than their wild-type littermates (PRESTI et al. 1998; YEOW et al. 1998). Numerous in vitro studies demonstrate that pretreating cells with IFN-α inhibits human HCMV replication by decreasing transcription of the immediate-early (IE) HCMV gene products (GRIBAUDO et al. 1993; MARTINOTTI et al. 1993). The antiviral effects of IFN also extend to clinical therapy, as IFN-α treatment significantly reduces the incidence of serious HCMV infections in seropositive renal transplant recipients (HIRSCH et al. 1983).

Studies of IFN-γ in MCMV models demonstrate that IFN-γ is a critical cytokine in controlling acute and chronic CMV infection (BIRON 1994; Presti et al. 1998). IFN-γ accounts for the majority of natural killer (NK) cell-mediated antiviral effects during acute infection. Neutralization of IFN-γ prevents MCMV clearance from the salivary gland and impairs control of MCMV infection, and IFN-γ depletion increases MCMV titers in the liver and spleen (JONJIC et al. 1989; LUCIN et al. 1992; ORANGE et al. 1995). CMV infection in IFN-γ receptor knockout mice leads to an uncontrolled persistent infection resulting in a severe vasculitis of large arterial vessels (PRESTI et al. 1998). Moreover, pretreatment of diverse cell types with IFN-γ inhibits HCMV replication (DAVIGNON et al. 1996; GRIBAUDO et al. 1993; HEISE and VIRGIN 1995; LUCIN et al. 1994; SCHUT et al. 1994) and restores HCMV antigen processing and presentation to HCMV-specific $CD8^+$ T lymphocyte clones (DAVIGNON et al. 1996; GRIBAUDO et al. 1993; HEISE and VIRGIN 1995; HENGEL et al. 1994, 1995; LUCIN et al. 1994; SCHUT et al. 1994).

Animal models demonstrate that the IFN-stimulated JAK/STAT signal transduction pathway is critical for controlling CMV infections. STAT1 knockout mice and IFN-α receptor/IFN-γ receptor double knockout mice, which are deficient in IFN-stimulated signal transduction and biological responses, are exquisitely sensitive to viral infection (DURBIN et al. 1996; PRESTI et al. 1998). In these mice, acute MCMV infection proceeds unchecked and rapidly leads to death.

4.2 IFN Signal Transduction

IFN-γ binds to the heterodimeric IFN-γ receptor (IFN-γR), which is associated intracellularly with the Janus kinases (JAKs) JAK1 and JAK2 (DARNELL et al. 1994). Binding of IFN-γ to its receptor initiates JAK1- and JAK2-mediated tyrosine phosphorylation of the cytoplasmic tail of the IFN-γR (KOTENKO et al. 1995; MULLER et al. 1993; PARGANAS et al. 1998). The phosphorylated IFN-γR serves as a docking site for latent cytoplasmic STAT-1 (DARNELL et al. 1994). After docking at the receptor, STAT-1 is phosphorylated by the JAKs and forms a homodimer through cognate phosphotyrosine-SH2 domain interactions (DARNELL 1997; DARNELL et al. 1994). STAT-1 homodimers translocate to the nucleus where they bind IFN-γ activation sequence (GAS) elements present in the promoters of IFN-γ inducible genes (Fig. 3) (CHEN et al. 1998b; DARNELL et al. 1994).

IFN-α signal transduction shares similar signaling molecules and basic signal transduction mechanisms as IFN-γ. IFN-α binds to its receptor, which stimulates

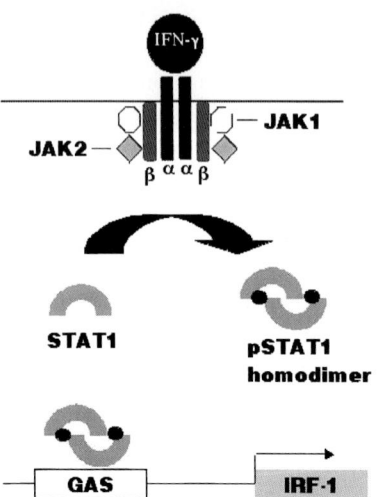

Fig. 3. IFN-γ signal transduction. IFN-γ binds to its receptor, triggering JAK1/JAK2-mediated tyrosine phosphorylation. The JAKs phosphorylate STAT-1 molecules, which form a homodimer through cognate phosphotyrosine-SH2 domain interactions. STAT-1 homodimers translocate to the nucleus, where they bind IFN-γ activation sequence (*GAS*) elements present in the promoters of IFN-γ-inducible genes

the activation of kinases JAK1 and TYK2 (a JAK family kinase specific for IFN-α signaling) (Lew et al. 1991). STAT1 and STAT2 molecules are tyrosine phosphorylated by JAK1 and TYK2. Phosphorylated STAT1/STAT2 heterodimers unite with p48 to form the transcription factor complex ISGF3, which binds to the ISRE sites in many IFN-α-responsive promoters (Ghislain and Fish 1996; Li et al. 1997; Schindler et al. 1992). Alternatively, phosphorylated STAT1 homodimers and STAT1/STAT2 heterodimers can translocate to the nucleus and bind to elements such as the inverted repeat (IR) element of the IFN regulatory factor-1 (IRF-1) gene to activate transcription in an ISGF3-independent manner (Fig. 4) (Haque and Williams 1994; Li et al. 1996).

4.3 Cytomegalovirus Inhibition of IFN-Stimulated Gene Expression and Antiviral Responses

HCMV infection blocks IFN-γ- and IFN-α-stimulated MHC expression. The defect in IFN-induced MHC expression occurs proximal to IFN-stimulated transcription factor activation and gene expression. Given that IFN-γ- and IFN-α-stimulated responses are disrupted at a similar level, it is likely that they share a common defect in their signal transduction pathways.

4.3.1 IFN-γ

Detailed investigation of the IFN-γ signal transduction revealed a lesion in this pathway that resulted in a complete block in IFN-γ signaling. EMSA experiments demonstrated that IFN-γ did not induce STAT-1 homodimer formation in HCMV-infected cells, suggesting that there was either a disruption in the phosphorylation of STAT-1 molecules or a decrease in STAT-1 levels or another component of

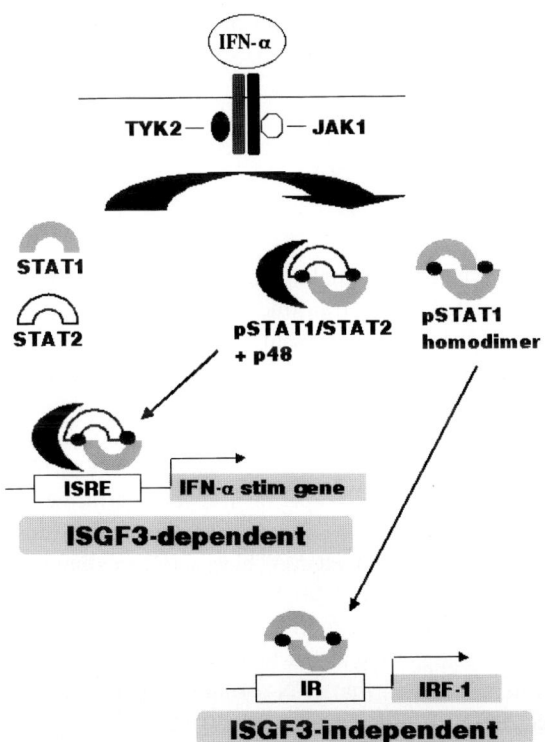

Fig. 4. IFN-α signal transduction. IFN-α binds to its receptor, which stimulates the activation of JAK1 and TYK2 and the JAK1/TYK2-mediated phosphorylation of STAT-1 and STAT-2. STAT-1 and STAT-2 form heterodimers and unite with p48, thereby forming ISGF3, which binds to ISRE sites in many ISGF3- dependent IFN-α-responsive promoters. Alternatively, phosphorylated STAT-1 homodimers and STAT-1/STAT-2 heterodimers can move to the nucleus and bind to elements such as the IR element of the IRF-1 gene to activate transcription in an ISGF3-independent manner

JAK/STAT signal transduction. Immunoprecipitation experiments revealed that IFN-γ-stimulated tyrosine phosphorylation of JAK1, JAK2, IFNγR, and STAT-1 was blocked in infected cells. Moreover, JAK1 protein was not detected in HCMV-infected cells (MILLER et al. 1998). JAK1 is a Janus family protein tyrosine kinase that is required for IFN-γ-stimulated tyrosine phosphorylation, STAT-1 homodimer formation, and IFN-γ-stimulated gene expression and biological responses (MULLER et al. 1993). The HCMV-mediated decrease of JAK1 protein levels is capable of disrupting the IFN-γ-stimulated activation of the signal transduction pathway and the subsequent downstream responses such as STAT-1 homodimer formation and MHC induction. The decrease of JAK1 protein may be the lesion responsible for the lack of IFN-γ-stimulated MHC class I, MHC class II, β_2-microglobulin, LMP2, LMP7, TAP1, and TAP2 expression in HCMV-infected cells.

Further analyses of JAK1 expression in HCMV-infected cells revealed that JAK1 levels are decreased posttranslationally by a subset of HCMV genes. Steady-state JAK1 mRNA levels are not altered in the course of HCMV infection despite decreased JAK1 protein at 48 and 72h after infection (MILLER et al. 1998). Moreover, treatment of HCMV-infected cells with carboxybenzyl-leucyl-leucyl-leucine vinyl sulfone (Z-L$_3$VS), a specific inhibitor of the proteasome, reverses the decrease of JAK1 protein in HCMV-infected cells, suggesting that JAK1 protein levels are decreased by a proteasome-dependent mechanism (MILLER et al. 1998). In addition, treatment of infected cells with phosphonoacetic acid (PFA), which inhibits HCMV late gene expression, has no effect on the HCMV-mediated block in IFN-γ signal transduction, suggesting that HCMV IE and/or E genes decrease JAK1 expression (MILLER et al. 1998).

4.3.2 IFN-α

JAK1 is an essential component of type I IFN signaling; therefore, decreased JAK1 protein would inhibit IFN-α-stimulated signal transduction, transcription factor activation, and gene expression and is likely to represent a common lesion in IFN signal transduction that blocks both IFN-γ- and IFN-α-stimulated responses in HCMV-infected cells. However, differential display experiments in HCMV-infected cells at 8h after infection have shown that a constituent of the viral particle, independent of the IFN signal transduction pathway, upregulates a subset of IFN-responsive genes (ZHU et al. 1997). Therefore, experiments were performed to directly test the integrity of IFN-α-stimulated responses in HCMV-infected cells.

Gene expression experiments revealed that HCMV blocks IFN-α-stimulated ISGF3-dependent (MHC class I, 2′,5′-OAS, and MxA) and ISGF3-independent (IRF-1) gene expression in infected fibroblasts and ECs (MILLER et al. 1999). EMSA analyses, utilizing a DNA-binding element that binds STAT-1 homodimers and STAT-1/STAT-2 heterodimers, revealed that IFN-α-stimulated transcription factor activation was blocked in HCMV-infected cells (MILLER et al. 1999). Immunoprecipitation experiments demonstrated that IFN-α-stimulated tyrosine phosphorylation was also blocked in HCMV-infected cells, a cellular phenotype that correlates with decreased JAK1 protein (MILLER et al. 1999). Thus the HCMV-mediated decrease of JAK1 protein may mediate disruption of IFN-α-stimulated responses in a manner analogous to the defect in the IFN-γ pathway.

However, analyses of the individual components of the IFN-α signal transduction pathway uncovered an additional lesion in this signal transduction system. There is decreased expression of p48 protein (the DNA-binding component of ISGF3) in HCMV-infected fibroblasts and ECs (MILLER et al. 1998). Moreover, HCMV IE and/or E genes mediate the decrease in p48 protein expression.

These two defects in IFN-α signal transduction block distinct arms of the IFN-α signal transduction system. There are significant data demonstrating that IFN-α activates genes, such as IRF-1, independent of the p48 component of ISGF3 (HAQUE and WILLIAMS 1994; LI et al. 1996). Thus the decrease of JAK1 protein may inhibit IFN-α-stimulated ISGF3-independent gene expression. In addition, p48 is the principal mediator of ISGF3 binding to ISRE DNA elements in IFN-α-stimulated promoters (VEALS et al. 1992). Studies in p48-negative cell lines and in p48 knockout mice show that IFN-α-stimulated ISGF3-dependent gene expression is inhibited in cells lacking p48 (KIMURA et al. 1996). Moreover, p48 is critical to the establishment of IFN-α- and IFN-γ-induced antiviral states (KIMURA et al. 1996). Thus the HCMV-mediated decrease in JAK1 and p48 may block IFN-α-stimulated ISGF3-dependent gene expression.

5 Summary

These studies demonstrate that HCMV has evolved multiple mechanisms for inhibiting MHC transcriptional control and globally blocking IFN-stimulated responses. Specifically, HCMV may disrupt IFN-γ-stimulated MHC class II expression through two distinct mechanisms, decreased JAK1 protein and disrupted IFN-γ signal transduction as well as a direct repression of CIITA expression. Moreover, IFN-γ-stimulated MHC class I, $β_2$m, proteasome subunit, and TAP expression are inhibited by the decrease of JAK1 protein. IFN-α-stimulated MHC class I expression and upregulation of direct antiviral effector molecules such as 2′,5′-OAS and MxA are blocked by disruption of ISGF3-dependent gene expression secondary to the HCMV decrease of JAK1 and p48 proteins. Finally, IFN-α-stimulated ISGF3-independent gene expression is inhibited, which may be secondary to the decrease in JAK1.

Viral disruption of MHC transcriptional control is not a unique phenomenon. Murine hepatitis virus and Kirsten murine sarcoma virus block MHC class II transcription within infected cells (JOSEPH et al. 1991). Infection of murine macrophages with the parasite *Leishmania donovani* or with mycobacteria inhibits IFN-γ-stimulated MHC class II induction (KWAN et al. 1992). Recently, it was shown that chlamydia inhibits IFN-γ-inducible MHC class II expression by degrading USF-1 (ZHONG et al. 1999).

Divergent viruses have evolved means of disrupting IFN-stimulated JAK/STAT signal transduction and thus IFN-stimulated antiviral and immunoregulatory responses. For example, adenovirus E1A gene products decrease p48, and overexpression of p48 can restore IFN-α signal transduction in E1A-transfected cells (LEONARD and SEN 1997). In addition, in HeLa cells E1A also decreases STAT-1 levels, further preventing IFN-α- and IFN-γ-induced signaling (LEONARD and SEN 1996, 1997; TAKEDA et al. 1994). The human papillomavirus (HPV) E7 protein inhibits the induction of IFN-α-inducible genes through decreased ISGF3 formation and p48 nuclear translocation, without decreasing p48 levels (BARNARD and MCMILLAN 1999). Mumps disrupts IFN-induced gene expression in infected cells through decreased STAT-1, mediated by a posttranscriptional mechanism (YOKOSAWA et al. 1998). Ebola inhibits the induction of IFN-induced genes including MHC class I, IRF-1, and 2′,5′-OAS through decreased formation of STAT-1 homodimers and ISGF3 complexes (HARCOURT et al. 1999). The hepatitis B virus (HBV) terminal protein disrupts the signaling response to IFN-α and IFN-γ by interfering with the formation of active ISGF3 complexes (FOSTER et al. 1991). Moreover, the Epstein-Barr virus (EBV) nuclear antigen 2 (EBNA-2) gene similarly prevents activation of IFN-α by a mechanism downstream of the activation of ISGF3 (KANDA et al. 1992).

These critical immunoevasive strategies are important for understanding viral pathogenesis and persistence. The ability of CMV to evade the type I and II IFN-stimulated antiviral and immunoregulatory responses in ECs and, moreover, to disrupt IFN-stimulated MHC class I and class II expression and the antigen processing machinery contributes to reservoirs of persistent virus (FISH et al. 1995; KOFFRON et al. 1998; ORTIZ et al. 1997; PRESTI et al. 1998).

Understanding the direct responses of IFNs in virally infected cells may also explain the limited benefit of IFN therapies. For example, it is hypothesized that HBV terminal protein decreases the efficacy of IFN therapy and antagonizes infected hepatocyte antigen presentation to cytotoxic T lymphocytes (FOSTER et al. 1991). Moreover, IFN-α monotherapy for HCV infection clears persistent virus in only a small fraction of patients.

Future studies are needed to determine the CMV gene products mediating these processes and the precise molecular mechanisms for these effects. These gene products will undoubtedly be unique tools for studying JAK/STAT signal transduction, MHC expression and antigen presentation, T lymphocyte function, and tissue transplantation.

References

Barnard P, McMillan NA (1999) The human papillomavirus E7 oncoprotein abrogates signaling mediated by interferon-α. Virology 259:305–313
Biron CA (1994) Cytokines in the generation of immune responses to, and resolution of, virus infection. Curr Opin Immunol 6:530–538
Boss JM (1997) Regulation of transcription of MHC class II genes. Curr Opin Immunol 9:107–113
Chen M, Shirai M, Liu Z, Arichi T, Takahashi H, Nishioka M (1998) Efficient class II major histocompatibility complex presentation of endogenously synthesized hepatitis C virus core protein by Epstein-Barr virus-transformed B-lymphoblastoid cell lines to CD4+ T cells. J Virol 72: 8301–8308
Chen X, Vinkemeier U, Zhao Y, Jeruzalmi D, Darnell JE Jr, Kuriyan J (1998) Crystal structure of a tyrosine phosphorylated Stat-1 dimer bound to DNA. Cell 93:827–839
Darnell JE Jr (1997) STATs and gene regulation. Science 277:1630–1635
Darnell JE Jr, Kerr IM, Stark GR (1994) Jak-Stat pathways and transcriptional activation in response to IFNs and other extracellular signaling proteins. Science 264:1415–1421
David M (1995) Transcription factors in interferon signaling. Pharmacol Ther 65:149–161
Davignon J-L, Castanie P, Yorke JA, Gautier N, Clement D, Davrinche C (1996) Anti-human cytomegalovirus activity of cytokines produced by CD4+ T-cell clones specifically activated by IE1 peptides in vitro. J Virol 70:2162–2169
Doherty P (1995) Immunity to viruses. The Immunologist 3:231–233
Durbin JE, Hackenmiller R, Simon MC, Levy DE (1996) Targeted disruption of the mouse STAT1 gene results in compromised innate immunity to viral disease. Cell 84:443–450
Fish KN, Stenglein SG, Ibanez C, Nelson JA (1995) Cytomegalovirus persistence in macrophages and endothelial cells. Scand J Infect Dis 99:34–40
Foster GR, Ackrill AM, Goldin RD, Kerr IM, Thomas HC, Stark GR (1991) Expression of the terminal protein region of hepatitis B virus inhibits cellular responses to interferons alpha and gamma and double stranded RNA. Proc Natl Acad Sci USA 88:2888–2892
Ghislain JJ, Fish EN (1996) Application of genomic DNA affinity chromatography identifies multiple interferon-alpha-regulated Stat2 complexes. J Biol Chem 271:12408–12413
Girdlestone J (1996) Transcriptional regulation of MHC class I genes. Eur J Immunogenet 23:393–413
Glimcher LH, Kara CJ (1992) Sequences and factors: a guide to MHC class II transcription. Annu Rev Immunol 10:13–49
Gobin SJ, Peijnenburg A, Keijsers V, van den Elsen P (1997) Site α is crucial for two routes of IFNγ-induced MHC class I transactivation: the ISRE-mediated route and a novel pathway involving CIITA. Immunity 6:601–611

Gobin SJ, van den Elsen PJ (1999) The regulation of HLA class I expression: is HLA-G the odd one out? Semin Cancer Biol 9:55–59

Gribaudo G, Ravaglia S, Caliendo A, Cavallo R, Gariglio M, Martinotti MG, Landolfo S (1993) Interferons inhibit onset of murine cytomegalovirus immediate-early gene transcription. Virology 197:303–311

Haque SJ, Williams BRG (1994) Identification and characterization of an interferon (IFN)-stimulated response element-IFN-stimulated gene factor 3-independent signaling pathway for IFN-α. J Biol Chem 269:19523–19529

Harcourt BH, Sanchez A, Offerman MK (1999) Ebola virus selectively inhibits responses to interferons, but not to interleukin-1β, in endothelial cells. J Virol 73:3491–3496

Heemels MT, Ploegh H (1995) Generation, translocation, and presentation of MHC class I-restricted peptides. Annu Rev Biochem 64:463–491

Heise MT, Connick M, Virgin HW (1998) Murine cytomegalovirus inhibits interferon gamma-induced antigen presentation to CD4 T cells by macrophages via regulation of expression of major histocompatibility complex class II-associated genes. J Exp Med 187:1037–1046

Heise MT, Virgin HW (1995) The T-cell-independent role of gamma interferon and tumor necrosis factor alpha in macrophage activation during murine cytomegalovirus and herpes simplex virus infections. J Virol 69:904–909

Hengel H, Eblinger C, Pool J, Goulmy E, Koszinowski UH (1995) Cytokines restore MHC class I complex formation and control antigen presentation in human cytomegalovirus-infected cells. J Gen Virol 76:2987–2997

Hengel H, Lucin P, Jonjic S, Ruppert T, Koszinowski UH (1994) Restoration of cytomegalovirus antigen presentation by gamma interferon combats viral escape. J Virol 68:289–297

Hirsch MS, Schooley RT, Cosimi AB, Russell PS, Delmonico FL, Tolkoff-Rubin NE, Herrin JT, Cantell K, Farrell ML, Rota TR, Rubin RH (1983) Effects of interferon-alpha on cytomegalo-virus reactivation syndromes in renal-transplant recipients. N Engl J Med 308:1489–1493

Hobart M, Ramassar V, Goes N, Urmson J, Halloran PF (1997) IFN regulatory factor-1 plays a central role in the regulation of the expression of class I and II MHC genes in vivo. J Immunol 158:4260–4269

Isaacs A, Lindenmann J (1957) Virus interference. The interferons. Proc R Soc Lond B 147:258–267

Johnsen A, France J, Sy M-S, Harding CV (1998) Downregulation of the transporter for antigen presentation, proteasome subunits and class I MHC in tumor cell lines. Cancer Res 58:3660–3667

Jonjic S, Mutter W, Weiland F, Reddehase MJ, Koszinowski UH (1989) Site-restricted persistent cytomegalovirus infection after selective long-term depletion of CD4 T lymphocytes. J Exp Med 169:1199–1212

Jonjic S, Pavic I, Lucin P, Rukavina D, Koszinowski UH (1990) Efficacious control of cytomegalovirus infection after long-term depletion of CD8+ T lymphocytes. J Virol 64:5457–5464

Joseph J, Knobler RL, Lublin FD, Hart MN (1991) Mouse hepatitis virus (MHV-4, JHM) blocks gamma interferon-induced major histocompatibility complex class II antigen expression on murine cerebral endothelial cells. J Neuroimmunol 33:181–190

Kanda K, Decker T, Aman P, Wahlstrom M, von Gabain A, Kallan B (1992) The EBNA2- related resistance towards alpha interferon (IFN-α) in Burkitt's lymphoma cells effects induction of IFN-induced genes but not the activation of transcription factor ISGF3. Mol Cell Biol 12:4930–4936

Kara CJ, Glimcher LH (1991) In vivo footprinting of MHC class II genes: bare promoters in the bare lymphocyte syndrome. Science 252:709–712

Kimura T, Kadokawa Y, Harada H, Matsumoto M, Sato M, Kashiwazaki Y, Tarutani M, Tan RS, Takasugi T, Matsuyama T, Mak TW, Noguchi S, Taniguchi T (1996) Essential and nonredundant roles of p48 (ISGF3-γ) and IRF-1 in both type I and type II interferon responses, as revealed by gene targeting studies. Genes Cells 1:115–124

Kimura T, Nakayama K, Penninger J, Kitagawa M, Harada H, Matsuyama T, Tanaka K, Kamijo R, Vilcek J, Mak TW (1994) Involvement of the IRF-1 transcription factor in antiviral responses to interferons. Science 264:1921–1924

Knight DA, Waldman WJ, Sedmak DD (1997) Human cytomegalovirus does not induce human leukocyte antigen class II expression on arterial endothelial cells. Transplantation 63:1366–1369

Koffron AJ, Hummel M, Patterson BK, Yan S, Kaufman DB, Fryer JP, Stuart J, Stuart FP (1998) Cellular localization of latent murine cytomegalovirus. J Virol 72:95–103

Kootstra G, Hammerling GJ, Momburg F (1997) Generation, intracellular transport and loading of peptides associated with MHC class I molecules. Curr Opin Immunol 9:80–88

Koromilas AE, Roy S, Barber GN, Katze MG, Sonenberg N (1992) Malignant transformation by a mutant of the interferon-inducible dsRNA-dependent protein kinase. Science 257:1685–1689

Kotenko SV, Izotova LS, Pollack BP, Mariano TM, Donnelly RJ, Muthukumaran G, Cook JR, Garotta G, Silvennoinen O, Ihle JN (1995) Interaction between the components of the interferon-γ receptor complex. J Biol Chem 270:20915–20921

Kwan WC, McMaster WR, Wong N, Reiner NE (1992) Inhibition of expression of major histocompatibility complex class II molecules in macrophages infected with *Leishmania donovani* occurs at the level of gene transcription via a cyclic AMP-independent mechanism. Infect Immun 60:2115–2120

Le Roy E, Muhlethaler-Mottet A, Davrinche C, Mach B, Davignon J (1999) Escape of human cytomegalovirus from HLA-DR-restricted CD4+ T-cell response is mediated by repression of γ interferon-induced class II transactivator expression. J Virol 73:6582–6589

Lehtonen A, Matikainen S, Julkunen I (1997) Interferons upregulate Stat1, Stat2, and IRF family transcription factor gene expression in human peripheral blood mononuclear cells and macrophages. J Immunol 159:794–803

Leonard GT, Sen GC (1996) Effects of adenovirus E1A protein on interferon-signaling. Virology 224:25–33

Leonard GT, Sen GC (1997) Restoration of interferon responses of adenovirus E1A-expressing HT1080 cell lines by overexpression of p48 protein. J Virol 71:5095–5101

Li X, Leung S, Kerr IM, Stark GR (1997) Functional subdomains of STAT2 required for preassociation with the alpha interferon receptor and for signaling. Mol Cell Biol 17:2048–2056

Li X, Leung S, Qureshi S, Darnell JE, Stark GR (1996) Formation of STAT1-STAT2 heterodimers and their role in the activation of IRF-1 gene transcription by interferon-α. J Biol Chem 271:790–794

Lu HT, Riley JL, Babcock GT, Huston M, Stark GR, Boss JM, Ransohoff RM (1995) Interferon (IFN) β acts downstream of IFN-γ-induced class II transactivator messenger RNA accumulation to block major histocompatibility complex class II gene expression and requires the 48-kD DNA-binding protein, ISGF3-γ. J Exp Med 182:1517–1525

Lucin P, Jonjic S, Messerle M, Polic B, Hengel H, Koszinowski UH (1994) Late phase inhibition of murine cytomegalovirus replication by synergistic action of interferon-gamma and tumour necrosis factor. J Gen Virol 75:101–110

Lucin P, Pavic I, Polic B, Jonjic S, Koszinowski UH (1992) Gamma interferon-dependent clearance of cytomegalovirus infection in salivary glands. J Virology 66:1977–1984

Martin BK, Chin KC, Olsen JC, Skinner CA, Dey A, Ozato K, Ting JP (1997) Induction of MHC class I expression by the MHC class II transactivator CIITA. Immunity 6:591–600

Martinotti MG, Gribaudo G, Gariglio M, Caliendo A, Lembo D, Angeretti A, Cavallo R, Landolfo S (1993) Effect of interferon-α on immediate early gene expression of murine cytomegalovirus. J Interferon Res 13:105–109

Meurs EF, Galabru J, Barber GN, Zatze MG, Hovanessian AG (1992) Tumor suppressor function of the interferon-induced double stranded RNA-activated protein kinase. Proc Natl Acad Sci USA 90:232–236

Miller DM, Rahill BM, Boss JM, Laimore MD, Durbin JE, Waldman WJ, Sedmak DD (1998) Human cytomegalovirus inhibits MHC class II expression by disruption of the Jak/Stat pathway. J Exp Med 187:675–683

Miller DM, Zhang Y, Rahill BM, Kazor K, Rofagha S, Eckell JJ, Sedmak DD (2000) Human cytomegalovirus blocks interferon-γ stimulated up-regulation of major histocompatibility complex class I expression and the class I antigen processing machinery. Transplantation 69:687–690

Miller DM, Zhang Y, Rahill BM, Waldman WJ, Sedmak DD (1999) Human cytomegalovirus inhibits IFN-alpha-stimulated antiviral and immunoregulatory responses by blocking multiple levels of IFN-alpha signal transduction. J Immunol 162:6107–6113

Muhlethaler-Mottet A, Di Berardino W, Otten LA, Mach B (1998) Activation of the MHC class II transactivator CIITA by interferon-γ requires cooperative interaction between Stat1 and USF-1. Immunity 8:157–166

Muller D, Koller BH, Whitton JL, LaPan KE, Brigman KK, Frelinger JA (1992) LCMV-specific, class II-restricted cytotoxic T cells in β-2 microglobulin deficient mice. Science 255:1576–1578

Muller M, Briscoe J, Laxton C, Guschin D, Ziemiecki A, Silvennoinen O, Harpur AG, Barbieri G, Witthuhn BA, Schindler C (1993) The protein tyrosine kinase JAK1 complements defects in interferon-alpha/beta and -gamma signal transduction. Nature 366:129–135

Orange JS, Wang B, Terhorst C, Biron CA (1995) Requirement for natural killer cell-produced interferon γ in defense against murine cytomegalovirus infection and enhancement of this defense pathway by interleukin 12 administration. J Exp Med 182:1045–1056

Ortiz BD, Nelson PJ, Krensky AM (1997) Switching gears during T-cell maturation: RANTES and late transcription. Immunol Today 18:468–471

Otten LA, Steimle V, Bontron S, Mach B (1998) Quantitative control of MHC class II expression by the transactivator CIITA. Eur J Immunol 28:473–478

Parganas E, Wang D, Stravopodis D, Topham DJ, Marine JC, Teglund S, Vanin EF, Bodner S, Colamonici OR, van Duersen JM, Grosveld G, Ihle JN (1998) Jak2 is essential for signaling through a variety of cytokine receptors. Cell 93:385–395

Pollock JL, Presti RM, Paetzold S, Virgin HW (1997) Latent murine cytomegalovirus infection in macrophages. Virology 227:168–179

Presti RM, Pollock JL, Dal Canto AJ, O'Guin AK, Virgin HW (1998) Interferon-γ regulates acute and latent murine cytomegalovirus infection and chronic disease of the great vessels. J Exp Med 188:577–588

Ramassar V, Goes N, Hobart M, Halloran PF (1996) Evidence for the in vivo role of class II transactivator in basal and IFN-g induced class II expression in mouse tissue. Transplantation 62:1901–1907

Riley JL, Westerheide SD, Price JA, Brown JA, Boss JM. (1995) Activation of class II MHC genes requires both the X box region and the class II transactivator (CIITA). Immunity 2:533–543

Samuel CE (1991) Antiviral actions of interferon-regulated cellular proteins and their surprisingly selective antiviral activities. Virology 183:1–11

Schindler C, Darnell JE (1995) Transcriptional responses to polypeptide ligands: The jak-stat pathway. Annu Rev Biochem 64:621–651

Schindler C, Shuai K, Prezioso VR, Darnell JE (1992) Interferon-dependent tyrosine phosphorylation of a latent cytoplasmic transcription factor. Science 257:809–813

Scholz M, Hamann A, Blaheta RA, Auth MK, Encke A, Markus BH (1992) Cytomegalovirus- and interferon-related effects on human endothelial cells: cytomegalovirus infection reduces upregulation of HLA class II antigen expression after treatment with interferon-gamma. Hum Immunol 35:230–238

Schut RL, Gekker G, Hu S, Chao CC, Pomeroy C, Jordan MC, Peterson PK (1994) Cytomegalovirus replication in murine microglial cell cultures: suppression of permissive infection by interferon-gamma. J Infect Dis 169:1092–1096

Sedmak DD, Chaiwiriyakul S, Knight DA, Waldman WJ (1995) The role of interferon β in cytomegalovirus-mediated inhibition of HLA-DR induction. Arch Virol 140:111–126

Sedmak DD, Guglielmo AM, Knight DA, Birmingham DJ, Huang EH, Waldman WJ (1994) Cytomegalovirus inhibits major histocompatibility class II expression on infected endothelial cells. Am J Pathol 144:683–692

Shirayoshi Y, Burke PA, Appella E, Ozato K (1988) Interferon induced transcription of a major histocompatibility class I gene accompanies binding of inducible nuclear factors to the interferon consensus sequence. Proc Natl Acad Sci USA 85:5884–5888

Silacci P, Mottet A, Steimle V, Reith W, Mach B (1994) Developmental extinction of major histocompatibility complex class II gene expression in plasmocytes is mediated by silencing of the transactivator gene CIITA. J Exp Med 180:1329–1336

Steimle V, Otten LA, Zufferey M, Mach B (1993) Complementation cloning of an MHC class II transactivator mutated in hereditary MHC class II deficiency (or bare lymphocyte syndrome). Cell 75:135–146

Steimle V, Siegrist CA, Mottet A, Lisowska-Grospierre B, Mach B (1994) Regulation of MHC class II expression by interferon-gamma mediated by the transactivator gene CIITA. Science 265:106–109

Takeda T, Nakajima K, Kojima H, Hirano T (1994) E1A repression of IL-6-induced gene activation by blocking the assembly of IL-6 response element binding complexes. J Immunol 153:4573–4582

Ting JPY, Baldwin AS (1993) Regulation of MHC gene expression. Curr Opin Immunol 5:8–16

van den Elsen PJ, Gobin SJP, van Eggermond MCA, Peijnenburg A (1998) Regulation of MHC class I and II gene transcription: differences and similarities. Immunogenetics 48:208–221

Varga SM, Welsh RM (1998) Detection of a high frequency of virus-specific CD4+ T cells during acute infection with lymphocytic choriomenigitis virus. J Immunol 161:3215–3218

Veals SA, Schindler C, Leonard D, Fu XY, Aebersold R, Darnell JE Jr, Levy DE (1992) Subunit of an alpha-interferon-responsive transcription factor is related to interferon regulatory factor and Myb families of DNA-binding proteins. Mol Cell Biol 12:3315–3324

Vilcek J, Sen GC (1996) Interferons and other cytokines. In: Fields BN, Knipe DM, Howley PM (eds) Fields Virology, 3rd Edition, Lippencott-Raven Publishers, Philadelphia

Vogel J, Kress M, Khoury G, Jay G (1986) A transcriptional enhancer and an interferon-responsive sequence in major histocompatibility complex class I genes. Mol Cell Biol 6:3550–3554

Yeow W-S, Lawson CM, Beilharz MW (1998) Antiviral activities of individual murine IFN-(alpha) subtypes in vivo: intramuscular injection of IFN expression constructs reduces cytomegalovirus replication. J Immunol 160:2932–2939

Yokosawa N, Kubota T, Fujii N (1998) Poor induction of interferon-induced 2′,5′-oligoadenylate synthetase (2-5 AS) in cells persistently infected with mumps virus is caused by decrease of STAT-1α. Arch Virol 143:1985–1992

Zhong G, Fan T, Liu L (1999) Chlamydia inhibits interferon gamma-inducible major Histocompatibility complex class II expression by degradation of upstream stimulatory factor 1. J Exp Med 189: 1931–1938

Zhou H, Glimcher LH (1995) Human MHC class II gene transcription directed by the carboxyl terminus of CIITA, one of the defective genes in type II MHC combined immune deficiency. Immunity 2: 545–553

Zhou H, Su H, Zhang X, Douhan J, Glimcher L (1997) CIITA-dependent and independent class II MHC expression revealed by a dominant negative mutant. J Immunol 158:4741–4749

Zhu H, Cong JP, Shenk T (1997) Use of differential display analysis to assess the effect of human cytomegalovirus infection on the accumulation of cellular RNAs: Induction of interferon-responsive RNAs. Proc Natl Acad Sci USA 94:13985–13990

Counteraction of Interferon-Induced Antiviral Responses by Herpes Simplex Viruses

D.A. Leib

The outcome of a viral infection of a host involves the complex interplay of viral determinants of virulence and host resistance factors. Among the first lines of defense for the host in attempts to control viral infection are the interferons (IFNs). A large body of work has now shown that the IFNs are a family of soluble proteins that serve to mediate antiviral effects, to regulate cell growth, and to modulate the activation of immune responses. The innate antiviral activities of IFNs are exceedingly potent and rapid. It is, therefore, not surprising that so many viruses have evolved ways to either preclude the synthesis of IFNs or evade downstream antiviral events. Such evasion allows for the virus to spread before the development of a specific adaptive immune response and likely represents a pivotal determinant of virulence for the invading virus. This review describes some of the research on herpes simplex virus (HSV) that has elucidated genes involved in evasion of the IFN response. In particular, the roles of specific viral genes in resistance to the antiviral effects of PKR and RNaseL are described, along with other HSV genes and loci associated with resistance to IFN for which mechanisms have yet to be described.

1	Introduction	171
2	Interferon Signaling and Innate Antiviral Activities	173
2.1	Induction of the Antiviral State Through PKR	174
3	Inhibition of the Effects of PKR by HSV	175
3.1	Infected Cell Protein 34.5	175
3.2	US11	177
4	Other Inhibitory Mechanisms	178
4.1	Inhibition of the 2'–5'/RNaseL System	178
4.2	HSV Genes and Loci Associated with IFN Resistance	179
4.2.1	ICP0	179
4.2.2	Virion Host Shutoff Protein	180
5	Conclusions	181
References		182

1 Introduction

The outcome of a viral infection of a host involves the complex interplay of viral determinants of virulence and host resistance factors. Among the first lines of defense for the host in attempts to control viral infection are the interferons (IFNs).

Departments of Ophthalmology and Visual Sciences and Molecular Microbiology, Washington University School of Medicine Box 8096, 660 South Euclid Avenue, St. Louis, MO 63110, USA

Interferon was first described by Isaacs and Lindenmann as a soluble factor that interfered with viral replication in cultured cells (ISAACS and LINDENMANN 1957). A large body of work has now shown that the IFNs are a family of soluble proteins that serve to mediate antiviral effects, to regulate cell growth, and to modulate the activation of immune responses. The innate antiviral activities of IFNs are exceedingly potent and rapid. It is, therefore, not surprising that so many viruses have evolved ways to either preclude the synthesis of IFNs or evade downstream antiviral events. Such evasion allows for the virus to spread before the development of a specific adaptive immune response and likely represents a pivotal determinant of virulence for the invading virus.

The two most common outcomes for viral infection are recovery of the host coupled with complete clearance of the pathogen and death of the host. This is the so-called "get better or die" paradigm of microbial infection. In some cases, however, the life cycle is more complex ,with the virus being capable of establishing infections that neither are cleared nor lead to death of the host, with the establishment of lifelong persistent or latent infections. For such viruses, evasion of innate immunity during acute replication before escape into the immunologically safe haven of latency represents a battle that must be won. Indeed, for a number of persistent viruses, such as hepatitis B, hepatitis C, and human immunodeficiency virus (HIV) multiple mechanisms have now been described by which these viruses can counteract the antiviral effects of interferons (see GOODBOURN et al. 2000 for a review). The herpesviruses are among the most studied examples of viruses that exhibit latency, and herpes simplex virus (HSV) is considered one of several prototypic examples of this family.

Studies using animal models and observations of human infections have resulted in the classic theory of HSV pathogenesis (WILDY 1982). According to this theory there are four stages that characterize an HSV infection. *Entry* into the host occurs at the time of a primary infection, and HSV replicates at peripheral sites such as the eyes, skin, or mucosae. *Spread* to the axonal terminae of sensory neurons is followed by retrograde intra-axonal transport to neuronal cell bodies in sensory ganglia, where further viral replication may occur. *Establishment of latency* then occurs, at which time lytic gene expression is repressed. At this stage no infectious virus can be detected but the viral genome persists in the neuron in a transcriptionally active state, but with little or no expression of protein. *Reactivation* occurs when certain poorly defined stimuli such as stress, immunosuppression, or UV light cause the controls responsible for maintaining latency to break down. This leads to the production of infectious virus in the ganglion followed by anterograde transport to the periphery, which, after further replication, may be manifest as lesions at or near the site of primary infection.

During primary acute infection, and during reactivation, the actively replicating HSV is susceptible to immune attack. During latency in the nervous system, in contrast, the virus is relatively sheltered. During the early events of acute infection HSV is especially vulnerable to the effects of innate immunity such as IFNs. Such innate responses are capable of shaping the future adaptive response to the virus, and it is likely therefore that evasion of innate immunity is least as important to the infective success of the virus as evasion of adaptive immune attack. The IFNs are the major components of innate immunity to viruses because they serve to directly induce an antiviral state in infected cells as well as promoting the adaptive response through activation of natural killer (NK) cells, cytotoxic T cells, and macrophages and inducing MHC antigens (STARK et al. 1998). Together, the interferons elicit a potent series of responses to which viruses have developed a variety of countermeasures. In this review the counteraction of interferon-induced innate immune responses by HSVs will be discussed.

2 Interferon Signaling and Innate Antiviral Activities

Several excellent reviews have been written on this subject, and the reader is referred to them for details, the salient points of which are summarized below (DAVID 1995; GOODBOURN et al. 2000; STARK et al. 1998). The IFNs belong to two principal classes, namely type I (predominantly IFN-α and IFN-β) and type II (IFN-γ). IFN-α and IFN-β are synthesized in many cell types, especially leukocytes and fibroblasts in response to viral infection. IFN-γ, in contrast, is produced by activated NK and T cells, after recognition of infected cells and initiation of the adaptive immune response. IFN-α is a family of closely related proteins, products of a multigene family, whereas IFN-β is the product of a single gene. The unifying property of the type I IFNs is that they bind to a common cellular receptor and signal via a common pathway. Although IFN-γ binds to a distinct receptor, both types of IFN signal through related pathways and lead to activation of some common genes, resulting in generation of an antiviral state in stimulated cells. Binding to cognate receptors leads to activation of cytoplasmic tyrosine kinases of the Janus family of kinases (JAKs). Once activated, the JAKs in turn phosphorylate transcription factors known as signal transducers and activators of transcription (STATs). The STATs then dimerize and translocate to the nucleus, where they bind specific sequences within the promoters of a number of interferon-inducible genes and upregulate their expression. Many of these genes contribute to the direct inhibition of viral replication via induction of the an antiviral state (Fig. 1). In particular, the induction of expression of double-stranded RNA-dependent protein kinase (PKR) and 2–5(A) oligoadenylate synthetases serves to block the growth and virulence of many viruses. These two pathways are described in further detail below. A third pathway exists, involving the induction of the Mx proteins that mediate the inhibition of several types of RNA virus, and will not be discussed further (ARNHEITER et al. 1996). In addition, it should be noted that further alternative IFN-mediated antiviral pathways may exist (ZHOU et al. 1999). This evidence comes from triply deficient mice (lacking PKR, RNaseL, and Mx1) showing substantial antiviral response to IFN-α. The nature of these other pathways has yet to be determined, but current use of microarrays and mouse genomic

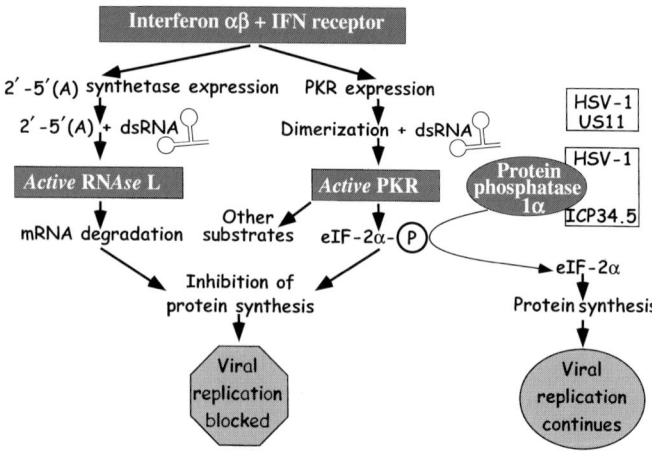

Fig. 1. Interferon-mediated induction of the antiviral state through RNaseL and PKR. For RNaseL, interferons interacting with their receptors lead to expression of 2–5A synthetase, which, after double-stranded RNA binding, produces 2′,5′-oligoadenylates that lead to RNaseL activation and blockage of viral replication. Interferon also induces expression of PKR, which is also activated through double-stranded RNA binding. One of the activities of PKR is to phosphorylate eIF2α, leading to cessation of protein synthesis and blockage of viral replication. To evade this antiviral mechanism, HSV-1 ICP34.5 binds to protein phosphatase 1α and dephosphorylates eIF2α, lifting the block on viral replication. HSV-1 US11 is known to be involved in the downregulation of PKR activity by either preventing the accumulation of activated PKR or precluding the phosphorylation of eIF2α

technologies will likely define these other pathways and their perturbation by viral infection.

2.1 Induction of the Antiviral State Through PKR

Double-stranded RNA-dependent PKR is a 68-kDa serine-threonine protein kinase belonging to the eukaryotic initiation factor-2 α-subunit (eIF-2α) protein kinase family (LEBLEU et al. 1976; MEURS et al. 1990; ROBERTS et al. 1976). PKR also has numerous effects, including the phosphorylation of IκB leading to activation of NF-κB, regulation of growth factor and calcium-mediated signal transduction, and induction of apoptosis (DER et al. 1997; KUMAR et al. 1994; TAN and KATZE 1999; WILLIAMS 1999). The transcription of PKR is potently induced by IFNs although it is ubiquitously expressed to low levels at all times in an inactive form. On binding of double-stranded RNA (dsRNA) or single-stranded RNA with regions of lengthy and extensive secondary structure to its N terminus, PKR undergoes a conformational change (BISCHOFF and SAMUEL 1985; NANDURI et al. 1998). This change exposes the catalytic domain of the kinase. This in turn leads to dimerization and autophosphorylation on specific critical residues and activation of PKR (THOMIS and SAMUEL 1995). The requirement for binding of dsRNA, normally rare in nature but very common in virally infected cells, provides an important checkpoint for the regulation of potentially destructive PKR activity.

As indicated above, activated PKR has many critical functions that contribute to the host's initial response to viral infection. Perhaps the best-understood antiviral function, and the one most targeted by viruses, is the ability of PKR to phosphorylate the α subunit of the basal eukaryotic translation initiation factor eIF2 (Meurs et al. 1992). This results in an inability of initiation factors to be recycled. The net result is a lack of the ternary (eIF2-GTP-MtRNA) complex, which is required for the initiation of translation, leading to a profound inhibition of both host and viral protein synthesis (Ramaiah et al. 1994). In addition, PKR is emerging as an important player in the regulation of apoptosis through FADD-mediated and other cell death signaling pathways (Tan and Katze 1999). This gives further teleological rationale to PKR being such an important target for inhibition by viruses.

3 Inhibition of the Effects of PKR by HSV

3.1 Infected Cell Protein 34.5

Work from several laboratories identified a locus in the internal repeat sequences of HSV-1 that was associated with profound changes in neurovirulence. Indeed, mutations in this region were found to cause up to one million-fold increases in the LD_{50} after intracerebral injection in mice (Bolovan et al. 1994; Chou et al. 1990). Sequencing and mapping revealed the presence of a 263-amino acid open reading frame in this region, termed infected cell protein (ICP)34.5 (also known as RL1), and targeted mutation of this gene yielded similarly dramatic alterations in neurovirulence (Chou and Roizman 1990; Dolan et al. 1992). For some time, however, the mechanism of action of ICP34.5 remained obscure. An early clue as to the possible function of ICP34.5 came from the observation that in a human neuroblastoma cell line, two viruses lacking ICP34.5 induced a premature and total shutoff of protein synthesis that was associated with viral DNA synthesis (Chou and Roizman 1994). Such shutoff was not observed in nonneuronal cultured cells. Another clue came from the fact that the carboxy-terminal domain of ICP34.5 shares significant homology with the murine MyD116 and the hamster GADD34 proteins, which are 71% identical to each other (Zhan et al. 1994). These two genes belong to the growth arrest and DNA damage (GADD) gene family. As their name suggests, GADD genes are induced to be expressed in growth-arrested or DNA-damaged cells. The function of GADD34 is presently unknown, but it has been shown that its overexpression can lead to apoptosis (Hollander et al. 1997). The carboxyl terminus of GADD34 can functionally substitute for the carboxyl terminus of ICP34.5 in terms of precluding shutoff of protein synthesis, thereby showing their functional homology (He et al. 1996). A homolog of ICP34.5, termed I14L, was also identified in the African swine fever virus genome, although its role in virulence currently remains unknown (Sussman et al. 1992).

Perhaps the key breakthrough in this field was the discovery that PKR was activated in both wild-type and ICP34.5 mutant-infected cells, but eIF2α was phosphorylated only in cells infected with the mutants (Chou et al. 1995). This was believed to correlate with the observed shutoff of protein synthesis seen in cells infected with viruses lacking the carboxyl terminus of ICP34.5. This idea was further confirmed by experiments showing that ICP34.5 forms a complex with protein phosphatase 1α, or PP1, and redirects its activity to cause the dephosphorylation of eIF2α, thereby negating one of the antiviral effects of activated PKR and allowing protein synthesis to continue in HSV-infected cells (He et al. 1997). Further analysis has shown that ICP34.5 has structural and functional characteristics similar to those of a PP1 regulatory subunit, perhaps explaining how ICP34.5 is able to regulate the activities of PP1 (He et al. 1997).

Is there a direct connection between the avirulence of ICP34.5 mutants and the ability of ICP34.5 to interfere with the activities of PKR? Recent data have shown that an HSV-1 null mutant in gene ICP34.5 grew to normal levels and showed wild-type virulence in mice lacking both type 1 (IFN-α/β) and type 2 (IFN-γ) IFN receptors (Leib et al. 1999). This was despite the fact that in normal mice the mutant exhibited a 10,000-fold reduction in replication and neurovirulence. Deletion of interferon receptors, however, also resulted in increased replication in mice of HSV recombinants with null mutations in several other genes. Increased virulence in the IFN receptor-deleted mice was therefore not a specific property of viruses lacking ICP34.5 but was consistent with the idea that ICP34.5 targeted an IFN-dependent pathway. A subsequent study showed that an ICP34.5-deleted virus exhibits wild-type replication and virulence in a host from which the PKR gene has been deleted (Leib et al. 2000; Tan and Katze 2000). The restoration of virulence was shown to be specific to HSV ICP34.5-null viruses infecting mice lacking PKR. For example, there was no restoration of virulence to a virus lacking thymidine kinase in PKR-deficient mice. Similarly, there was no restoration of virulence to an ICP34.5-null virus in mice lacking RNaseL, an independent IFN-inducible antiviral pathway. Furthermore, it was demonstrated that the lack of PKR in the host did not result in significantly enhanced replication or virulence of wild-type virus. This implies that ICP34.5 is a highly efficient inactivator of the antiviral effects of PKR. This is consistent with the observation that the dephosphorylation of eIF2 in wild-type HSV-infected cells is more than 1,000-fold higher than cells infected with an ICP34.5 mutant (Cassady et al. 1998b). Together, these data formally demonstrate that the PKR pathway is a specific target for ICP34.5 in vivo.

Recently, it was shown that enhanced permissivity of transformed cells to HSV infection, in the presence or absence of ICP34.5, is associated with inhibition of PKR activation through activation of the host Ras signaling pathway (Farassati et al. 2001). This lends further credence to the idea that ICP34.5-deleted viruses may have utility as oncolytic agents. Some important issues, however remain unresolved. First, deletions in the amino-terminal domain of ICP34.5 also reduce virulence even though there is no discernable effect on protein synthesis or the ability of ICP34.5 to bind to PP1. Second, substitution of the ICP34.5 C-terminus with the GADD34 C-terminus also restores protein synthesis in neoplastic cells but not virulence in mice (ANDREANSKY et al. 1996). It is clear, therefore, that for an ICP34.5-altered virus, there is no direct relationship between its ability to sustain protein synthesis in neoplastic cells and its virulence in vivo, suggestive of multiple functions and interactions for ICP34.5.

3.2 US11

US11 is an abundant protein that is expressed as a late gene and is packaged into the tegument of the virus (JOHNSON et al. 1986). It associates with ribosomes, binds RNA in a sequence-specific fashion, and accumulates in the nucleoli of infected cells (MACLEAN et al. 1987; ROLLER and ROIZMAN 1990, 1992). US11 can also functionally substitute for Rev in HIV infection and can allow restoration of protein synthesis in HeLa cells after heat shock (DIAZ-LATOUD et al. 1997; DUC DODON et al. 2000). Genetic evidence has demonstrated that US11 can compensate for the loss of ICP34.5 function in culture. On serial passage of some ICP34.5 mutants, variants were isolated that had regained the ability to grow in neoplastic neuronal cells (MOHR and GLUZMAN 1996). Further characterization of these suppressor mutants revealed a deletion that extends from the adjacent immediate-early gene ICP47 into a repetitive region of the viral genome. This rearrangement allows US11 to be expressed from an immediate-early RNA that initiates from the ICP47 promoter, thus altering its expression profile from a late gene to an immediate-early gene. The alteration of the kinetics of US11 expression was sufficient to allow ICP34.5 mutants to sustain protein synthesis on nonpermissive cells (CASSADY et al. 1998a; MOHR and GLUZMAN 1996). US11 expression as an immediate-early or early gene is capable of either reducing the accumulation of activated PKR or precluding the phosphorylation of eIF2 even in the presence of activated PKR (CASSADY et al. 1998a; MULVEY et al. 1999). These two functions are clearly not mutually exclusive, and either will result in the continuation of protein synthesis.

The above properties of US11 and of the ICP34.5 suppressor mutants are consistent with the idea that US11 could partner with ICP34.5 in aiding HSV to evade the effects of IFNs and PKR. A number of anomalies, however, exist. First, if US11 is such an abundant tegument protein, why is there a requirement for its disregulation to suppress the protein synthesis shutoff seen in ICP34.5 mutants? Tegument proteins are introduced into the cell before any de novo protein synthesis, so a requirement for US11 to be regulated as an early gene should be superfluous. Second, and perhaps most significantly, although the ICP34.5 second site suppressor mutants have regained the ability to sustain protein synthesis and replication in neoplastic cells, they remain avirulent in mice (MOHR et al. 2001). Such viruses are also capable of inhibiting tumor growth in vivo, suggesting that they may have utility in the treatment of certain cancers (TANEJA et al. 2001). This is analogous to the situation mentioned above in which substitution of the ICP34.5 C-terminus with the GADD34 C-terminus also restores protein synthesis in neoplastic cells but not virulence (ANDREANSKY et al. 1996). Furthermore, US11 is apparently dispensable for virulence in mice (NISHIYAMA et al. 1993). Why then would HSV carry US11 as an apparently redundant function? It has been suggested that perhaps US11 represents an ancient and vestigial function in the virus that has been superseded in HSV by the acquisition of the domain of ICP34.5 capable of dephosphorylating eIF2 (CASSADY et al. 1998a). Alternatively, perhaps US11 provides an additional pathway for the virus to inhibit PKR in cells in which the GADD34-related function of ICP34.5 is absent or rendered ineffective (MULVEY et al. 1999). US11 may have other presently unknown functions, and the precise role of this gene in the pathogenesis and replication of HSV has yet to be determined.

4 Other Inhibitory Mechanisms

Evasion of the effects of IFN is clearly a critical determinant of pathogenesis, and several viruses exhibit multiple mechanisms to evade IFN (GOODBOURN et al. 2000). HSV is no exception, although it is clear that compared to the inhibition of PKR by ICP34.5 and US11 these other mechanisms are less well defined. These other inhibitory mechanisms are described below, categorized either by the IFN-responsive pathway inhibited or by the viral gene product or products believed to encode resistance to IFN-mediated antiviral activity. A number of HSV genes and loci have been associated with resistance to the antiviral effects of IFNs, although to date the molecular basis of resistance remains unknown, and in some cases the actual genes involved have not been mapped.

4.1 Inhibition of the 2′–5′/RNaseL System

IFNs induce the expression of 2′–5′(A)oligoadenylate synthetases that, after stimulation by viral dsRNA, synthesize oligomers of adenosine known as 2′5′-oligoadenylates (2–5A) (KERR and BROWN 1978). The synthesized 2–5A binds to the N-terminus of the inactive monomeric form of ribonuclease L (RNaseL), leading to its dimerization and activation (DONG and SILVERMAN 1995). Once activated, RNaseL causes the nonspecific cleavage of single-stranded RNA, such as mRNA, as well as the site-specific cleavage of 28S rRNA. In contrast to PKR-mediated inhibition, however, RNaseL appears to preferentially degrade viral over cellular RNA. This is because the synthesized 2–5A is highly unstable and RNaseL activation is dependent on the localized activity of 2–5(A) oligoadenylate synthetases. This activity is highest at intracellular sites proximal to its activator, namely viral dsRNA (NILSEN and BAGLIONI 1979).

One report in the literature has noted that there is minimal activation of RNaseL in HSV-infected IFN-treated human conjunctival (Chang) cells (CAYLEY et al. 1984). At least two possible explanations exist for this observation. The first, favored by the authors, is that 2'5'derivatives are synthesized in HSV-infected cells which are only weak activators of RNaseL, relative to authentic 2'5'A. These derivatives therefore antagonize RNaseL activation. An alternative explanation is that HSV infection has a direct inhibitory effect on RNaseL activity. There has been surprisingly little follow-up on this observation. One study of HSV infection of RNaseL-deficient mice using the mouse ocular model revealed little or no alterations in growth or virulence (LEIB et al. 2000). This is consistent with the ideas that RNaseL plays no role in limiting HSV replication in this infection model, or more likely, that HSV has some mechanism to overcome the effect of RNaseL. Such a mechanism would render the effect of RNaseL-deficiency inapparent in this system. Another study showed that there was a higher yield of infectious HSV from PKR-deficient than from PKR/RNaseL-deficient fibroblasts (KHABAR et al. 2000). This suggests the possibility that RNaseL may degrade certain mRNAs that actually inhibit HSV replication in the absence of PKR. Given that a number of other viruses such as HIV, encephalomyocarditis virus, and vaccinia virus can inhibit the activation of RNaseL, it would not be surprising that HSV does the same. What is currently lacking is a knowledge of the precise molecular basis for the inhibition of this important pathway.

4.2 HSV Genes and Loci Associated with IFN Resistance

4.2.1 ICP0

Some recent studies have suggested that ICP0 is especially important for viral resistance to IFN (MOSSMAN et al. 2000). ICP0 is an immediate-early gene whose deletion results in viruses that are able to grow, albeit at reduced levels, both in cell culture and in vivo as well as playing a critical role in reactivation from neurons (LEIB et al. 1989; SACKS and SCHAFFER 1987; STOW and STOW 1986). ICP0 also has multiple effects on cell metabolism with reported interactions with a ubiquitin-specific protease and colocalization with and subsequent disruption of nuclear domains known as ND10 (EVERETT et al. 1997; MAUL et al. 1993). In the study by Mossman et al. four viruses constructed in different laboratories, all deficient in ICP0, were found to be hypersensitive to IFN-α in cell culture. Viruses with mutations in five other genes of disparate function did not display such hypersensitivity. Moreover, it was of interest that the ICP0 mutants, which were hypersensitive to IFN in Vero cells, were all normally sensitive to IFN-α on U2OS cells, which can restore normal growth to ICP0 mutants. Although the precise mechanism for this hypersensitivity remains to be determined, it is intriguing that abundant ND10 proteins such as PML and Sp100 are strongly IFN regulated (MAUL 1998). Given that ND10s are major sites of viral DNA transcription and replication, their disruption by ICP0 may represent a mechanism for the virus to inhibit an IFN-induced repressor protein (EVERETT et al. 1998). The IFN sensitivity of ICP0 is also consistent with other studies that showed that an ICP0-deleted virus grew poorly in wild-type mice but robustly in IFN receptor-deleted mice (LEIB et al.

1999). It should be noted, however, that growth of the ICP0 mutant was not restored completely to wild-type virus levels in the IFN-deficient mice, supporting the idea that ICP0 has multiple functions for promotion of growth and virulence in vivo.

4.2.2 Virion Host Shutoff Protein

HSV causes the rapid shutoff of macromolecular synthesis through the destabilization of mRNA in infected cells through the action of a viral tegument factor known as the virion host shutoff (*vhs*) protein (KWONG and FRENKEL 1989; READ and FRENKEL 1983). Although *vhs* is dispensable for replication in cell culture, it plays a key role in the promotion of replication in peripheral epithelia and the nervous system (STRELOW et al. 1997; STRELOW and LEIB 1995). The precise reason for the very profound attenuation of *vhs*-deficient viruses is unknown, although a number of possibilities have been suggested. One study showed that cells infected with an HSV-2 mutant that lacks ICP47 [the transporters associated with antigen presentation (TAP) inhibitor] and *vhs* are sensitive to CTL lysis (TIGGES et al. 1996). This suggests that the combined effects of the HSV-2 *vhs* and ICP47 gene products are to block antigen presentation by MHC class I.

More relevant to this review are studies that have examined the possible role of *vhs* in viral resistance to IFN. The idea of *vhs* mediating IFN resistance is attractive for several reasons. First, IFN-α profoundly suppresses immediate-early gene expression (MITTNACHT et al. 1988; OBERMAN and PANET 1988). The presence of *vhs* in the tegument allows it to function before any de novo viral gene expression, consistent with it counteracting IFNs before immediate-early gene expression. Second, the ability of *vhs* to destabilize mRNA rapidly in infected cells might render IFNs ineffective by causing the rapid degradation of the many IFN-induced transcripts that are pivotal to the establishment of the antiviral state.

In a study of HSV-1 strain VR-3, the 50% plaque inhibitory concentrations of IFNs were measured in a plaque assay in human embryo lung cells (HELs) infected with a *vhs*-null virus or a marker-rescued (*vhs* repaired) virus (SUZUTANI et al. 2000). The 50% inhibitory concentration of IFN-α for the *vhs* mutant at 26.6IU/ml was reported to be 20-fold lower than that for the repaired virus. Incubation with 1000IU/ml resulted in a 30–50-fold inhibition of plaque formation of the *vhs* mutant, with a threefold inhibition of the repair virus relative to untreated controls. The authors concluded from this work that *vhs* mutants show increased sensitivity to IFN relative to control virus. The study of Mossman et al. using strain KOS showed that 1000IU/ml of IFN-α in infected Vero cells gave only a sixfold reduction in plaque formation for a *vhs*-null virus and that this reduction was equivalent to that seen for wild-type virus (MOSSMAN et al. 2000). These authors concluded, therefore, that there was no difference in sensitivity to IFN-α between the *vhs* mutant and control virus. The discrepancy may have arisen from any number of technical differences between the studies, such as sources of IFN, cell types used for the assays, and wild-type strain of virus used, making the meaningful comparisons of these studies difficult. The pathogenesis of a *vhs*-null mutant has been examined in type I and type II IFN receptor-deficient mice (LEIB et al. 1999). Although there was significant enhancement of the growth and virulence of a strain KOS *vhs* mutant in IFN receptor-deficient compared with wild-type mice, the enhancement was less than that seen for other mutants (ICP0 and ribonucleotide reductase). In fact, the enhancement of the *vhs* mutant in the cornea was comparable to that seen for wild-type virus in IFN receptor-deficient mice, suggesting that this enhancement was not specific for lack of *vhs* activity.

It has been noted that HSV-2 is in general significantly more resistant to IFNs than HSV-1 (LAUSCH et al. 1991), so studies of IFN resistance using HSV-2 are of particular interest. It is also notable that *vhs* of HSV-2 is about 50- to 100-fold more active than that of HSV-1 (EVERLY and READ 1997; FENWICK and EVERETT 1990). It has been shown that resistance of HSV-2 to IFN-β was mediated by a 7.4-kb *Bgl*II-N cloned region of the viral genome (NARITA et al. 1998). This 7.4-kb restriction fragment not only contains the *vhs* gene but also portions of two genes that flank *vhs*, so the resistance to IFNβ could not be unequivocally attributed to the *vhs* gene. Another HSV-2 locus (*Bam*HI D) has been identified that confers resistance to IFN-α and IFN-β (SU et al. 1993). This locus is distinct from that described by NARITA and colleagues and is more than 60kb away. At first glance, this appears to represent an independent IFN resistance locus mechanistically distinct from *vhs*. It is possible, however, that these observations are linked. The *Bam*HI D fragment contains, along with two other genes, the UL13 protein kinase, which has been shown to be necessary for *vhs* activity, *vhs* synthesis, or both (NG et al. 1997; OVERTON et al. 1994). It is possible that both groups have identified loci that regulate *vhs* activity and that these in turn control resistance to IFNs. Alternatively, other genes, included in the respective genomic fragments and distinct from *vhs*, may be controlling IFN resistance. Clearly, further work with defined mutations, rather than use of marker-transfer, will be needed to elucidate this important issue. It is possible, however, that all of these disparate observations are pointing to a role for *vhs* in mediating resistance to IFN, at least in the context of HSV-2 infection.

5 Conclusions

In this review, data have been presented that demonstrate that HSV has at least two genes (ICP34.5 and US11), and most likely an additional two or more genes (ICP0, *vhs*), which render the virus resistant to the antiviral effects of IFNs. As is common in virology, these genes have multiple functions, and we are only just beginning to understand the complex interplay of viral immunomodulatory proteins with their host resistance counterparts. The use of mice deficient in defined components of the IFN antiviral pathway in combination with defined HSV mutants will likely elucidate the roles of these and other yet to be discovered viral IFN evasion functions in pathogenesis. Moreover, such work has and will continue to prove useful in the development of potential vaccines, gene delivery vectors, and antitumor agents

(FARASSATI et al. 2001; MARKERT et al. 2000; MCMENAMIN et al. 1998; RAMPLING et al. 2000; SPECTOR et al. 1998; TANEJA et al. 2001; WALKER et al. 1998).

Acknowledgements. David Leib is supported by National Institutes of Health Grants EY-10707 and EY-09083 and is a recipient of a Robert McCormick Scholarship from Research to Prevent Blindness. Thanks to members of the Leib laboratory and Ian Mohr for comments on the manuscript.

References

Andreansky SS, He B, Gillespie GY, Soroceanu L, Markert J, Chou J, Roizman B, Whitley RJ (1996) The application of genetically engineered herpes simplex viruses to the treatment of experimental brain tumors. Proc Natl Acad Sci USA 93:11313–11318

Arnheiter H, Frese M, Kambadur R, Meier E, Haller O (1996) Mx transgenic mice – animal models of health. Curr Top Microbiol Immunol 206:119–147

Bischoff JR, Samuel CE (1985) Mechanism of interferon action. The interferon-induced phosphoprotein P1 possesses a double-stranded RNA-dependent ATP-binding site. J Biol Chem 260:8237–8239

Bolovan CA, Sawtell NM, Thompson RL (1994) ICP34.5 mutants of herpes simplex virus type 1 strain 17syn+ are attenuated for neurovirulence in mice and for replication in confluent primary mouse embryo cell cultures. J Virol 68:48–55

Cassady KA, Gross M, Roizman B (1998a) The herpes simplex virus US11 protein effectively compensates for the gamma1(34.5) gene if present before activation of protein kinase R by precluding its phosphorylation and that of the alpha subunit of eukaryotic translation initiation factor 2. J Virol 72:8620–8626

Cassady KA, Gross M, Roizman B (1998b) The second-site mutation in the herpes simplex virus recombinants lacking the gamma134.5 genes precludes shutoff of protein synthesis by blocking the phosphorylation of eIF-2alpha. J Virol 72:7005–7011

Cayley PJ, Davies JA, McCullagh KG, Kerr IM (1984) Activation of the ppp(A2'p)nA system in interferon-treated, herpes simplex virus-infected cells and evidence for novel inhibitors of the ppp(A2'p)nA-dependent RNase. Eur J Biochem 143:165–174

Chou J, Chen JJ, Gross M, Roizman B (1995) Association of a M(r) 90,000 phosphoprotein with protein kinase PKR in cells exhibiting enhanced phosphorylation of translation initiation factor eIF-2 alpha and premature shutoff of protein synthesis after infection with gamma 134.5-mutants of herpes simplex virus 1. Proc Natl Acad Sci USA 92:10516–10520

Chou J, Kern ER, Whitley RJ, Roizman B (1990) Mapping of herpes simplex virus-1 neurovirulence to gamma 134.5, a gene nonessential for growth in culture. Science 250:1262–1266

Chou J, Roizman B (1990) The herpes simplex virus 1 gene for ICP34.5, which maps in inverted repeats, is conserved in several limited-passage isolates but not in strain 17syn+. J Virol 64:1014–1020

Chou J, Roizman B (1994) Herpes simplex virus 1 gamma(1)34.5 gene function, which blocks the host response to infection, maps in the homologous domain of the genes expressed during growth arrest and DNA damage. Proc Natl Acad Sci USA 91:5247–5251

David M (1995) Transcription factors in interferon signaling. Pharmacol Ther 65:149–161

Der SD, Yang YL, Weissmann C, Williams BR (1997) A double-stranded RNA-activated protein kinase-dependent pathway mediating stress-induced apoptosis. Proc Natl Acad Sci USA 94:3279–3283

Diaz-Latoud C, Diaz JJ, Fabre-Jonca N, Kindbeiter K, Madjar JJ, Arrigo AP (1997) Herpes simplex virus Us11 protein enhances recovery of protein synthesis and survival in heat shock treated HeLa cells. Cell Stress Chaperones 2:119–131

Dolan A, McKie E, MacLean AR, McGeoch DJ (1992) Status of the ICP34.5 gene in herpes simplex virus type 1 strain 17. J Gen Virol 73:971–973

Dong B, Silverman RH (1995) 2–5A-dependent RNase molecules dimerize during activation by 2–5A. J Biol Chem 270:4133–4137

Duc Dodon M, Mikaelian I, Sergeant A, Gazzolo L (2000) The herpes simplex virus 1 US11 protein cooperates with suboptimal amounts of human immunodeficiency virus type 1 (HIV-1) Rev protein to rescue HIV-1 production. Virology 270:43–53

Everett RD, Meredith M, Orr A, Cross A, Kathoria M, Parkinson J (1997) A novel ubiquitin-specific protease is dynamically associated with the PML nuclear domain and binds to a herpesvirus regulatory protein [corrected and republished article originally printed in EMBO J 1997 Feb 3;16(3): 566–577]. EMBO J 16:1519–1530

Everett RD, Orr A, Preston CM (1998) A viral activator of gene expression functions via the ubiquitin-proteasome pathway. EMBO J 17:7161–7169

Everly DN, Jr., Read GS (1997) Mutational analysis of the virion host shutoff gene (UL41) of herpes simplex virus (HSV): characterization of HSV type 1 (HSV-1)/HSV-2 chimeras. J Virol 71:7157–7166

Farassati F, Yang AD, Lee PW (2001) Oncogenes in Ras signalling pathway dictate host-cell permissiveness to herpes simplex virus 1. Nat Cell Biol 3:745–750

Fenwick ML, Everett RD (1990) Transfer of UL41, the gene controlling virion-associated host cell shutoff, between different strains of herpes simplex virus. J Gen Virol 71:411–418

Goodbourn S, Didcock L, Randall RE (2000) Interferons: cell signalling, immune modulation, antiviral response and virus countermeasures. J Gen Virol 81:2341–2364

He B, Chou J, Liebermann DA, Hoffman B, Roizman B (1996) The carboxyl terminus of the murine MyD116 gene substitutes for the corresponding domain of the gamma(1)34.5 gene of herpes simplex virus to preclude the premature shutoff of total protein synthesis in infected human cells. J Virol 70:84–90

He B, Gross M, Roizman B (1997) The gamma(1)34.5 protein of herpes simplex virus 1 complexes with protein phosphatase 1alpha to dephosphorylate the alpha subunit of the eukaryotic translation initiation factor 2 and preclude the shutoff of protein synthesis by double-stranded RNA-activated protein kinase. Proc Natl Acad Sci USA 94:843–848

Hollander MC, Zhan Q, Bae I, Fornace AJ, Jr (1997) Mammalian GADD34, an apoptosis- and DNA damage-inducible gene. J Biol Chem 272:13731–13737

Isaacs A, Lindenmann J (1957) Virus interference. I. The interferon. Proceedings of the Royal Society of London 147:258–267

Johnson PA, MacLean C, Marsden HS, Dalziel RG, Everett RD (1986) The product of gene US11 of herpes simplex virus type 1 is expressed as a true late gene. J Gen Virol 67:871–883

Kerr IM, Brown RE (1978) pppA2′p5′A2′p5′A: an inhibitor of protein synthesis synthesized with an enzyme fraction from interferon-treated cells. Proc Natl Acad Sci USA 75:256–260

Khabar KS, Dhalla M, Siddiqui Y, Zhou A, Al-Ahdal MN, Der SD, Silverman RH, Williams BR (2000) Effect of deficiency of the double-stranded RNA-dependent protein kinase, PKR, on antiviral resistance in the presence or absence of ribonuclease L: HSV-1 replication is particularly sensitive to deficiency of the major IFN-mediated enzymes [In Process Citation]. J Interferon Cytokine Res 20:653–659

Kumar A, Haque J, Lacoste J, Hiscott J, Williams BR (1994) Double-stranded RNA-dependent protein kinase activates transcription factor NF-κ B by phosphorylating I kappa B. Proc Natl Acad Sci USA 91:6288–6292

Kwong AD, Frenkel N (1989) The herpes simplex virus virion host shutoff function. J Virol 63:4834–4839

Lausch RN, Su YH, Ritchie M, Oakes JE (1991) Evidence endogenous interferon production contributed to the lack of ocular virulence of an HSV intertypic recombinant. Curr Eye Res 10:39–45

Lebleu B, Sen GC, Shaila S, Cabrer B, Lengyel P (1976) Interferon, double-stranded RNA, and protein phosphorylation. Proc Natl Acad Sci USA 73:3107–3111

Leib DA, Coen DM, Bogard CL, Hicks KA, Yager DR, Knipe DM, Tyler KL, Schaffer PA (1989) Immediate-early regulatory gene mutants define different stages in the establishment and reactivation of herpes simplex virus latency. J Virol 63:759–768

Leib DA, Harrison TE, Laslo KM, Machalek MA, Moorman NJ, Virgin HW (1999) Interferons regulate the phenotype of wild-type and mutant herpes simplex viruses in vivo. J Exp Med 189:663–672

Leib DA, Machalek MA, Williams BR, Silverman RH, Virgin HW (2000) Specific phenotypic restoration of an attenuated virus by knockout of a host resistance gene [see comments]. Proc Natl Acad Sci USA 97:6097–6101

MacLean CA, Rixon FJ, Marsden HS (1987) The products of gene US11 of herpes simplex virus type 1 are DNA- binding and localize to the nucleoli of infected cells [published erratum appears in J Gen Virol 1988 Mar;69(Pt 3):763]. J Gen Virol 68:1921–1937

Markert JM, Medlock MD, Rabkin SD, Gillespie GY, Todo T, Hunter WD, Palmer CA, Feigenbaum F, Tornatore C, Tufaro F, Martuza RL (2000) Conditionally replicating herpes simplex virus mutant, G207 for the treatment of malignant glioma: results of a phase I trial [see comments]. Gene Ther 7:867–874

Abbreviations

KSHV	Kaposi's sarcoma-associated herpesvirus
HHV8	human herpesvirus 8
HVS	Herpesvirus saimiri
EBV	Epstein Barr virus
NK	natural killer
KS	Kaposi's sarcoma
PEL	primary effusion lymphoma
IFNs	Interferons
IRFs	IFN regulatory factors
vIRF	viral interferon regulatory factor
HAT	histone acetyltransferase
IL-6	interleukin-6
vGPCR	viral G protein-coupled receptor
VEGF	vascular endothelial growth factor
vCCP	viral complement control protein
KIRs	killer immunoglobulin-related receptors
FLICE	Fas-associated death domain-like interleukin 1β-converting enzyme
FLIP	FLICE inhibitory protein
CTL	cytotoxic T lymphocytes
MHC	major histocompatibility complex

1 Introduction

In the early stage of viral infection there is an intensive race between the virus and the host's defense systems, which are mediated through (1) nonspecific or innate immune defenses by interferon, natural killer (NK) cells, complement cascade, apoptosis, and macrophages; (2) specific or adaptive immune defenses by cytotoxic T lymphocytes (CTL), helper T lymphocytes, and anti-viral antibodies (PLOEGH 1998). The conclusion of this race will depend on how rapidly the virus can infect and/or replicate before it is cleared by an active immune system.

For more than a century, Kaposi's sarcoma (KS) had been an interesting neoplasm only to a few clinicians because of its rare occurrence and very localized outbreaks, primarily in Mediterranean white men. In addition, this complex lesion progresses slowly and presents with only mild clinical manifestation. The explosion of the AIDS epidemic around the mid-1980s brought increased attention to KS as it is the major malignancy among AIDS patients, occurring in approximately 20% of male HIV patients. Significant efforts were put forth to increase our understanding of its etiology and pathology. The hunt for the causative agent came to an end when Drs. Chang and Moore discovered fragments of DNA specific to KS lesions that were similar to other known gamma herpesvirus including Herpesvirus saimiri (HVS) and Epstein-Barr virus (EBV) (CHANG et al. 1994). This new herpesvirus, named KS-associated herpesvirus (KSHV) or human herpesvirus 8 (HHV8), was soon found associated with all epidemiological forms of KS (ANTMAN and CHANG 2000; CHANG 1997; GANEM 1997; NEIPEL et al. 1999; NEIPEL and FLECKENSTEIN 1999; REITZ et al. 1999; SCHULZ et al. 1998). KSHV has also been identified in primary effusion lymphoma (PEL) and an immunoblast variant of Castleman's disease, both of which are of B cell origin (CESARMAN et al. 1995; RENNE et al. 1996). Extensive studies have demonstrated that KSHV is present in spindle-shaped endothelial cells and infiltrating lymphocytes in KS tumors (BLASIG et al. 1997; BROWNING et al. 1994; OTSUKI et al. 1996; SIRIANNI et al. 1997). Serological and immunohistochemical assays also demonstrate that anti-KSHV antibodies against lytic and latent KSHV antigens are consistently present in groups at high risk for KS (GAO et al. 1996; KEDES et al. 1996). Progression to KS has been shown to be preceded by KSHV seroconversion (PARRAVICINI et al. 1997). Thus, the presently available data provide compelling evidence that KSHV is a necessary cause for KS in HIV-infected or other immunocompromised patients. However, this finding is not much of a surprise when one considers, as an analogy, the pathogenesis associated with EBV and HVS. In healthy individuals, EBV remains predominantly in a latent state with only a handful of genes being expressed without apparent symptoms (KIEFF 1996). However, individuals with congenital or acquired immunosuppression have a higher risk for developing polyclonal lymphoproliferative disease or malignant B lymphomas.

Despite the diversity of their biology and disease associations, a common theme of herpesviruses is a lifelong infection in the presence of an active host immune system. To establish this, herpesviruses have been shown to encode a variety of gene products that modulate host immune surveillance (JOHNSON and HILL 1998). KSHV has also evolved to contain a battery of genes whose products play an important role in the escape from host immune surveillance (Table 1). In this review, we will discuss our understanding of KSHV strategies to escape host immune surveillance, which ultimately contributes to lifelong infection and pathogenesis associated with this virus.

Table 1. KSHV immune evasion strategies

	KSHV Gene
Innate immunity	
Interferons	vIRF
Complement cascade	vCCP
Natural killer cell lysis	K5
Cytokines/chemokines	vIL-6, vMIP, vGPCR
Apoptosis	vBcl-2, vFLIP
Adaptive immunity	
Cytotoxic T lymphocyte lysis	K3, K5

2 Modulation of Host Innate Immunity

The microorganisms that are encountered daily in the life of a normal healthy individual only occasionally cause perceptible disease. Most are detected and destroyed within hours by defense mechanisms that are not antigen-specific and do not require a prolonged period of induction: these are the mechanisms of innate immunity. The primary components of innate immunity are chemokines, cytokines, NK cells, the complement cascade, and apoptosis (Table 1). KSHV contains a series of genes that downregulate host innate immune responses at various levels.

2.1 Viral Interferon Regulatory Factor

Interferons (IFNs) are a family of cytokines that exhibit such diverse biological effects as the inhibition of cell growth and protection against viral infection. Viruses have evolved a variety of mechanisms to counteract the inhibitory effects of IFN (VILCEK and SEN 1996). E1A of adenovirus inhibits IFN-induced signaling by downregulating the expression of STAT-1 and ISGF3γ. The terminal protein of hepatitis B virus also blocks signaling by IFNs (FOSTER et al. 1991) and EBNA-2 of EBV inhibits IFN signaling by abolishing the induction of IFN-stimulated genes (KANDA et al. 1992). The major secreted protein (M-T7) of myxoma virus blocks IFN signaling in a unique manner by binding IFN-α and neutralizing its activity (VILCEK and SEN 1996). The KSHV K9 open reading frame exhibits significant sequence homology with cellular IFN regulatory factors (IRFs). Expression of K9 dramatically represses transcriptional activation induced by IFN-$\alpha/\beta/\gamma$ (GAO et al. 1997; LI et al. 1998; ZIMRING et al. 1998). Furthermore, K9 expression leads to transformation of rodent fibroblast cells, resulting in morphological change, focus formation, growth at reduced serum concentration, and tumor induction in nude mice (GAO et al. 1997; LI et al. 1998). Thus the K9 gene of KSHV encodes the first viral interferon regulatory factor (vIRF), which functions as a repressor of cellular IFN-mediated signal transduction and as an oncogene to induce cell growth transformation.

Recent detailed studies have demonstrated that these functional activities of vIRF appear to be attributed in part to an interaction with and inhibition of p300 (BURYSEK et al. 1999; JAYACHANDRA et al. 1999; LI et al. 2000). Interaction of vIRF with p300 inhibits the histone acetyltransferase (HAT) activity of p300 in vitro and induces a dramatic hypoacetylation of nucleosomal histone H3 and H4 in vivo, resulting in global alteration of nucleosomal chromatin structure (LI et al. 2000). As a consequence, vIRF downregulates the transcriptional activity of an early inflammatory gene and the macrophage inhibitory factor gene (BURYSEK et al. 1999; LI et al. 2000), whereas it upregulates transcription of the cellular myc gene (JAYACHANDRA et al. 1999). Thus the modulation of p300 HAT activity is likely part of the mechanisms that vIRF employs to block cellular IFN-mediated antiviral activity.

2.2 Virocrines

Chemokines are molecules that interact with G protein-coupled chemokine receptors and play a key role in orchestrating leukocytic recruitment during inflammatory responses, including those to viral infections (ADAMS and LLOYD 1997; MURDOCH and FINN 2000). They have also been implicated in hematopoiesis, angiogenesis, and lymphocyte development (LOCATI and MURPHY 1999). In addition, cytokines play a critical role in the regulation of the immune response and constitute important targets for virus immune evasion strategies. One strategy employed by KSHV is to encode virocrines composed of virokines and viral receptors. Virokines are viral proteins that mimic cellular cytokines or chemokines, whereas viral receptors mimic cellular cytokine receptors. These virocrines are important players in the continuing encounter between viruses and their hosts.

2.2.1 Virokines

2.2.1.1 vMIP-I, vMIP-II, and vMIP-III

The migration of leukocytes from blood vessels to sites of infection and inflammation is an important part of host defenses (TAUB 1996). Many viruses, particularly herpesviruses, have captured and modified cellular chemokine and chemokine receptor genes to decoy the host immune response (LALANI et al. 2000; PARRY et al. 2000). Three open reading frames of KSHV, called vMIP-I, vMIP-II, and vMIP-III, share 25%–40% homology at the amino acid level with a CC chemokine, macrophage inflammatory protein (MIP-1) (MOORE et al. 1996; NICHOLAS et al. 1997b). Among these, vMIP-II exhibits an unusually broad spectrum of receptor binding activities (KLEDAL et al. 1997). Competition assays have demonstrated that vMIP-II efficiently competes with cellular chemokines for binding to cellular chemokine receptors, including CC chemokine receptor (CCR) 1, CCR2, CCR5, CXCR4, and CX3CR1 (KLEDAL et al. 1997). However, unlike cellular chemokines, vMIP-II binding to these receptors does not elicit Ca^{++} influx, suggesting that viral chemokine may function as a competitive antagonist on a wide variety of chemokine receptors (BOSHOFF et al. 1997; CHEN et al. 1998; KLEDAL et al. 1997; MOORE et al. 1996) and also significantly reduces the infiltration of inflammatory leukocytes and suppresses the onset of the host inflammatory response in a rat model (CHEN et al. 1998). In contrast, vMIP-II can function as CCR3 agonist to drive a Th2-type immune response in vivo (BOSHOFF et al. 1997; CHEN et al. 1998; KLEDAL et al. 1997; MOORE et al. 1996). Unlike vMIP-II, which is promiscuous in its binding profile, vMIP-I and vMIP-III selectively engage CCR8 and CCR4 and function as agonists for the cellular chemokine receptors (DAIRAGHI et al. 1999; ENDRES et al. 1999; STINE et al. 2000). In addition, all three, vMIP-I, vMIP-II, and vMIP-III, induce angiogenesis in the chorioallantoic membrane of chicken eggs (BOSHOFF et al. 1997; LOCATI and MURPHY 1999; MOORE et al. 1996; NICHOLAS et al. 1997b; STINE et al. 2000). This evidence indicates that viral chemokines may contribute to the deregulation of host immune responses and the development of angioneoplasms associated with KSHV.

2.2.1.2 vIL-6

Previously known as B-cell differentiating factor, interleukin-6 (IL-6) is expressed in lymphocytes, macrophages, and endothelial cells. Although it acts on most cells, it is particularly important in inducing B cells to differentiate into antibody-producing plasma cells and is considered to be an important growth factor for multiple myeloma, lymphoma, and leukemia. KSHV ORF K2, which has homology to human IL-6 at the amino acid level, is constitutively expressed at the latent/lytic stage of viral lifecycle in BCP-1 cells, but only at the lytic stage in BC-1 cells. Immunohistochemistry has shown that only a minor population of virus-infected cells in KS lesions express vIL-6, whereas a high level of vIL-6 expression is detected in tissues from multicentric Castleman's disease, indicating a confined role of vIL-6 in KSHV-associated lymphoproliferative disorders (PARRAVICINI et al. 2000). Functional studies with a cloned vIL-6 gene have demonstrated that it is able to support proliferation of an IL-6-dependent mouse myeloma cell line and to promote the growth of KSHV-infected PEL cells (AOKI et al. 1999; MOORE et al. 1996; NICHOLAS et al. 1997b). Despite their similarities in sequence and function, cellular IL-6 and vIL-6 display differences in receptor usage. Although cellular IL-6 requires both IL-6R and gp130 for intracellular signaling, vIL-6 appears to bind only gp130 and this binding is sufficient to elicit its signal transduction (HOISCHEN et al. 2000; MOLDEN et al. 1997; OSBORNE et al. 1999). Thus vIL-6 is a multifunctional cytokine that potentially contributes to KSHV-associated lymphoproliferative diseases by constitutively activating cellular signal transduction pathways and by preventing apoptosis of virus-infected lymphoid cells.

2.2.2 Viral Receptor

2.2.2.1 vGPCR

The G protein-coupled receptors belong to a class of cellular receptors that are characterized by a uniform molecular architecture of seven transmembrane α-helices linked by extra- and intracellular peptide loops (SCHONEBERG et al. 1999). Binding of diverse agonists to these heptahelical receptors leads to a reversible activation of a limited repertoire of heterotrimeric guanine nucleotide-binding proteins (G proteins), forwarding the signal to intracellular effectors such as enzymes and ion channels. KSHV encodes a G protein-coupled receptor (vGPCR) homologous to the human receptor for the angiogenic chemokine interleukin-8 (IL-8R) (CESARMAN et al. 1996; NEIPEL et al. 1997; RUSSO et al. 1996). Competitive binding assays show that KSHV vGPCR has binding affinity for various CXC chemokines (IL-8, MGSA, NAP-2, and PF-4) and limited members of the CC chemokines (I-309 and RANTES) (ARVANITAKIS et al. 1997; CESARMAN et al. 2000;

Ho et al. 1999). However, despite binding to a broad spectrum of chemokines, vGPCR has been shown to be agonist-independent, constitutively active, and capable of eliciting signals to induce the phosphoinositide-inositol triphosphate/ protein kinase C pathway (ARVANITAKIS et al. 1997). Along with this finding, expression of vGPCR in rodent kidney and fibroblast cells has been shown to elicit a constitutively active signal that strongly stimulates both proliferation and angiogenesis (ARVANITAKIS et al. 1997; BAIS et al. 1998). Angiogenic responses induced by vGPCR are mediated by upregulation of vascular endothelial growth factor (VEGF) (BAIS et al. 1998). In addition, KSHV GPCR expression within the hematopoietic cell lineage of transgenic mice has been shown to result in the development of angioproliferative lesions in multiple organs that morphologically resemble KS lesions (ARVANITAKIS et al. 1997; YANG et al. 2000). Thus, alteration of cellular signal transduction by vGPCR may contribute to enhanced viral propagation in the presence of an active immune system and deregulated angiogenesis of virus-containing cells for tumor development.

2.3 Viral Complement Control Protein

The complement system is the primary humoral defense mechanism against microorganisms (COOPER 1991). Many large DNA viruses including vaccinia virus, Herpesvirus simplex, and EBV are believed to have mechanisms of evading or exploiting the complement cascade (FRIEDMAN et al. 1996; HARRIS et al. 1990; ISAACS et al. 1992; MOLD et al. 1988). All γ-2 herpesviruses, including KSHV, HVS, RRV, and murine herpesvirus 68 (MHV68), contain an orf4 that has high homology with cellular complement control protein. Until now, only HVS orf4 has been studied for its functional activity. Two different forms of HVS orf4, called viral complement control protein (vCCP), have been detected, a membrane-bound and a secreted form. Both forms of vCCP have been shown to inhibit the complement cascade at the level of C3 and are associated with the virion particle (FODOR et al. 1995). In addition, HVS contains an additional gene, orf15, which strongly resembles the gene for human CD59, the membrane-attacking complex inhibitory factor (ROTHER et al. 1994). HVS vCD59 has been shown to inhibit lysis at the level of C9 in the cascade with less species restriction compared with cellular CD59. Like cellular CD59, it is also a GPI-linked membrane protein (ROTHER et al. 1994). Recently, cellular CD59 has been shown to be present in the HIV envelope, and it can protect HIV from complement-mediated lysis (SAIFUDDIN et al. 1995). The presence of complement control homologous genes in all γ-2 herpesviral genomes speaks to the importance of complement cascade to their lifecycle.

2.4 Viral Immune Modulator for NK Cell Killing

NK cells are a subpopulation of lymphoid cells that do not require prior sensitization to recognize and kill various targets including viruses, intracellular bacteria, parasites, and neoplastic cells (BIRON et al. 1999; KARRE and WELSH 1997; LANIER 1998b; LONG 1999; UNANUE 1997). NK cell-mediated cytotoxicity is regulated by opposing signals from receptors that activate or inhibit effector function (BOTTINO et al. 2000; LANIER 1998b; LONG 1999). It is well established that the interaction between killer immunoglobulin-related receptors (KIRs) on NK cells and their specific HLA class I allotypes on target cells provides the major turn-off signal that regulates NK cell-mediated cytotoxicity (BLERY et al. 2000; LANIER 1998a; LONG 1999). Thus, the loss of MHC class I molecules from the cell surface renders target cells susceptible to NK cell cytotoxicity. In contrast, the interaction between lymphocyte surface receptors on NK cells and their specific ligands on target cells is involved in providing the signals responsible for NK cell activation (BAKKER et al. 2000). Specifically, the B7 family can function as a ligand for CD28 or its derivative activating receptors to induce NK cell-mediated cytotoxicity (AZUMA et al. 1992; DANG et al. 1990; GALEA-LAURI et al. 1999; WILSON et al. 1999). Recently, we have discovered a novel strategy of KSHV to escape NK cell cytotoxicity. The KSHV K5 zinc finger membrane protein has been shown to downregulate surface expression of ICAM-1 and B7-2 (COSCOY and GANEM 2001; ISHIDO et al. 2000a). As a consequence, K5 expression drastically inhibits NK cell-mediated cytotoxicity in several experimental settings (ISHIDO et al. 2000a). Furthermore, we have demonstrated that B7-2 costimulatory molecule plays the major role in activation of NK cytotoxicity and ICAM-1 acts synergistically to enhance NK cell-mediated cytotoxicity (ISHIDO et al. 2000a). This is a novel viral immune evasion strategy by which KSHV achieves immune avoidance by downregulation of cellular ligands for NK cell-mediated cytotoxicity receptors.

2.5 Antiapoptotic Genes

Upon viral infection, infected cells can become the target of host immune responses or can go through a programmed cell death process, called apoptosis, as a defense mechanism to limit the ability of the virus to replicate. To prevent this, viruses have evolved elaborate mechanisms to subvert the apoptotic process and facilitate a persistent infection or prolong the survival of lytically infected cells to maximize the production of viral progeny. KSHV is also genetically equipped to prevent cellular apoptosis by harboring two antiapoptotic genes, vFLIP and vBcl-2.

2.5.1 vFLIP

K13 of KSHV encodes a viral FLIP (vFLIP), a homologue of the cellular FLICE (Fas-associated death domain-like interleukin 1β-converting enzyme) inhibitory protein (FLIP). KSHV vFLIP has been shown to protect cells from Fas/APO1-mediated apoptosis by inhibiting activation of caspase-3, -8, and -9 and also permits clonal growth in the presence of death stimuli in vitro (DJERBI et al. 1999). Moreover, vFLIP has also been shown to modulate the NF-κB signaling pathway through association with TRAFs and downstream signaling proteins (CHAUDHARY et al. 1999). Transcripts from the locus comprising vFLIP, vCyclin, and orf73 (LANA) appear to be differentially spliced in PEL cell lines, KS lesions, and lymph nodes. vFLIP is present at a very low level in early KS lesions, with expression dramatically increasing in late-stage lesions (STURZL et al. 1999). In parallel, the increase in the amount of vFLIP transcripts is associated with a reduction in apoptosis in KS lesions, indicating an in vivo antiapoptotic activity of vFLIP (STURZL et al. 1999).

2.5.2 vBcl-2

Although the overall amino acid sequence identity of KSHV orf16 to cellular Bcl-2 is only 15%–20%, it contains the BH1 and BH2 regions required for heterodimerization of Bcl-2 and for its death-repressor activity. vBcl-2 is expressed in both classic and AIDS-KS lesions and in cell lines derived from primary effusion lymphomas (SARID et al. 1997). Overexpression of vBcl-2 has been shown to block apoptosis as efficiently as cellular Bcl-2, Bcl-xL, or another viral Bcl-2 homolog encoded by EBV, BHRF1 (CHENG et al. 1997). In addition, vBcl-2 has been shown to block Bax-mediated toxicity in yeast (SARID et al. 1997). Although the ability of vBcl-2 to heterodimerize with the cellular Bcl-2 family is controversial, it is now clear that KSHV has pirated the cellular antiapoptotic Bcl-2 gene to escape negative regulatory effects on viral infection and replication.

3 Modulation of Host Adaptive Immunity

When an infectious agent breaches the early lines of host defense, an adaptive immune response will ensue, with the generation of antigen-specific effector cells that specifically target the pathogen and the memory cells that prevent subsequent infection with the same infectious agents. Adaptive immunity consists of both a humoral immune response and a cell-mediated immune response. Not surprisingly, KSHV also circumvents host cell-mediated immune response by harboring genes that inhibit antigen presentation process (Table 1).

3.1 Viral Immune Modulators for Cytotoxic T Lymphocyte Killing

A major immune defense against viral infection is mediated by cytotoxic T lymphocytes (CTLs), which recognize and lyse infected cells on engagement of the T cell receptor with major histocompatibility complex (MHC) class I molecules presenting viral peptides (HENGEL and KOSZINOWSKI 1997; PLOEGH 1998; MILLER and SEDMAK 1999; WIERTZ et al. 1996b). To achieve persistent infection, herpesvirus encodes a variety of proteins that function to lower MHC I display by several mechanisms (PLOEGH 1998). These include binding and retention of MHC I chains in the endoplasmic reticulum (ER), dislocation of class I chains from the ER, inhibition of the peptide transporter TAP involved in antigen presentation, and shunting of newly assembled chains to lysosomes (AHN et al. 1997; FRUH et al. 1995; HILL et al. 1995; JONES et al. 1996; LEVITSKAYA et al. 1995; PLOEGH 1998; WIERTZ et al. 1996a; YORK et al. 1994; ZIEGLER et al. 1997). KSHV K3 and K5 proteins, which exhibit 40% amino acid identity to each other (NICHOLAS et al. 1997a; RUSSO et al. 1996), dramatically downregulate surface expression of MHC class I molecules (COSCOY and GANEM 2000; ISHIDO et al. 2000b). Biochemical analyses have demonstrated that although K3 and K5 do not affect expression and intracellular transport of class I molecules, their expression induces rapid endocytosis of class I molecules (COSCOY and GANEM 2000; ISHIDO et al. 2000b). Despite their similarity in sequence and function, K3 and K5 differ in their specificity as K3 drastically downregulates HLA-A, -B, -C, and -E, whereas K5 exclusively downregulates HLA-A and -B (ISHIDO et al. 2000b). This selective downregulation of HLA allotypes by K5 is partly due to differences in the amino acid sequences of the HLA transmembrane regions. Because the magnitude of K3 downregulation of class I molecules is more pronounced than that of K5 (ISHIDO et al. 2000b), K3 may play a major role in the inhibition of CTL cytotoxicity. Although MHC class I downregulation may protect KSHV-infected cells from CTL recognition, indiscriminate downregulation of HLA allotypes by K3 invites NK cell susceptibility (ISHIDO et al. 2000b). To prevent this, K5 has been shown to downregulate ICAM-1 and B7-2, which are NK cytotoxic receptors (ISHIDO et al. 2000b), as discussed in Section 2.4. Thus KSHV uses two genes, K3 and K5, with similar but distinct activities to ensure comprehensive protection from host immune effectors.

4 Conclusions

Soon after viral infection, there is an intensive race between the virus to spread to uninfected cells and the host's immune system to eliminate infected cells (PLOEGH 1998). Thus the success of a persistent virus infection lies in its capacities for evasion of host defense mechanisms by inhibiting the function of cellular proteins that are important components of a host immune response. Like other herpesviruses, KSHV encodes a diverse array of viral genes that contribute to circumventing host immune surveillance mechanisms. By using unique viral genes and counterparts to cellular genes, KSHV deregulates normal cellular pathways that, otherwise, lead to apoptosis of infected cells and activation of host immune responses. These include the K3 and K5 genes, which comprehensively inhibit host immune effectors, the vIRF gene, which suppresses IFN-mediated antiviral activity, vFLIP and vBcl-2, which inhibit the apoptosis-mediated host defense, vMIP, vIL-6, and vGPCR, which modulate cytokine-mediated immune responses, and vCCP, which inhibits the complement cascade. On the other hand, the capability of KSHV to express vIRF, the vMIP, and vIL-6 along with vGPCR may also allow the virus to induce autocrine/paracrine mechanisms of cellular signal transduction to promote deregulated cell growth.

Extensive studies of individual KSHV gene products have provided us with a better understanding of the molecular mechanisms that underlie the immune evasion and pathogenesis associated with this virus. However, because of the lack of a permissive culture system, we are still deficient in our comprehension of the contribution of these gene products to viral replication, persistent infection, and pathogenesis in the context of viral genome and infected host. Fortunately, several animal homologues of KSHV that have an efficient cell culture system have been identified and well characterized for the study of viral replication and pathogenesis. These are RRV, HVS, and MHV68 (BERGQUAM et al. 1999; DESROSIERS et al. 1997; JUNG et al. 1999; MANSFIELD et al. 1999; NASH and SUNIL-CHANDRA 1994; SPECK and VIRGIN 1999; WONG et al. 1999). Thus, combined studies of KSHV gene products in cell culture and in defined in vivo experimental settings will provide comprehensive understanding of the molecular mechanisms of viral immune evasion and pathogenesis.

Acknowledgements. This work was supported by U.S. Public Health Service Grants CA-31363, CA-82057, CA-86841, CA-91819, AI-38131, and RR-00168 and ACS Grant RPG001102. R.E. Means is a Cancer Research Institute Fellow. J. Jung is a Leukemia & Lymphoma Society Scholar.

References

Adams DH, Lloyd AR (1997) Chemokines: leucocyte recruitment and activation cytokines. Lancet 349:490–495
Ahn K, Gruhler A, Galocha B, Jones TR, Wiertz EJ, Ploegh HL, Peterson PA, Yang Y, Fruh K (1997) The ER-luminal domain of the HCMV glycoprotein US6 inhibits peptide translocation by TAP. Immunity 6:613–621
Antman K, Chang Y (2000) Kaposi's sarcoma. N Engl J Med 342:1027–1038
Aoki Y, Jaffe ES, Chang Y, Jones K, Teruya-Feldstein J, Moore PS, Tosato G (1999) Angiogenesis and hematopoiesis induced by Kaposi's sarcoma-associated herpesvirus-encoded interleukin-6 [see comments]. Blood 93:4034–4043
Arvanitakis L, Geras-Raaka E, Varma A, Gershengorn MC, Cesarman E (1997) Human herpesvirus KSHV encodes a constitutively active G-protein-coupled receptor linked to cell proliferation. Nature 385:347–350
Azuma M, Cayabyab M, Buck D, Phillips JH, Lanier LL (1992) Involvement of CD28 in MHC-unrestricted cytotoxicity mediated by a human natural killer leukemia cell line. J Immunol 149:1115–1123
Bais C, Santomasso B, Coso O, Arvanitakis L, Raaka EG, Gutkind JS, Asch AS, Cesarman E, Gershengorn MC, Mesri EA (1998) G-protein-coupled receptor of Kaposi's sarcoma-associated herpesvirus is a viral oncogene and angiogenesis activator. Nature 391:86–89
Bakker AB, Wu J, Phillips JH, Lanier LL (2000) NK cell activation: distinct stimulatory pathways counterbalancing inhibitory signals. Hum Immunol 61:18–27
Bergquam EP, Avery N, Shiigi SM, Axthelm MK, Wong SW (1999) Rhesus rhadinovirus establishes a latent infection in B lymphocytes in vivo. J Virol 73:7874–7876
Biron CA, Nguyen KB, Pien GC, Cousens LP, Salazar-Mather TP (1999) Natural killer cells in antiviral defense: function and regulation by innate cytokines. Annu Rev Immunol 17:189–220
Blasig C, Zietz C, Haar B, Neipel F, Esser S, Brockmeyer NH, Tschachler E, Colombini S, Ensoli B, Sturzl M (1997) Monocytes in Kaposi's sarcoma lesions are productively infected by human herpesvirus 8. J Virol 71:7963–7968

Blery M, Olcese L, Vivier E (2000) Early signaling via inhibitory and activating NK receptors. Hum Immunol 61:51–64
Boshoff C, Endo Y, Collins PD, Takeuchi Y, Reeves JD, Schweickart VL, Siani MA, Sasaki T, Williams TJ, Gray PW, et al. (1997) Angiogenic and HIV-inhibitory functions of KSHV-encoded chemokines [see comments]. Science 278:290–294
Bottino C, Biassoni R, Millo R, Moretta L, Moretta A (2000) The human natural cytotoxicity receptors (NCR) that induce HLA class I-independent NK cell triggering. Hum Immunol 61:1–6
Browning PJ, Sechler JM, Kaplan M, Washington RH, Gendelman R, Yarchoan R, Ensoli B, Gallo RC (1994) Identification and culture of Kaposi's sarcoma-like spindle cells from the peripheral blood of human immunodeficiency virus-1-infected individuals and normal controls [see comments]. Blood 84:2711–2720
Burysek L, Yeow WS, Lubyova B, Kellum M, Schafer SL, Huang YQ, Pitha PM (1999) Functional analysis of human herpesvirus 8-encoded viral interferon regulatory factor 1 and its association with cellular interferon regulatory factors and p300. J Virol 73:7334–7342
Cesarman E, Chang Y, Moore PS, Said JW, Knowles DM (1995) Kaposi's sarcoma-associated Herpesvirus-like DNA sequences in AIDS-related body-cavity-based lymphomas. N Engl J Med 332:1186–1191
Cesarman E, Mesri EA, Gershengorn MC (2000) Viral G protein-coupled receptor and Kaposi's sarcoma: a model of paracrine neoplasia? [comment]. J Exp Med 191:417–422
Cesarman E, Nador RG, Bai F, Bohenzky RA, Russo JJ, Moore PS, Chang Y, Knowles DM (1996) Kaposi's sarcoma-associated herpesvirus contains G protein-coupled receptor and cyclin D homologs which are expressed in Kaposi's sarcoma and malignant lymphoma. J Virol 70:8218–8223
Chang Y (1997) Kaposi's Sacoma and Kaposi's Sarcoma-associated Herpesvirus (human Herpesvirus 8): Where are we now? J Natl Cancer Inst 89:1829–1874
Chang Y, Cesarman E, Pessin MS, Lee F, Culpepper J, Knowles DM, Moore PS (1994) Identification of herpesvirus-like DNA sequences in AIDS-associated Kaposi's sarcoma [see comments]. Science 266:1865–1869
Chaudhary PM, Jasmin A, Eby MT, Hood L (1999) Modulation of the NF-κB pathway by virally encoded death effector domains-containing proteins. Oncogene 18:5738–5746
Chen S, Bacon KB, Li L, Garcia GE, Xia Y, Lo D, Thompson DA, Siani MA, Yamamoto T, Harrison JK, Feng L (1998) In vivo inhibition of CC and CX3 C chemokine-induced leukocyte infiltration and attenuation of glomerulonephritis in Wistar-Kyoto (WKY) rats by vMIP-II. J Exp Med 188:193–198
Cheng EH, Nicholas J, Bellows DS, Hayward GS, Guo HG, Reitz MS, Hardwick JM (1997) A Bcl-2 homolog encoded by Kaposi sarcoma-associated virus, human herpesvirus 8, inhibits apoptosis but does not heterodimerize with Bax or Bak. Proc Natl Acad Sci USA 94:690–694
Cooper NR (1991) Complement evasion strategies of microorganisms. Immunol Today 12:327–331
Coscoy L, Ganem D (2000) Kaposi's sarcoma-associated herpesvirus encodes two proteins that block cell surface display of MHC class I chains by enhancing their endocytosis. Proc Natl Acad Sci USA 97:8051–8056
Coscoy L, Ganem D (2001) A viral protein that selectively downregulates ICAM-1 and B7-2 and modulates T cell costimulation. J Clin Invest 107:1599–1606
Dairaghi DJ, Fan RA, McMaster BE, Hanley MR, Schall TJ (1999) HHV8-encoded vMIP-I selectively engages chemokine receptor CCR8. Agonist and antagonist profiles of viral chemokines. J Biol Chem 274:21569–21574
Dang LH, Michalek MT, Takei F, Benaceraff B, Rock KL (1990) Role of ICAM-1 in antigen presentation demonstrated by ICAM-1 defective mutants. J Immunol 144:4082–4091
Desrosiers RC, Sasseville VG, Czajak SC, Zhang X, Mansfield KG, Kaur A, Johnson RP, Lackner AA, Jung JU (1997) A herpesvirus of rhesus monkeys related to the human Kaposi's sarcoma-associated herpesvirus. J Virol 71:9764–9769
Djerbi M, Scrpanti V, Catrina AI, Bogen B, Biberfeld P, Grandien A (1999) The inhibitor of death receptor signaling, FLICE-inhibitory protein defines a new class of tumor progression factors [see comments]. J Exp Med 190:1025–1032
Endres MJ, Garlisi CG, Xiao H, Shan L, Hedrick JA (1999) The Kaposi's sarcoma-related herpesvirus (KSHV)-encoded chemokine vMIP- I is a specific agonist for the CC chemokine receptor (CCR)8. J Exp Med 189:1993–1998
Fodor WL, Rollins SA, Bianco-Caron S, Rother RP, Guilmette ER, Burton WV, Albrecht JC, Fleckenstein B, Squinto SP (1995) The complement control protein homolog of herpesvirus saimiri regulates serum complement by inhibiting C3 convertase activity. J Virol 69:3889–3892

Foster GR, Ackrill AM, Goldin RD, Kerr IM, Thomas HC, Stark GR (1991) Expression of the terminal protein region of hepatitis B virus inhibits cellular responses to interferons α and r and double-stranded RNA. Proc Natl Acad Sci USA 88:2888–2892

Friedman HM, Wang L, Fishman NO, Lambris JD, Eisenberg RJ, Cohen GH, Lubinski J (1996) Immune evasion properties of herpes simplex virus type 1 glycoprotein gC. J Virol 70:4253–4260

Fruh K, Ahn K, Djaballah H, Sempe P, van Endert PM, Tampe R, Peterson PA, Yang Y (1995) A viral inhibitor of peptide transporters for antigen presentation. Nature 375:415–418

Galea-Lauri J, Darling D, Gan SU, Krivochtchapov L, Kuiper M, Gaken J, Souberbielle B, Farzaneh F (1999) Expression of a variant of CD28 on a subpopulation of human NK cells: implications for B7-mediated stimulation of NK cells. J Immunol 163:62–70

Ganem D (1997) KSHV and Kaposi's sarcoma: The end of the beginning? Cell 91:157–160

Gao S-J, Boshoff C, Jayachandra S, Weiss RA, Chang Y, Moore PS (1997) KSHV *ORF K9* (vIRF) is an oncogene which inhibits the interferon signaling pathway. Oncogene 15:1979–1985

Gao S-J, Kingsley L, Hoover DR, Spira TJ, Rinaldo CR, Saah A, Phair J, Detels R, Parry P, Chang Y, Moore PS (1996) Seroconversion to antibodies against Kaposi's sarcoma-associated herpesvirus-related latent nuclear antigens before the development of Kaposi's sarcoma. N Engl J Med 335:233–241

Harris SL, Frank I, Yee A, Cohen GH, Eisenberg RJ, Friedman HM (1990) Glycoprotein C of herpes simplex virus type 1 prevents complement-mediated cell lysis and virus neutralization. J Infect Dis 162:331–337

Hengel H, Koszinowski UH (1997) Interference with antigen processing by viruses [see comments]. Curr Opin Immunol 9:470–476

Hill A, Jugovic P, York I, Russ G, Bennink J, Yewdell J, Ploegh H, Johnson D (1995) Herpes simplex virus turns off the TAP to evade host immunity. Nature 375:411–415

Ho HH, Du D, Gershengorn MC (1999) The N terminus of Kaposi's sarcoma-associated herpesvirus G protein-coupled receptor is necessary for high affinity chemokine binding but not for constitutive activity. J Biol Chem 274:31327–31332

Hoischen SH, Vollmer P, Marz P, Ozbek S, Gotze KS, Peschel C, Jostock T, Geib T, Mullberg J, Mechtersheimer S, et al. (2000) Human herpes virus 8 interleukin-6 homologue triggers gp130 on neuronal and hematopoietic cells. Eur J Biochem 267:3604–3612

Isaacs SN, Kotwal GJ, Moss B (1992) Vaccinia virus complement-control protein prevents antibody-dependent complement-enhanced neutralization of infectivity and contributes to virulence. Proc Natl Acad Sci USA 89:628–632

Ishido S, Choi JK, Lee BS, Wang C, DeMaria M, Johnson RP, Cohen GB, Jung JU (2000a) Inhibition of natural killer cell-mediated cytotoxicity by Kaposi's sarcoma-associated herpesvirus K5 protein. Immunity 13:365–374

Ishido S, Wang C, Lee BS, Cohen GB, Jung JU (2000b) Downregulation of major histocompatibility complex class I molecules by Kaposi's sarcoma-associated herpesvirus K3 and K5 proteins [In Process Citation]. J Virol 74:5300–5309

Jayachandra S, Low KG, Thlick AE, Yu J, Ling PD, Chang Y, Moore PS (1999) Three unrelated viral transforming proteins (vIRF, EBNA2, and E1A) induce the MYC oncogene through the interferon-responsive PRF element by using different transcription coadaptors. Proc Natl Acad Sci USA 96:11566–11571

Johnson DC, Hill AB (1998) Herpesvirus evasion of the immune system. Curr Top Microbiol Immunol 232:149–177

Jones TR, Wiertz EJ, Sun L, Fish KN, Nelson JA, Ploegh HL (1996) Human cytomegalovirus US3 impairs transport and maturation of major histocompatibility complex class I heavy chains. Proc Natl Acad Sci USA 93:11327–11333

Jung JU, Choi JK, Ensser A, Biesinger B (1999) Herpesvirus saimiri as a model for gammaherpesvirus oncogenesis. Semin Cancer Biol 9:231–239

Kanda K, Decker T, Aman P, Wahlstrom M, von Gabain A, Kallin B (1992) The EBNA2-related resistance towards a interferon (IFN-α) in Burkitt's lymphoma cells effects induction of IFN-induced genes but not the activation of transcription factor ISGF-3. Mol Cell Biol 12:4930–4936

Karre K, Welsh RM (1997) Viral decoy vetoes killer cell [news; comment]. Nature 386:446–447

Kedes D, Operaiski E, Busch M, Kohn R, Flood J, Ganem DE (1996) The seroepidemiology of human herpesvirus 8 (Kaposi's sarcoma-associated herpesvirus): distribution of infection in KS risk groups and evidence for sexual transmission. Nat Med 2:918–924

Kieff E (1996) Epstein-Barr Virus and Its Replication. In: Fields B, Howley PM, et al. (eds) Fields Virology. Lippincott-Raven, Philadelphia, pp 2343–2396

Kledal TN, Rosenkilde MM, Coulin F, Simmons G, Johnsen AH, Alouani S, Power CA, Lüttichau HR, Gerstoft J, Clapham PR, et al. (1997) A broad-spectrum chemokine antagonist encoded by Kaposi's sarcoma-associated herpesvirus. Science 277:1656–1659

Lalani AS, Barrett JW, McFadden G (2000) Modulating chemokines: more lessons from viruses. Immunol Today 21:100–106

Lanier LL (1998a) Follow the leader: NK cell receptors for classical and nonclassical MHC class I. Cell 92:705–707

Lanier LL (1998b) NK cell receptors. Annu Rev Immunol 16:359–393

Levitskaya J, Coram M, Levitsky V, Imreh S, Steigerwald-Mullen PM, Klein G, Kurilla MG, Masucci MG (1995) Inhibition of antigen processing by the internal repeat region of the Epstein-Barr virus nuclear antigen-1. Nature 375:685–688

Li M, Damania B, Alvarez X, Ogryzko V, Ozato K, Jung JU (2000) Inhibition of p300 histone acetyl-transferase by viral interferon regulatory factor [In Process Citation]. Mol Cell Biol 20:8254–8263

Li M, Lee H, Guo J, Neipel F, Fleckenstein B, Ozato K, Jung JU (1998) Kaposi's sarcoma-associated herpesvirus viral interferon regulatory factor. J Virol 72:5433–5440

Locati M, Murphy PM (1999) Chemokines and chemokine receptors: biology and clinical relevance in inflammation and AIDS. Annu Rev Med 50:425–440

Long EO (1999) Regulation of immune responses through inhibitory receptors. Annu Rev Immunol 17:875–904

Mansfield KG, Westmoreland SV, DeBakker CD, Czajak S, Lackner AA, Desrosiers RC (1999) Experimental infection of rhesus and pig-tailed macaques with macaque rhadinoviruses. J Virol 73:10320–10328

Miller DM, Sedmak DD (1999) Viral effects on antigen processing. Curr Opin Immunol 11:94–99

Mold C, Bradt BM, Nemerow GR, Cooper NR (1988) Epstein-Barr virus regulates activation and processing of the third component of complement. J Exp Med 168:949–969

Molden J, Chang Y, You Y, Moore PS, Goldsmith MA (1997) A Kaposi's sarcoma-associated herpes-virus-encoded cytokine homolog (vIL-6) activates signaling through the shared gp130 receptor subunit. J Biol Chem 272:19625–19631

Moore PS, Boshoff C, Weiss RA, Chang Y (1996) Molecular mimicry of human cytokine and cytokine response pathway genes by KSHV. Science 274:1739–1744

Murdoch C, Finn A (2000) Chemokine receptors and their role in inflammation and infectious diseases. Blood 95:3032–3043

Nash AA, Sunil-Chandra NP (1994) Interactions of the murine gammaherpesvirus with the immune system. Curr Opin Immunol 6:560–563

Neipel F, Albrecht JC, Fleckenstein B (1999) Human herpesvirus 8: is it a tumor virus? Proc Assoc Am Physicians 111:594–601

Neipel F, Albrecht J-C, Fleckenstein B (1997) Cell-homologous genes in the Kaposi's sarcoma-associated rhadinovirus human herpesvirus 8: determinants of its pathogenicity. J Virol 71:4187–4192

Neipel F, Fleckenstein B (1999) The role of HHV-8 in Kaposi's sarcoma. Semin Cancer Biol 9:151–164

Nicholas J, Ruvolo V, Zong J, Ciufo D, Guo H-G, Reitz MS, Hayward GS (1997a) A single 13-kilobase divergent locus in the kaposi sarcoma-associated herpesvirus (human herpesvirus 8) genome contains nine open reading frames that are homologous to or related to cellular proteins. J Virol 71:1963–1974

Nicholas J, Ruvolo VR, Burns WH, Sandford G, Wan X, Ciufo D, Hendrickson SB, Guo H-G, Hayward GS, Reitz MS (1997b) Kaposi's sarcoma-associated human herpesvirus-8 encodes homologues of macrophage inflammatory protein-1 and interleukin-6. Nat Med 3:287–292

Osborne J, Moore PS, Chang Y (1999) KSHV-encoded viral IL-6 activates multiple human IL-6 signaling pathways. Hum Immunol 60:921–927

Otsuki T, Kumar S, Ensoli B, Kingma DW, Yano T, Stetler-Stevenson M, Jaffe ES, Raffeld M (1996) Detection of HHV-8/KSHV DNA sequences in AIDS-associated extranodal lymphoid malignancies. Leukemia 10:1358–1362

Parravicini C, Chandran B, Corbellino M, Berti E, Paulli M, Moore PS, Chang Y (2000) Differential viral protein expression in Kaposi's sarcoma-associated herpesvirus-infected diseases: Kaposi's sarcoma, primary effusion lymphoma, and multicentric Castleman's disease. Am J Pathol 156:743–749

Parravicini C, Olsen SJ, Capra M, Poli F, Sirchia G, Gao SJ, Berti E, Nocera A, Rossi E, Bestetti G, et al. (1997) Risk of Kaposi's sarcoma-associated herpes virus transmission from donor allografts among Italian posttransplant Kaposi's sarcoma patients. Blood 90:2826–2829

Parry CM, Simas JP, Smith VP, Stewart CA, Minson AC, Efstathiou S, Alcami A (2000) A broad spectrum secreted chemokine binding protein encoded by a herpesvirus. J Exp Med 191:573–578

Ploegh HL (1998) Viral strategies of immune evasion. Science 280:248–253

Reitz MS, Jr, Nerurkar LS, Gallo RC (1999) Perspective on Kaposi's sarcoma: facts, concepts, and conjectures. J Natl Cancer Inst 91:1453–1458

Renne R, Zhong W, Herndier B, McGrath M, Abbey N, Ganem D (1996) Lytic growth of Kaposi's sarcoma-associate herpesvirus (human herpesvirus 8) in culture. Nat Med 2:342–346

Rother RP, Rollins SA, Fodor WL, Albrecht JC, Setter E, Fleckenstein B, Squinto SP (1994) Inhibition of complement-mediated cytolysis by the terminal complement inhibitor of herpesvirus saimiri. J Virol 68:730–737

Russo JJ, Bohenzky RA, Chien M-C, Chen J, Yan M, Maddalena D, Parry JP, Peruzzi D, Edelman IS, Chang Y, Moore PS (1996) Nucleotide sequence of the Kaposi's sarcoma-associated herpesvirus (HHV8). Proc Natl Acad Sci USA 93:14862–14867

Saifuddin M, Parker CJ, Peeples ME, Gorny MK, Zolla-Pazner S, Ghassemi M, Rooney IA, Atkinson JP, Spear GT (1995) Role of virion-associated glycosylphosphatidylinositol-linked proteins CD55 and CD59 in complement resistance of cell line-derived and primary isolates of HIV-1. J Exp Med 182:501–509

Sarid R, Sato T, Bohenzky RA, Russo JJ, Chang Y (1997) Kaposi's sarcoma-associated herpesvirus encodes a functional Bcl-2 homologue. Nature Medicine 3:293–298

Schoneberg T, Schultz G, Gudermann T (1999) Structural basis of G protein-coupled receptor function. Mol Cell Endocrinol 151:181–193

Schulz TF, Chang Y, Moore PS (1998) Kaposi's sarcoma-associated Herpesvirus (Human Herpesvirus 8), American Society for Microbiology

Sirianni MC, Uccini S, Angeloni A, Faggioni A, Cottoni F, Ensoli B (1997) Circulating spindle cells: correlation with human herpesvirus-8 (HHV-8) infection and Kaposi's sarcoma [letter]. Lancet 349:255

Speck SH, Virgin HW (1999) Host and viral genetics of chronic infection: a mouse model of gammaherpesvirus pathogenesis. Curr Opin Microbiol 2:403–409

Stine JT, Wood C, Hill M, Epp A, Raport CJ, Schweickart VL, Endo Y, Sasaki T, Simmons G, Boshoff C, et al. (2000) KSHV-encoded CC chemokine vMIP-III is a CCR4 agonist, stimulates angiogenesis, and selectively chemoattracts TH2 cells, Blood 95:1151–1157

Sturzl M, Hohenadl C, Zietz C, Castanos-Velez E, Wunderlich A, Ascherl G, Biberfeld P, Monini P, Browning PJ, Ensoli B (1999) Expression of K13/v-FLIP gene of human herpesvirus 8 and apoptosis in Kaposi's sarcoma spindle cells. J Natl Cancer Inst 91:1725–1733

Taub DD (1996) Chemokine-leukocyte interactions. The voodoo that they do so well. Cytokine Growth Factor Rev 7:355–376

Unanue ER (1997) Inter-relationship among macrophages, natural killer cells and neutrophils in early stages of Listeria resistance. Curr Opin Immunol 9:35–43

Vilcek J, Sen GC (1996) Interferons and other cytokines. In: Fields BN, Knipe DM, Howley PM (eds) Fields Virology. Lippincott-Raven Publishers, Philadelphia, pp 375–399

Wiertz EJ, Jones TR, Sun L, Bogyo M, Geuze HJ, Ploegh HL (1996a) The human cytomegalovirus US11 gene product dislocates MHC class I heavy chains from the endoplasmic reticulum to the cytosol. Cell 84:769–779

Wiertz EJ, Tortorella D, Bogyo M, Yu J, Mothes W, Jones TR, Rapoport TA, Ploegh HL (1996b) Sec61-mediated transfer of a membrane protein from the endoplasmic reticulum to the proteasome for destruction [see comments]. Nature 384:432–438

Wilson JL, Charo J, Martin-Fontecha A, Dellabona P, Casorati G, Chambers BJ, Kiessling R, Bejarano MT, Ljunggren HG (1999) NK cell triggering by the human costimulatory molecules CD80 and CD86. J Immunol 163:4207–4212

Wong SW, Bergquam EP, Swanson RM, Lee FW, Shiigi SM, Avery NA, Fanton JW, Axthelm MK (1999) Induction of B cell hyperplasia in simian immunodeficiency virus-infected rhesus macaques with the simian homologue of Kaposi's sarcoma-associated herpesvirus. J Exp Med 190:827–840

Yang TY, Chen SC, Leach MW, Manfra D, Homey B, Wiekowski M, Sullivan L, Jenh CH, Narula SK, Chensue SW, Lira SA (2000) Transgenic expression of the chemokine receptor encoded by human herpesvirus 8 induces an angioproliferative disease resembling Kaposi's sarcoma [see comments]. J Exp Med 191:445–454

York IA, Roop C, Andrews DW, Riddell SR, Graham FL, Johnson DC (1994) A cytosolic herpes simplex virus protein inhibits antigen presentation to CD8+ T lymphocytes. Cell 77:525–535

Ziegler H, Thale R, Lucin P, Muranyi W, Flohr T, Hengel H, Farrell H, Rawlinson W, Koszinowski UH (1997) A mouse cytomegalovirus glycoprotein retains MHC class I complexes in the ERGIC/cis-Golgi compartments. Immunity 6:57–66

Zimring JC, Goodbourn S, Offermann MK (1998) Human herpesvirus 8 encodes an interferon regulatory factor (IRF) homolog that represses IRF-1-mediated transcription. J Virol 72:701–707

Viral Chemokine Receptors and Chemokines in Human Cytomegalovirus Trafficking and Interaction with the Immune System
CMV Chemokine Receptors

P.S. Beisser[1], C.-S. Goh[2], F.E. Cohen[2], and S. Michelson[1]

The ubiquitous, opportunistic pathogen human cytomegalovirus (CMV) encodes several proteins homologous to those of the host organism. Four different CMV genes encode chemokine receptor-like peptides. These genes, UL33, UL78, US27, and US28, are expressed at various stages of infection in vitro. Their functions remain largely unknown. To date, chemokine binding and signalling has only been demonstrated for the US28 gene product. Putative ligands for the other CMV-encoded chemokine receptors are discussed on basis of phylogenetic analysis. The potential roles of these receptors in virus trafficking, persistence, and immune evasion are summarized. Similarly, modulation of expression of the host chemokines IL-8, MCP-1, MIP-1a and RANTES in relation to viral dissemination and persistence is reviewed.

1	Introduction.	205
2	Chemokine and Chemokine Receptor Interaction and Signaling	205
3	CMV-Encoded Chemokines.	207
4	CMV-Encoded Chemokine Receptors.	207
4.1	Sequence and Transcription Analysis of CMV Chemokine Receptors	209
4.2	Expression of CMV-Encoded Chemokine Receptors.	211
4.3	Chemokine Binding and Signaling Properties of CMV-Encoded Chemokine Receptors	213
4.4	Modulation of Host Cell Chemokine Production During CMV Infection	217
4.5	The Implication of US28 in Retroviral Infection In Vitro.	218
4.6	Adaptive Evolution of Human CMV Chemokines and Chemokine Receptors.	219
5	Putative CMV-Encoded Chemokine and Chemokine Receptor Functions.	220
5.1	The Role of CMV-Specific Chemokine and Chemokine Receptor in Viral Dissemination and Persistence.	222
5.2	Modulation of Host Cell Chemokine Production in Relation to CMV Dissemination and Persistence.	226
6	Conclusions.	228
	References.	229

Abbreviations

AA	arachidonic acid
BAL	bronchoalveolar lavage

---

[1] Unité d'Immunologie Virale, Institut Pasteur, 28 Rue du Docteur Roux, 75274 Paris Cedex 15, France
[2] Program in Medical Information Sciences, University of California, San Francisco CA 94143, USA

BM	bone marrow
CLTs	CMV latency-related transcripts
CMV	human cytomegalovirus
EGFP	enhanced green fluorescent protein
ENA 78	epithelial cell-derived neutrophil attractant 78 aas
FK	fractalkine or CX3C
GCP	granulocyte chemotactic protein
GPCR	G protein-coupled receptor
GRO	Growth-related gene
HA	hemagglutinin-specific peptide
HHV 6	human herpesvirus 6
HHV 8	human herpesvirus 8
HIV	human immunodeficiency virus
IFN	interferon
IP_3	inositol 3 phosphate
IP10	monokine induced by IFN-γ protein 10
I-TAC	IFN-inducible T cell chemoattractant
M	murine
MCK 1	murine CMV CC chemokine encoded by ORF M131
MCP	macrophage chemoattractant protein
MDC	macrophage-derived chemokine
MIE	major immediate-early
MIP	macrophage inflammatory protein
NAP	neutrophil activating protein
ORF	open reading frame
PCR	polymerase chain reaction
cPLA2	cytosolic phospholipase A2
PTX	*Bordetella pertussis* toxin
pUS27 and pUS28	gene products of US27 and of US28, respectively
R	rat
RANTES	regulated on activation, normal T expressed and secreted
ROS	reactive oxygen species
RT	reverse transcription
SDF	stromal cell-derived factor
SLC	Secondary lymphoid tissue chemokine
SMC	Smooth muscle cells
TECK	thymus-expressed chemokine
UL	long unique region of the genome
US	short unique region of the genome
UTR	untranslated region
vCXC 1	CMV UL146-encoded CXC chemokine
vMIP	HHV 8-encoded macrophage inflammatory protein
VSV-G	vesicular stomatosis virus protein G

1 Introduction

Human cytomegalovirus (CMV) has devised numerous means of getting around detection by the immune system. Many CMV-specific genes encode molecules that interfere with both innate and adaptive immunity (see other chapters in this volume). Some of these genes encode proteins that target antigen presentation, whereas others encode cytokines and chemokines, or cytokine or chemokine receptors. This review will focus on the human CMV homologs of chemokine receptors and induction of chemokines by CMV and will discuss some of the potentials that these molecules have in virus trafficking during CMV infection and immune evasion.

2 Chemokine and Chemokine Receptor Interaction and Signaling

Chemokines are soluble mediators implicated in infiltration, inflammation, and activation of leukocyte effector mechanisms. Many reviews have appeared recently, among which are those that cover the new nomenclature of chemokines and their receptors (MURPHY et al. 2000), coevolution of chemokine receptors and their ligands (GOH et al. 2000), chemokine-based lymphocyte trafficking (LOETSCHER et al. 2000), and viral antichemokines (MURPHY 2000). All chemokines have very similar overall structures, being composed of three beta sheets and an alpha helix, which separate the short N-terminal and the C-terminal domains. Chemokines are subdivided into four families based on the number and spacing of conserved cysteines: CXC with one amino acid (aa) separating the first two cysteines (C), CC with no intervening aa, CX3C with three intervening aa, and C with only one Cys residue. CXC chemokines can be further subdivided into "ERL^+", which are angiogenic, and ELR^-, which are usually angiostatic. Generally, CXC chemokines attract neutrophils and lymphocytes, whereas CC chemokines attract monocytes and macrophages (BAGGIOLINI et al. 1997). Almost all chemokines fall into either the CXC or the CC families, because only two chemokines have been described for the C family and one for the CX3C family. Cellular CXC chemokines bind only to CXC receptors (designated CXCR1–5) and CC chemokines bind only CCRs (designated CCR1–11) (reviewed in MURPHY et al. 2000). Several chemokines may bind to a given receptor, and, conversely, several receptors may bind the same chemokine (Fig. 1). For some CCRs and CXCRs, only one ligand has been found so far. These are often involved in homeostasis. Finally, Duffy antigen (reviewed in MURPHY et al. 2000), found on erythrocytes and endothelial cells, is thought to act as a "chemokine sink" and binds both CXC and CC chemokines but transmits no intracellular signal.

The structure, as well as the mechanism, of ligand binding and signal transduction by chemokine receptors is similar to those of other members of the G

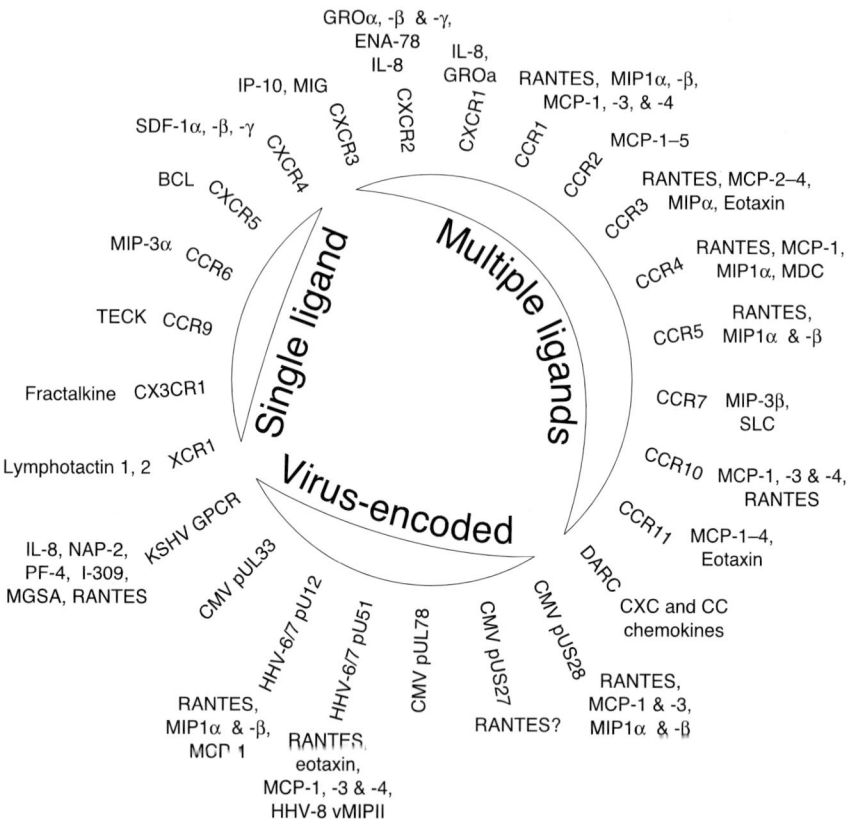

Fig. 1. Human cellular and human CMV-encoded chemokine receptors and their corresponding ligands. Cellular chemokine receptors that bind only one ligand (single ligand) or those that bind several ligands (multiple ligands) are shown. The human herpesvirus chemokine receptors (virus encoded) and their ligands are given where determined. The old nomenclature has been used here. For the new nomenclature see MURPHY et al. (2000)

protein-coupled receptor (GPCR) family (SELBIE and HILL 1998). Chemokine receptors have an extracellular N-terminal tail, seven transmembrane domains, and three extracellular loops, which are all important for chemokine binding. In addition, the receptors have three intracellular loops and an intracellular C-terminal tail, which are essential for G protein binding and activation. The mechanism of GPCR-mediated signaling is summarized in Table 1 and Fig. 2A (reviewed by GUTKIND 1998; HAMM 1998). Chemokine-chemokine receptor complexes can be internalized through clathrin-dependent receptor endocytosis into endosomes, where the bound chemokine is released and degraded and the receptor is rerouted to the plasma membrane (Fig. 2B) (reviewed in SIGNORET and MARSH 2000). Receptor stimulation (reviewed in BAGGIOLINI 1998) eventually leads to (a) differentiation, or inhibition of differentiation of leukocyte progenitors, (b) rolling and attachment to blood vessel endothelial cells, as well as transendothelial migration,

Table 1. Activation activities of G proteins

G_α subtypes (20)	Type of signal transduction	$G_{\beta\gamma}$ subunits $(6_{\beta\gamma}, 12_{\beta\gamma})$[a]
i[b]	Inhibition of adenylyl cyclases and activation of PI3K	−
	Activation of ion channels	+
q	Activation of GRK	+
	Activation of tyrosine kinases	+
	Activation of PLCβ[b]	+
s	Activation of adenylyl cyclases	−
12	Activation of ion channels	+
	Increase SAPK/JNK activity	−
	rho-dependent induction of actin polymerization	−
	Induction of NO synthase	−

[a] Numbers in parentheses denote the number of family members.
[b] $G_{\alpha i}$ proteins are sensitive to inhibition by pertussis toxin.
[c] Abbreviations: PI3K, phosphoinositol-3-kinase; GRK, G protein-coupled receptor kinase; PLCβ, phospholipase C-β; SAPK, stress-activated protein kinase; JNK, c-Jun N-terminal kinase.

(c) chemotaxis to inflamed tissue, or inhibition of chemotaxis in noninflamed tissue, and (d) induction of a variety of immunological responses such as cytotoxicity.

3 CMV-Encoded Chemokines

CMV produces a functional chemokine, encoded by UL146 and designated vCXC-1 (PENFOLD et al. 1999). This gene was found in the genome of the Toledo strain of CMV (CHA et al. 1996). UL146 encodes an ERL$^+$, CXC-type chemokine, which, like IL-8, probably does not bind to CMV GPCRs US28 or US27. The vCXC-1 chemokine attracts human peripheral blood neutrophils. It binds with high affinity to CXCR2-transfected, but not to CXCR1-transfected, mouse fibroblasts as well as to freshly isolated human neutrophils. In addition, the downstream open reading frame (ORF) UL147 also shows homology to CXC chemokines but lacks an ERL motif. The Towne strain of CMV carries a UL146-like gene (UL152) which is in the opposite orientation to that of Toledo UL146. Whether UL147 and UL152 encode functional chemokines remains to be investigated. A detailed review is given in the chapter by Saederup and Mocarski, this volume.

4 CMV-Encoded Chemokine Receptors

Potential human CMV chemokine receptors were first discovered when CHEE et al. (1990a) sequenced the genome of strain AD169. They described three receptor

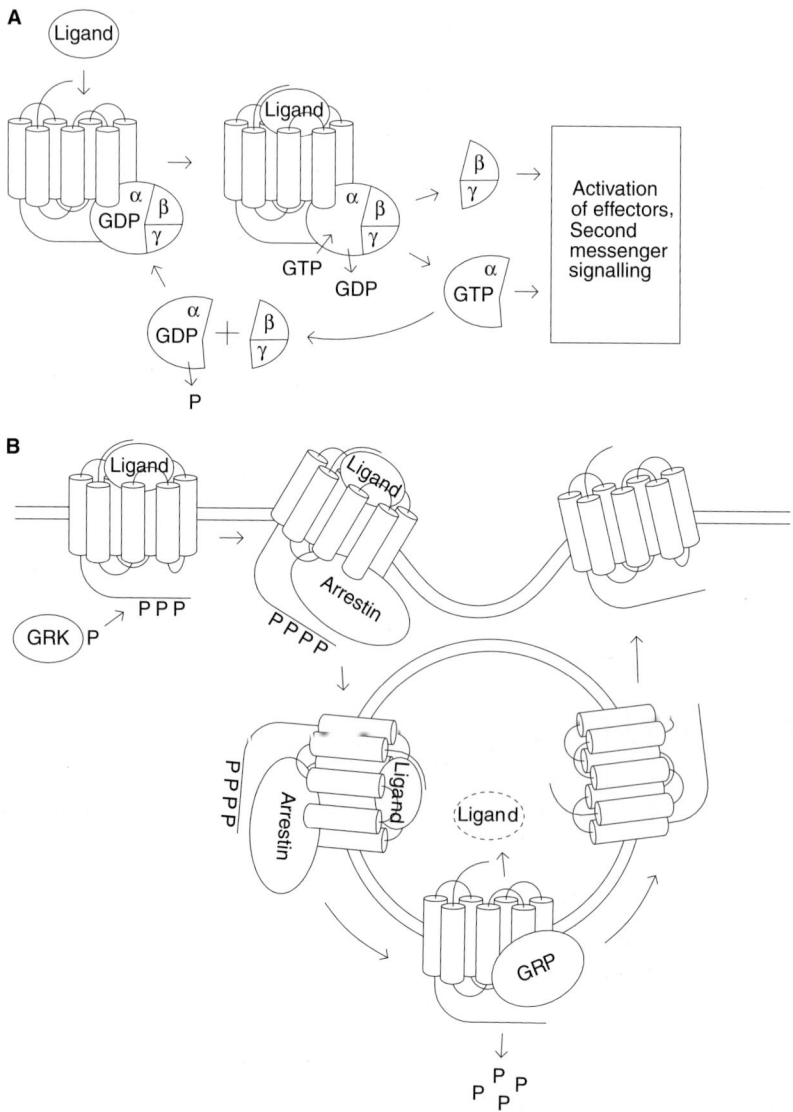

Fig. 2A,B. G protein-coupled receptor (*GPCR*-) mediated signaling. **A** The guanidine triphosphate (*GTP*) cycle in GPCR-mediated signaling. Note that the G proteins, as well as the effectors, are plasma membrane bound. Upon ligand binding, the GPCR undergoes a conformational change that enables it to interact with G proteins – the G protein-associated GDP is replaced by GTP, causing the G protein to dissociate into a G_α and a $G_{\beta\gamma}$ subunit, each of which are released from the receptor. Both G_α and $G_{\beta\gamma}$ subunits can activate signaling cascades that can result in the release of either inositol-3-phosphate (*IP₃*), cyclic AMP or Ca^{2+}. G_α subunits have an intrinsic hydrolyzing activity, resulting in dephosphorylation of the G_α-associated GTP. ON dephosphorylation, G_α can reassociate with $G_{\beta\gamma}$, thereby returning to an inactivated state. **B** Internalization of desensitized chemokine receptors. Chemokine receptors can be desensitized after an initial round of signaling, i.e., modified such that they can no longer be activated through successive chemokine binding events. This is established through phosphorylation of the intracellular C-terminus of the receptor by GPCR kinases (*GRK*) and subsequent binding with β-arrestins. On desensitization, the receptor can be internalized by inclusion into clathrin-coated endosome vesicles. On internalization, the chemokine is released and degraded, whereas the receptor is dephosphorylated by GPCR phosphatases and rerouted to the plasma membrane

homologs designated UL33, US28, and US27 according to their genomic locations within the long unique (UL) and short unique (US) regions of the genome (CHEE et al. 1990b). Subsequently, GOMPELS et al. (1995) defined another potential GPCR gene based on its similarity to a GPCR gene found in the human herpesvirus 6. (HHV-6) genome. The HCMV gene is designated UL78. CMVs from nonprimates carry positional and sequence equivalents of the UL33 and UL78 genes. However, it is important to note that so far only human CMV carries the GPCR genes US27 and US28, thus restricting in vivo study of the latter receptors.

In the following sections, the sequence, transcription, and expression properties of the CMV chemokine receptor genes and their gene products will be outlined, as will their known and anticipated chemokine-binding capacities.

4.1 Sequence and Transcription Analysis of CMV Chemokine Receptors

The first characterization of a virus-encoded GPCR was reported in conjunction with the cloning and characterization of the human chemokine receptor CCR1 by NEOTE et al. (1993) and GAO et al. (1993). They found that the amino acid (aa) sequence derived from US28 was 33% identical to that of CCR1 but also shared 32% identity to the sequences of both CXCR1 and CXCR2. When they cloned US28, they found that, because of an error in the original GenBank sequence, the predicted C-terminal of the protein was actually 65 aa, rather than 23 aa, in length, resulting in an overall length of 365 aa for the US28 gene product (pUS28). This was ultimately confirmed by several other research groups (BILLSTROM et al. 1998; GAO and MURPHY 1994; KUHN et al. 1995). Cloning of US27 and US28 has been done with genomic DNA from CMV. This approach does not take into consideration putative splicing of the US27- or US28-specific transcripts. To generate US27 and US28 expression constructs, we obtained US27-specific and US28-specific cDNAs from Toledo CMV-infected fibroblasts rather than clones of genomic DNA. By sequencing these cDNA clones, the potential 5' and 3' ends of the US27 and US28 mRNAs were determined (Fig. 3A). The US28 gene does not harbor any introns. Surprisingly, the US27-specific transcript was found to be spliced within the 5' untranslated region (UTR) (Fig. 3A). The relevance of this splicing event has not been investigated, but it suggests that US27 expression might be regulated at the posttranscriptional level. The UL33-specific transcript was also shown to be spliced (Fig. 3A) (DAVIS-POYNTER et al. 1997). This splicing, however, results in a transcript containing an UL33 ORF that is different from the UL33 ORF predicted by CHEE et al. (1990b). US27 and US28 share a common polyadenylation signal (AATAAA), of which the first two adenosines are also part of the US28 stop codon (TAA) (Fig. 3A). The aa coding content within the cDNA sequences of the Toledo strain was compared with the US27 and US28 aa sequences of the AD169 and Towne strains. The aa sequence derived from Toledo US28 differed by only two residues compared with the US28 sequence of both AD169 and Towne. However, higher sequence variability was found for the US27-derived aa sequence. The

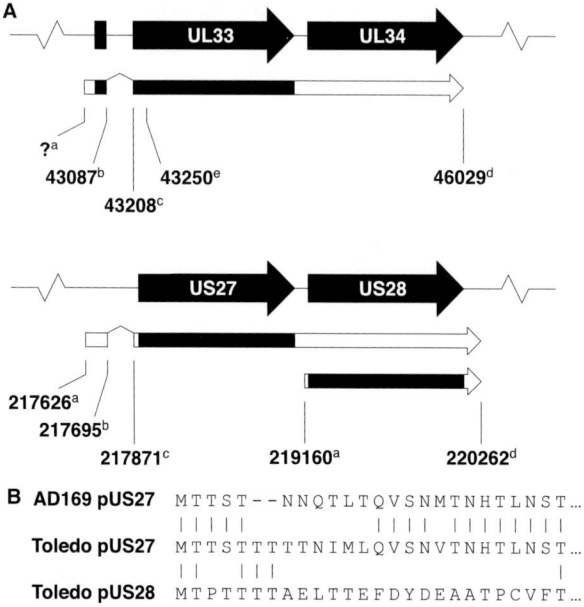

Fig. 3A,B. Sequence analysis and transcription of the CMV UL33, US27-, and US28-encoded chemokine receptors. **A** Splicing of UL33- and US27-specific transcripts. In the diagram, the chemokine receptor ORFs on the CMV genome (*black lines*) are indicated as *black arrows* and the transcripts (*white arrows*) as *black boxes*. The positions of the transcription start (*a*), splice donor (*b*), splice acceptor (*c*), and transcription termination (*d*) are indicated by *numbers* that correspond to the nucleotide positions of the CMV AD169 genomic sequence deposited under GenBank number NC001347. The number indicated by (*e*) denotes the initially predicted start codon of the UL33 ORF. **B** Alignment of the predicted aa sequences corresponding to the extracellular N-termini of the chemokine receptors encoded by CMV AD169 US27, Toledo US27, and Toledo US28

AD169 US27-derived aa sequence differs by 14 residues from that of Towne US27 and by 15 residues from that of Toledo US27. Moreover, both the potential Toledo and Towne US27-specific N-termini have two additional aa residues compared with the AD169 US27-specific N-terminus. The highest sequence variability between AD169 US27, on the one hand, and Towne and Toledo US27, on the other, was found within the potential N-terminal region of the receptor (Fig. 3B). An important component in binding of chemokines to their receptors is the interaction of the chemokine with the N-terminal of the chemokine receptor (reviewed in BAGGIOLINI et al. 1997). Whether the differences in N-terminal sequences among pUL27s of AD169, Towne, and Toledo reflect differences in chemokine binding affinity among the different CMV strains remains to be investigated.

Transcription of each of the UL33, UL78, US27, and US28 genes is initiated at different times postinfection (pi). Transcripts of UL33, 3.3kb in length, are detected by Northern blot analysis as early as 4h pi and become more abundant during the late phase of infection (BODAGHI et al. 1998; DAVIS-POYNTER et al. 1997; DAVIS-POYNTER et al. 1995). However, inhibition of viral replication with phosphonoacetic acid (PAA) for 2 or 7 days pi prevented detection of UL33 transcripts by

Northern blot (DAVIS-POYNTER et al. 1997; WELCH et al. 1991). Although it was not determined for UL33 transcription, UL33-specific transcripts of the murine CMV counterpart M33 could not be detected in infected cells treated with cycloheximide (CHX) (DAVIS-POYNTER et al. 1997). This indicates that M33, and possibly UL33, are not expressed as immediate-early genes. UL78-specific transcripts were detected in fibroblasts exclusively at the early stage of infection, as determined by microarray analysis (CHAMBERS et al. 1999). However, similar genes found in the mouse and rat CMV genomes, M78 and R78, respectively, were shown to be transcribed at early times, whereas R78 is also transcribed late after infection, as demonstrated by Northern blot analysis (BEISSER et al. 1999; OLIVEIRA and SHENK 2001). The US27 gene is transcribed as a 2.9-kb mRNA only at late times (>48 h) after infection (BODAGHI et al. 1998; WELCH et al. 1991). The US28 gene is transcribed both at early (8h pi) and late times after infection by Northern blot analysis (BODAGHI et al. 1998) and at immediate-early times (2h) pi as detected by reverse transcription (RT) followed by amplification by polymerase chain reaction (PCR) (ZIPETO et al. 1999). In contrast to UL33 transcription, it was found that US28 transcription was not inhibited by CHX treatment. Furthermore, US28-specific transcripts can be found in peripheral blood mononuclear cells in vivo (PATTERSON et al. 1998), as well as in a CMV-infected pre-monocyte cell line THP-1 in vitro (ZIPETO et al. 1999). Both US28- and CMV latency-related transcripts (CLTs) from the major immediate-early (MIE) locus (KONDO and MOCARSKI 1995) were detected by RT-PCR in CMV Toledo-infected, THP-1 monocytic cells 7 days pi. Infectious virus could not be recovered from supernatants of these cells, but virus could be reactivated after 2 weeks of coculture with MRC-5 fibroblasts (BEISSER et al. 2001). These findings suggest that, like MIE-derived CLTs, US28 is transcribed in latently infected cells. Because transcripts from UL33 were found at very early time points pi (DAVIS-POYNTER et al. 1997), similar to detection of immediate-early US28 transcripts, it might be worthwhile to determine whether UL33 and other immediate-early genes are transcribed during latency. This could eventually lead to a better understanding of gene regulation during latent CMV infection.

4.2 Expression of CMV-Encoded Chemokine Receptors

The investigation of CMV-specific chemokine receptor detection at different times postinfection (p.i.) has been frustrated by a lack of specific antibodies to these proteins, with one exception – polyclonal antibodies were developed against a UL33 C-terminal peptide by MARGULIES et al. (1996). With these antibodies, UL33-encoded receptors (pUL33) were detected in CMV virions, dense bodies, and noninfectious enveloped particles, as well as in intracytoplasmic inclusions. The presence of pUL33 in virions and dense bodies led to several speculations: (a) pUL33 could participate in viral adsorption by attaching to its natural ligand(s) expressed by specific cell types; (b) pUL33 may be disposed at the cell surface on virus adsorption and penetration, where it could play a role in very early cell activation, which would augment viral infection; and (c) other CMV-specific chemokine receptors, if similarly incorporated into the envelopes of virions and dense bodies, could also participate in viral entry and/or host cell activation.

Expression of the putative UL78 gene product (pUL78) has not yet been reported. However, the murine cytomegalovirus (MCMV) M78 gene product colocalizes with Golgi markers and is incorporated into viral particles, where on fusion with cells, it plays a role in activating accumulation of viral IE RNAs (OLIVEIRA and SHENK 2001). A UL78-like receptor, encoded by the HHV-6 gene U51 (pU51), was shown to accumulate in the ergastoplasm of HEK 293 and 143tk$^-$ cells after transfection (MENOTTI et al. 1999). This localization appeared to be cell type dependent. The pU51 receptor localizes to plasma membranes in T cells, which is a permissive cell type for HHV-6 (MENOTTI et al. 1999).

Determination of the localization of the gene product of US27 (pUS27) and of pUS28 within infected and transfected cells has relied on the adjunction of different peptide tags such as N-terminal c-myc (PLESKOFF et al. 1997), N-terminal FLAG (STREBLOW et al. 1999), N-terminal immunoglobulin domains of CD4 (FRAILE-RAMOS et al. 2001), C-terminally tagged enhanced green fluorescent protein (EGFP), or an N-terminally-tagged hemagglutinin-specific peptide (HA) (FRAILE-RAMOS et al. 2001; Bodaghi and Beisser, unpublished results). With expression vectors containing either HA- or GFP-tagged US27 or US28 genes, it has been possible to localize these receptors in both transiently and stably transfected cells. Cell types used include HeLa and Cos 7 cells (FRAILE-Ramos et al. 2001), an astrocytoma cell line (U373 MG), HEK 293, and an erythrocytoma cell line (K562) (Bodaghi and Beisser, unpublished results). The receptors have a marked tendency to be localized predominantly within the perinuclear cell center of HeLa, Cos 7, and U373 MG cells. When U373 MG cells were cotransfected with a chemokine receptor gene tagged either at the N- or the C-terminus, confocal microscopy showed that US27-EGFP and HA-US27 expression constructs resulted in colocalization of their respective gene products (Fig. 4A–C). Similar results were obtained for US28 expression (not shown), indicating that the presence of either a C- or N-terminal tag does not differentially affect localization of the US27- and US28-encoded receptors. When cells transfected with either a tagged US27 or US28 expression vector were subsequently infected with Toledo CMV (Fig. 4D–F), several observations were made: (a) there was enhanced expression of the transfected receptor, which is not surprising in light of their being driven by the MIE CMV promoter/enhancer; (b) there was no change in the subcellular location of tagged receptors after infection; and (c) transfection of receptors did not render astrocytoma cells resistant to infection. Similarly, on cotransfection of astrocytoma cells with expression vectors containing either HA-US27 and US28-EGFP, or vice versa, US27-EGFP and HA-US28, the respective gene products colocalize (Fig. 4G–H). This suggests that both pUS27 and pUS28 are expressed in the same subcellular compartments in astrocytoma cells. Finally, it was reported that the pUS28 receptor could be expressed in aorta smooth muscle cells (SMC) by recombinant adenovirus containing an N-terminal FLAG-tagged US28 gene (STREBLOW et al. 1999). In these cells, the receptor adopted a polarized distribution, and it is presumed that the receptor appears at the cell membrane. Recent immunofluorescent and electron microscope studies (FRAILE-RAMOS et al. 2001) in HeLa and Cos cells demonstrated that the majority of pUS28 is within multivesicular endosomes, whereas only 20% localizes to the cell surface.

Although many chemokine binding and signaling studies have been performed with pUS28, and transcription of the US28 gene has been confirmed in their respective expression systems (BILLSTROM et al. 1998; BODAGHI et al. 1998; GAO and MURPHY 1994; NEOTE et al. 1993; VIEIRA et al. 1998), direct evidence for cell surface expression of pUS27 and/or pUS28 has been reported only by PLESKOFF et al. (1997) and FRAILE-RAMOS et al. (2001) in transfected HeLa, Cos-7, and HEK 293 cells. The cell surface expression of both pUS27 and pUS28 is significantly lower compared to that of human cellular chemokine receptors. A comparative example is shown in Fig. 4I, in which HEK 293 cells were transfected with vectors containing either US27 or US28, each tagged with an N-terminal, HA-encoding sequence, or a vector containing the CCR5 receptor. Stabilization of HA-US27 and HA-US28 in U373 MG or K562 cells with a selective agent and subsequent cell sorting of cells expressing HA epitopes failed to result in an enrichment of HA-US27- or HA-US28-expressing cells (Beisser et al., unpublished data). In addition, HEK 293 cells expressing myc-tagged pUS28 could not be stabilized (Pleskoff et al., personal communication). However, US27 and US28 could be stably expressed in U373 MG cells that stably express CMV IE1 (Beisser et al., unpublished data). This suggests that (a) both pUS27 and pUS28 inhibit cell growth and might even be toxic to the cell and (b) this possible growth inhibitory effect or toxicity can be compensated for by the presence of IE1 proteins. Currently, possible relationships between pUS28 expression and induction of cell death are under investigation.

4.3 Chemokine Binding and Signaling Properties of CMV-Encoded Chemokine Receptors

Binding of chemokines to the gene products of either UL33 or UL78 has not yet been reported. Moreover, fibroblasts infected with a CMV mutant, from which both US27 and US28 are deleted, failed to internalize RANTES or deplete extracellular MCP-1, whereas wild-type (wt) CMV was able to internalize both chemokines (BODAGHI et al. 1998). This suggests that neither UL33 nor UL78 is involved in RANTES internalization or macrophage chemoattractant protein (MCP)-1 depletion. In contrast, similar receptors encoded by the HHV-6 genes U12 and U51 were shown to bind several CC chemokines. Cells transfected with U12 were shown to bind RANTES, macrophage inflammatory protein (MIP)-1α, MIP-1β, and MCP-1 (ISEGAWA et al. 1998), whereas cells transfected with U51 bind RANTES, eotaxin, MCP-1, -3, and -4, as well as human herpesvirus 8 vMIP-II (PENFOLD et al. 1999). Additionally, the receptor encoded by U12 was shown to induce Ca^{2+} signaling on stimulation by the aforementioned chemokines. Thus, although the genomic positions of the CMV UL33 and UL78 genes and the HHV-6 U12 and U51 genes are

Fig. 4A–I. Subcellular localization of the CMV Toledo US27- and US28-encoded chemokine receptors (pUS27 and pUS28, respectively) in the astrocytoma cell line U373 MG. **A** An immunofluorescence micrograph (rhodamine staining) of astrocytes expressing HA-tagged pUS27. **B** The same field showing the expression of EGFP-tagged pUS27. **C** The same field combined with the corresponding bright-field micrograph. **D** Astrocytoma cells expressing EGFP-tagged pUS27. **E** Human CMV Toledo-infected astrocytoma cells expressing EGFP-tagged pUS27. **F** The same field showing CMV-infected cells expressing major immediate early antigens. **G** Astrocytoma cells expressing HA-tagged pUS28. **H** The same field showing the expression of EGFP-tagged pUS27. All magnifications are ×640. **I** Cell surface expression of HA-tagged US27 and pUS28 in HEK 392 cells determined by FACS analysis

conserved, respectively, it is possible that the corresponding gene products of the respective betaherpesviruses have different functional properties.

The chemokine-binding property of pUS27 is not well characterized. However, it was shown that cells infected with a US28 deletion mutant of CMV could bind and internalize RANTES (BODAGHI et al. 1998). In contrast, RANTES binding and internalization could not be detected in cells infected with a mutant CMV strain from which both US27 and US28 were deleted. This suggests that pUS27 can bind RANTES. However, this has not yet been confirmed by conventional ligand binding studies.

The US28-encoded receptor is at present one of the most extensively studied viral chemokine receptors. It binds RANTES, MIP-1α and -β, and MCP-1 and 3 (BILLSTROM et al. 1998; BODAGHI et al. 1998; GAO and MURPHY 1994; NEOTE et al. 1993; VIEIRA et al. 1998) but not the CXC chemokine IL-8 (BILLSTROM et al. 1998; GAO and MURPHY 1994; NEOTE et al. 1993). Table 2 gives binding affinities of CC chemokines as determined in both US28-transfected and CMV-infected cells. It appears that, in general, RANTES and MIP-1α have higher affinities for US28 than do the chemokines MIP-1β, MCP-1, and MCP-3 (see references in Table 2). In addition, US28 displays high affinity for the soluble form, and possibly also for the membrane-bound form, of the CX3C chemokine fractalkine (HASKELL et al. 2000; KLEDAL et al. 1998). pUS28 expressed in Cos-7 and HeLa cells is constitutively active (CASAROSA et al. 2001; FRAILE-RAMOS et al. 2001), and in Cos cells (CASAROSA et al. 2001) it increases inositol-3-phosphate (IP_3) production by activating phospholipase C via $G\alpha q/11$. RANTES and MCP-1 stimulate IP_3 production further, but this activity is partially inhibited by fractalkine, which therefore acts as a partial inverse agonist. Additionally, US28-transfected Cos-7 cells show constitutive activation of NF-κB via $G\alpha q/11$ and $G\beta/\gamma$ subunits, which is again partially inhibited by fractalkine. Neither IP_3 production nor NF-κB activation could be inhibited by pertussis toxin (PTX), confirming their $G_{\alpha i}$-independent activation.

In human cells, some CC chemokines that can bind to US28 (MCP-1 and -3, MIP-1α) stimulate arachidonic acid (AA) release in association with phosphorylation of cytosolic phospholipase A2 (cPLA2) (LOCATI et al. 1996). Some of the very early metabolic changes in fibroblasts infected with active CMV involve stimulation of AA release (reviewed in ALBRECHT et al. 1989), which depends on a PTX-sensitive, phosphorylated cPLA2 chain of events (SHIBUTANI et al. 1997). This chain of events consists of (a) phosphorylation, membrane mobilization, and activation of cPLA2, (b) concomitant increase in AA release and increase of

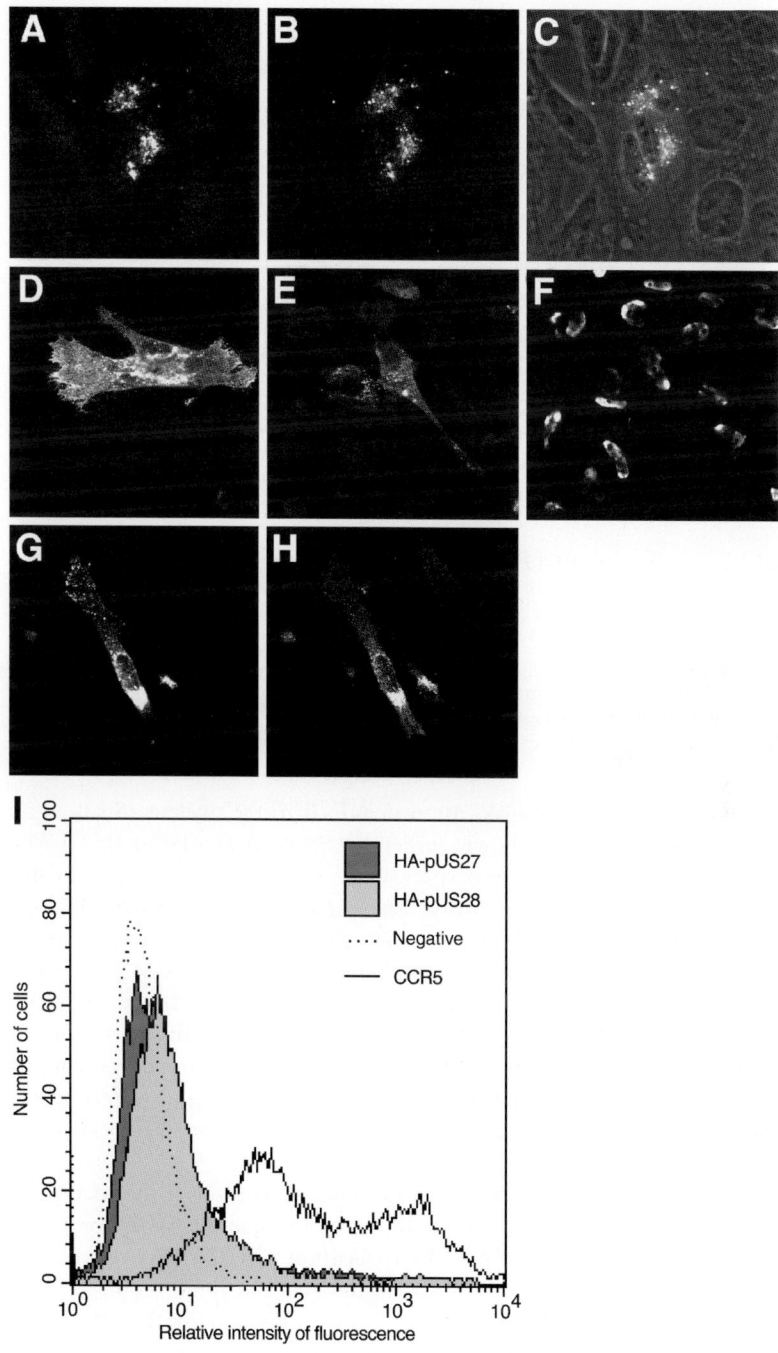

Table 2. Chemokine binding to CMV US28

Cell system	Ligand(s)	K_d (nM)	Reference
HEK 293 cells (transiently expressing)	MIP-1α	≈1[a]	NEOTE et al. 1993
K562 cells (stably expressing)	MCP-1 MIP-1α MIP-1β RANTES	6.1 2.5 5.1 3.4	GAO and MURPHY 1994
Cos 7 cells (transiently expressing)	MCP-1 RANTES	0.46 0.17	KUHN et al. 1995
HEK 293	RANTES	~10	BILLSTROM et al. 1998
Cos 7 cells (transiently expressing)	Soluble CX3C Soluble CX3C with mucin stalk MCP-1 MIP-1α MIP-1β RANTES	0.29–0.51[b] 2.8 0.748[b] 0.608[b] 0.708[b] 0.49[b]	KLEDAL et al. 1998
CMV-infected HUVEC	RANTES	10	BILLSTROM et al. 1998
CMV-infected fibroblasts	MIP-1α MIP-1β RANTES MCP-1 and 3	0.75[b] 0.75[b] 0.75[b] 5x[c]	BODAGHI et al. 1998

[a] NEOTE et al. (1993) report 2 binding affinities for MIP-1α, the second being ≈380nM.
[b] These were given as IC_{50}. Conversion to K_d was done with the formula $K_d = IC_{50}$ – concentration of radioactive ligand reported by the authors.
[c] The authors merely say that five times-higher concentrations of MCP-1 and -3 were required to compete the same amount of ^{125}I-MIP-1α.

cyclooxygenase levels, and (c) translocation of NFκB to the nucleus (SPEIR et al. 1998; ZHU et al. 1997). It was shown earlier by SPEIR et al. (1996) that CMV infection also induces reactive oxygen species (ROS), which are involved in this cPLA2 to NF-κB translocation pathway. The early induction of RANTES by CMV infection could stimulate these events in cells bearing CCRs responsive to RANTES. If pUS27 and pUS28 are structural components of the CMV envelope, similar to what has been shown for the UL33 gene product (MARGULIES et al. 1996), these receptors could be deposited by the viral envelope on the cell membrane at the time of viral entry. US28, deposited on the cell membrane by incoming viral elements or expressed at immediate-early times (ZIPETO et al. 1999), might play a role in NF-κB translocation and subsequent gene activation (YUROCHKO and HUANG 1999).

CMV infection of fibroblasts results in sustained activation of the MAP kinases ERK1, ERK2, and p38, which presumably play a role in the phosphorylation of transcription factors important for CMV replication (CREB, AP-1, etc.) (BRUENING et al. 1998; RESCHKE et al. 1999; RODEMS and SPECTOR 1998). In this respect, it is interesting that RANTES stimulation of US28 stably expressed in HEK 293 cells resulted in activation of ERK2, which was sensitive to inhibition with PTX (BILLSTROM et al. 1998); this activity was greater in HEK 293 cells cotransfected with $G_{α16}$ protein. Activation of MAP kinases can be stimulated through chemokine receptors coupled to α subunits of G_s, G_q and G_i families, as well as via βγ subunits (FAURE et al. 1994; SELBIE and HILL 1998). Although RANTES induction appears to be concomitant to MAP kinase activation, MCP-1 production is often constitutive in uninfected cell cultures (BODAGHI et al. 1998; STREBLOW et al. 1999) and can also activate US28. Finally, in CMV-infected cells, endogenous Ca^{2+} levels increase with time after infection (GARNETT 1979). One could wonder whether the continuous stimulation of US28 by MCP-1 and high concentrations of RANTES, which has been shown to mobilize calcium in infected and transfected cells (BILLSTROM et al. 1998; GAO and MURPHY 1994; NEOTE et al. 1993; VIEIRA et al. 1998), might not contribute to this elevation of Ca^{2+} levels. This could be additional to Ca^{2+} signaling associated with IP_3 production mediated by the constitutive activity of pUS28 (CASAROSA et al. 2001).

The consequences of chemokine-mediated activation of host cells would depend on the extent of viral replication within a given cell. Abortive infection would presumably lead to induction of CC and CXC chemokines, whereas full viral replication would be more likely to decrease ambient CC chemokine concentrations.

4.4 Modulation of Host Cell Chemokine Production During CMV Infection

The production of chemokines of the host organism is regulated at both transcriptional and posttranscriptional levels. This occurs on stimulation with cytokines in an inflammatory situation, such as during viral infection. In addition to the production of virus-encoded chemokines and chemokine receptors, it was shown that CMV infection also modulates the expression of cellular chemokines of both the CXC and CC families. Infection of human fibroblasts with any laboratory strain of CMV, as well as clinical isolates, upregulates constitutive production of IL-8 (CRAIGEN and GRUNDY 1996; CRAIGEN et al. 1997; MURAYAMA et al. 1997). We have studied IL-8 production after infection of bone marrow (BM) myofibroblasts isolated from human BM. Constitutive IL-8 production by uninfected cells was high (ranging from 4 to 57 ng/ml) and was not modified in 12/13 BM myofibroblast cultures infected with either AD169 or Toledo strains of CMV (Michelson and Charbord, unpublished results). In contrast, AD169 strain and endothelial cell-adapted clinical isolates of CMV upregulate IL-8 production in endothelial cells (ALMEIDA-PORADA et al. 1997; GRUNDY et al. 1998). CMV infection of fibroblasts has also been shown to increase extracellular production of RANTES (MICHELSON et al. 1997), as well as MCP-1 secretion (HIRSCH and SHENK 1999), at early times of infection. MIP-1α production increases in supernatants of CMV-infected, global BM stroma cultures (LAGNEAUX et al. 1996).

These modulations of chemokine production after CMV infection may be indirect, through induction of inflammatory cytokines [TNF-α, IL-1β, interferon (IFN)-γ and -β]. Prior cytokine induction was partially controlled for in some studies. The induction of IL-8 expression in infected fibroblasts was not the result of the presence of TNF-α or IL-1 (CRAIGEN and GRUNDY 1996), whereas stimulation of IL-8 production in endothelial cells might have been related to IL-1 and IL-6 (ALMEIDA-PORADA et al. 1997). Induction of RANTES in fibroblasts could not be attributed to the presence of the TNF-α or IL-1β (MICHELSON et al. 1997). However, in subsequent studies, RANTES secretion by infected fibroblasts was reduced by 60% in the presence of IFN-β-neutralizing antibodies (Bodaghi et al., unpublished results).

In contrast to their upregulation during the early phase, at later times after CMV infection of fibroblasts and endothelial cells, CC chemokine excretion is drastically reduced. Ligand binding to chemokine receptors leads to internalization of the ligand-receptor complex, destruction of the bound ligand, and subsequent recirculation of the receptor to the surface. Through this process, pUS28 has been shown to withdraw chemokines from the supernatants of infected fibroblasts (BODAGHI et al. 1998; VIEIRA et al. 1998) and endothelial (BILLSTROM et al. 1998, 1999; RANDOLPH-HABECKER et al. 1997) and astrocytoma (Michelson et al., unpublished results) cells. Infection of fibroblasts with laboratory or clinical CMV isolates results in the disappearance of RANTES from culture supernatants starting 16–24h after infection (MICHELSON et al. 1997). RANTES can be seen to accumulate intracellularly concomitant to its disappearance from supernatants. Exogenous, biotinylated RANTES added to infected cells 48–72h pi can be detected within cells after a 3-h adsorption when cells are infected with either the US27 or US28 null mutants of CMV (BODAGHI et al. 1998) but not when they are infected with a combined US27-US28 null mutant. The pUS28 receptor appears to have a considerable capacity for chemokine internalization, for it can simultaneously deplete RANTES and constitutively produced MCP-1 from supernatants of infected cells.

It was reported recently that CMV infection downregulates transcription of the gene encoding MCP-1 in fibroblasts, as detected by Northern blot analysis (HIRSCH and SHENK 1999). However, in our laboratory (BODAGHI et al. 1998), infection of fibroblasts with a mutant CMV deleted of both US28 and US27 did not affect constitutive production of MCP-1 in fibroblasts, suggesting that the downregulation of MCP-1 gene transcription is associated with either pUS27 or pUS28 expression. Somehow, simultaneous binding of RANTES and MCP-1 to pUS28 may have a feedback effect on the transcription of chemokine genes. This notion is supported by the findings of MILNE et al. (2000), who studied RANTES binding to HHV-6 U51. In this system, RANTES-specific transcripts were reduced tenfold in cells transfected with U51 expression vectors, whereas transcripts of β-actin and IL-8 were not affected. A similar feedback mechanism has not been described for cellular GPCRs to our knowledge.

4.5 The Implication of US28 in Retroviral Infection In Vitro

The US28-encoded chemokine receptor can serve as a coreceptor for human immunodeficiency virus (HIV) entry and play a role in cell-to-cell fusion between cells expressing HIV envelopes and those expressing pUS28 (CHOE et al. 1998; OHAGEN et al. 2000; PLESKOFF et al. 1997; RUCKER et al. 1997). The US27-encoded receptor promotes neither cell fusion nor HIV infection. HeLa, U373 MG, and neuroblastoma (U87) cells coexpressing pUS28 and CD4 can be infected by some monocyte-tropic and dual-tropic HIV strains but not by T lymphocyte-tropic HIV strains. Fusion of cells expressing pUS28 with cells expressing monocyte-tropic and, much less efficiently, T-cell tropic HIV envelopes also occurs. Thus pUS28 behaves much like the CC chemokine receptors CCR3 and CCR5 as concerns coreceptor activity for HIV but is much less efficient.

Co-expression of US28 with retroviral proteins other than those of HIV, such as human T cell lymphoma-leukemia virus-1 gp46 and gp21, as well as vesicular stomatosis virus (VSV)-G proteins, also leads to increased cell-to-cell fusion (PLESKOFF et al. 1998). Various mutations within the US28 gene affect fusion with cells expressing HIV envelopes or VSV-G. Deletion of N-terminal aa 2–22 abolishes fusion with HIV envelope-expressing cells but leads to increased fusion with cells coexpressing VSV-G. Removal of the C-terminus (aa residues 296–355) has no effect on HIV-coinduced fusion but again increases fusion mediated by VSV-G. In contrast, a point mutation in the second extracellular domain of US28 decreases its capacity to mediate cell-to-cell fusion. US28-mediated fusion was seen with human, macaque, and feline cells but not with murine or rat cells. Thus US28 expression may contribute to transfer of CMV, HIV, and perhaps other viruses from cell to cell via fusion.

4.6 Adaptive Evolution of Human CMV Chemokines and Chemokine Receptors

Human chemokines and their receptors have coevolved in a correlated manner, as evidenced by the correlated patterns of clustering between evolutionary trees of well-characterized chemokines and chemokine receptors (GOH et al. 2000). Consequently, through computation, we can augment our experimental understanding of cytokine ligand-receptor preferences. By analyzing the potential coevolution of chemokines and chemokine receptors of both human and viral origin, inferences can be made about the human protein-binding partners of the orphan CMV chemokines and receptors. Here, phylogenetic trees were constructed from the multiple sequence alignment of both chemokines (Fig. 5A) and their receptors (Fig. 5B), according to a method described by GOH et al. (2000), to predict the probable interaction of CMV-encoded chemokines and chemokine receptors with chemokines and chemokine receptors of the host. For this purpose, both CXC chemokines encoded by HCMV UL146 and UL147 and a murine cytomegalovirus (MCMV) ORF m131-encoding CC chemokine, MCK-1 (SAEDERUP et al. 1999), were included in the chemokine tree to determine their binding specificities. In addition, human CMV chemokine receptor sequences derived from HCMV UL78, UL33, US27, and US28, as well as both R33 and R78 from rat cytomegalovirus (RCMV) and both M33 and M78 from MCMV, were added to the chemokine receptor trees.

In the chemokine tree (Fig. 5A), MCMV MCK-1 clusters next to the macrophage-derived chemokine group and to the MIP-3α, MIP-3β, secondary lymphoid tissue chemokine and thymus-expressed chemokine (TECK) groups. This implies that MCK-1 is a CC chemokine-like protein that can potentially bind to CCR4, CCR6, CCR7, and/or CCR9. These receptors are predominantly expressed on macrophages and dendritic cells. The chemokines encoded by HCMV UL146 and UL147 cluster with CXC-type chemokines. The UL146 gene product, vCXC–1, clusters with ligands of the CXCR5 receptor. Although the vCXC-1 chemokine was found only to bind to CXCR2 out of an array of CCR1–CCR8, CXCR1–CXCR4, CX3CR1, and the US28-encoded receptor (PENFOLD et al. 1999), it could also be a potential binding partner for CXCR5. The UL147-encoded chemokine groups together with monokine induced by IFN-γ protein 10 (IP10) and IFN-inducible T cell chemoattractant (I-TAC) – all ligands for the CXCR3 receptor. Therefore, we can predict that the gene product of UL147 will bind to CXCR3 or a closely related receptor.

In the chemokine receptor tree (Fig. 5B), CMV US28, US27, and UL78 cluster together very closely within the CX3CR group. Among the chemokine receptors, human CMV US28 has the highest similarity with CX3CR1 – the receptor for CX3C or fractalkine. This corresponds to experimental findings that human CMV US28 binds CX3C (KLEDAL et al. 1998). Although US27 and UL78 are in the same cluster as US28, they appear to branch away from the rest of the chemokine receptor tree. It is possible that they can bind other CC chemokines, but it would be difficult to assign binding partners to these proteins. Finally, human CMV UL33, another orphan viral chemokine receptor, clusters quite closely to CXCR4. This suggests that CMV UL33 could bind stromal cell-derived factor (SDF)-1 or a CXC chemokine that is closely related. Together, these inferences on ligand-receptor specificity can possibly aid in the characterization of binding preferences of the CMV proteins.

5 Putative CMV-Encoded Chemokine and Chemokine Receptor Functions

The CMV-encoded chemokine and chemokine receptors could have diverse and combined functions. These include activation of the host cell (discussed above),

Fig. 5. Phylogenetic trees of chemokines (**A**) and chemokine receptors (**B**). By employing a linear regression analysis on the evolutionary pairwise distances among all the proteins in the multiple sequence alignment, a correlation coefficient was calculated based on the known binding partners in the chemokine and the chemokine receptor trees. Because of the similarity of the clustering patterns between the trees, a correlation coefficient of 0.57 with a *P* value less than 10^{-4} was obtained for the non-CMV chemokines and their receptors. The encircling of groups is based on the branching of the chemokine receptor tree. The *Roman numeral* that indicates each chemokine group refers to the corresponding receptor group, each of which has been numbered accordingly

Viral Chemokine Receptors and Chemokines 221

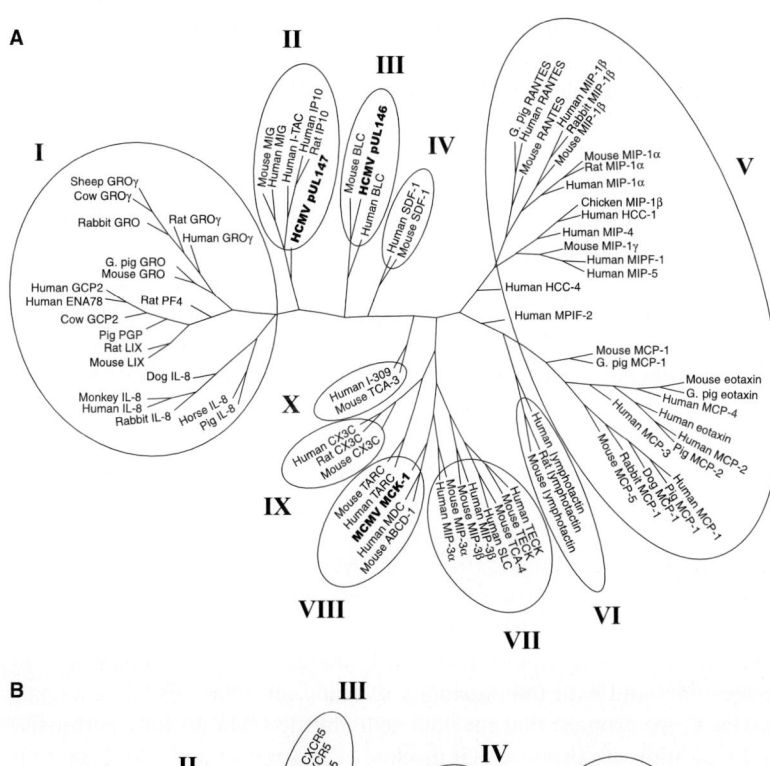

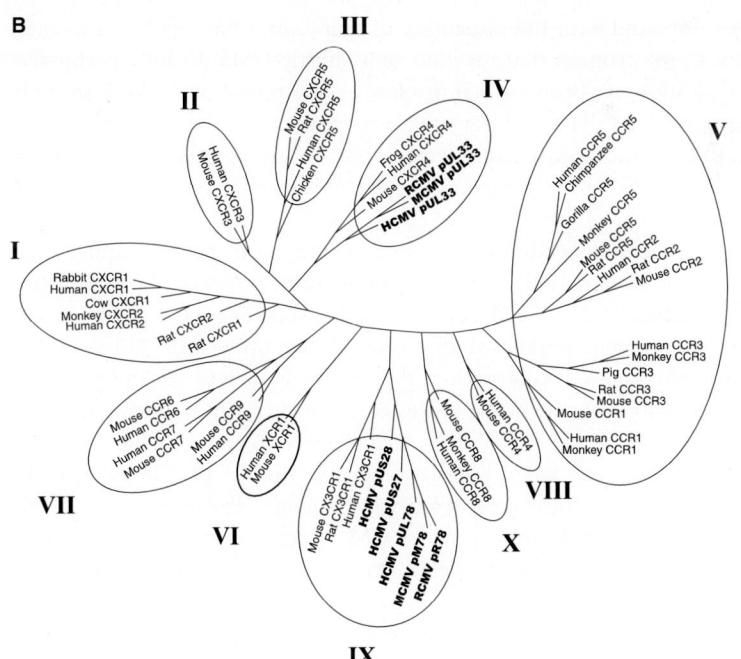

with or without subsequent stimulation of viral replication, viral dissemination by chemokine-regulated trafficking of infected cells, and modulation of the behavior and trafficking of cells involved in hematopoiesis and immune responses. These functions may have effects at both cellular and systemic levels.

5.1 The Role of CMV-Specific Chemokine and Chemokine Receptor in Viral Dissemination and Persistence

Cell types that are fully permissive for CMV infection, i.e., allow full viral replication leading to excretion of new infectious particles and cell lysis, include fibroblasts, SMC, endothelial cells, and epithelial cells of the retina and excretory organs, such as salivary glands. Infection of most of these cell types is associated with immunosuppression and CMV disease. However, infection of epithelial cells from excretory organs is probably essential for virus transmission between healthy individuals. In contrast to its ability to replicate in the afore-mentioned cell types, CMV remains latent, i.e., in a nonreplicative state, in myeloid cells, such as granulocyte/monocyte progenitors and mature monocytes. The possible mechanisms of trafficking of CMV in vivo between fully permissive cells on the one hand and cells that are latently infected with CMV on the other are not well understood. Considering that trafficking of myeloid cells toward inflammatory sites is mediated by chemokine receptors and with the possibility of many myeloid cells being latently infected by CMV, we propose that myeloid cells shuttle CMV to fully permissive target cells. In addition, we propose that myeloid cells can take up CMV from fully permissive, infected cells. The resulting two-way traffic may be orchestrated by virus-encoded chemokines and chemokine receptors. The putative roles of these chemokines and chemokine receptors in in vivo CMV trafficking are illustrated in Fig. 6.

Many of the viral dissemination pathways suggested below require that transmission of virus from the infected cell to an adjacent target cell be established either via CMV-induced cell-to-cell contacts or by shedding of virus from the carrier cell and subsequent uptake of the virus by the target cell. Some CMV-encoded proteins, predominantly structural glycoproteins such as gB and gH, were shown to play an important role in the cell-to-cell spread of CMV during infection in vitro (BALDWIN et al. 2000; BOLD et al. 1996; NAVARRO et al. 1993; RESCHKE et al. 1999). In addition, we have discussed the potential of pUS28 to provoke cell-to-cell fusion in association with retroviral proteins (PLESKOFF et al. 1997, 1998). Consequently, not only trafficking of CMV excreting cells, but also subsequent transmission of the virus through cell-to-cell contacts, may be mediated by pUS28.

CMV infection induces the production of cellular chemokines. In particular, IL-8 production is increased on infection of fibroblasts (CRAIGEN and GRUNDY 1996; CRAIGEN et al. 1997), endothelial cells (ALMEIDA-PORADA et al. 1997; GRUNDY et al. 1998) and monocytic THP-1 cells (MURAYAMA et al. 1997). Recently, GRUNDY et al. (1998) illustrated how a CMV-induced increase of IL-8 production, and perhaps growth-related gene (GRO)α, could assist viral dissemination. Supernatants from CMV-infected endothelial cells contained elevated levels of IL-8 and GROα, relative to supernatants of uninfected cells; these supernatants promoted neutrophil migration across an endothelial cell barrier. Neutrophils that are either cocultivated with, or migrated across, infected endothelial cells take up viral products, in particular pp65 (GRUNDY et al. 1998; REVELLO et al. 1998). CMV could be reactivated by subsequent coculture of pp65$^+$ neutrophils with fibroblasts. These observations were confirmed by GERNA et al. (2000), who showed cell fusion between neutrophils and infected endothelial cells by electron microscopy. They also reported that CMV replicated abortively in neutrophils (GERNA et al. 2000). Thus it is likely that CMV is shuttled between fully permissive cells by neutrophils. Recently, a CMV-encoded chemokine (vCXC-1) was identified. This chemokine was shown to be a potent chemotractant of neutrophils (PENFOLD et al. 1999). Therefore, neutrophil-mediated shuttling of CMV might be initiated by the attraction of neutrophils to infected cells expressing vCXC-1, as well as by upregulation of IL-8 and GROα, (Fig. 6).

Many CXC chemokines that can bind specifically to CXCR2 function as inhibitors of myelopoiesis (reviewed in BROXMEYER and KIM 1999). The CMV chemokine vCXC-1 desensitizes the cellular receptor CXCR2 expressed at the surface of neutrophils to further stimulation by neutrophil activating protein (NAP)-2, GROα, -β, or -γ, epithelial cell-derived neutrophil attractant 78 aas (ENA-78), or granulocyte chemotactic protein (GCP)-2 (PENFOLD et al. 1999). The majority of these chemokines (NAP-2, GROβ, ENA-78, and GCP-2) are inhibitory to hematopoiesis. Thus vCXC-1 can potentially interact with chemokine receptor(s) involved in myelopoiesis, although it is not yet known whether this would be stimulatory or inhibitory. It is also not known whether vCXC-1 is expressed by CMV-infected hematopoietic progenitors. If so, it could serve an autoregulatory function in which vCXC-1 would stimulate the release of CMV-harboring, differentiated myeloid cells into the circulation for further dissemination. Alternatively, it could autosuppress the differentiation of CMV-infected progenitors in the absence of other inhibitory chemokines to preserve latency. The putative stimulatory/suppressive effect of vCXC-1 on myelopoiesis is indicated in Fig. 6.

Previously, cells expressing pUS28 were shown to bind the CX3C chemokine fractalkine (KLEDAL et al. 1998), interacting with many of the same epitopes of fractalkine as does CX3CR1 (MIZOUE et al. 2001). Fractalkine exists in a soluble and a membrane-bound version. In its membrane-bound form it consists of a chemokine-like domain, a mucin stalk, a transmembrane domain, and a cytoplasmic tail. KLEDAL et al. (1998) proposed a role for pUS28 in the adhesion of leukocytes latently infected by CMV to the surface of CX3C-expressing endothelial cells. Recent studies by HASKELL et al. (2000) supported this proposal. They constructed chimeras of RANTES, MIP-1α, MCP-1, and IL-8 bound to the fractalkine mucin stalk and anchored these chimeric proteins, as well as native fractalkine, to glass slides. Using these immobilized chimeras and fractalkine, they showed that 300-19 cells transfected with US28 can adhere to antibody-tethered fractalkine and become immobilized under shear flow conditions. Although cells adhered to CC

Fig. 6. Proposed mechanisms of chemokine- and chemokine receptor-dependent trafficking and persistence of CMV. The chemokines or chemokine receptors suggested to play a role in CMV trafficking are indicated in the *figure adjacent* to each of the cells that express these molecules. CMV-infected monocytes (*Mo*) and macrophages (Mϕ expressing either pUL33 or pUS28 could infect bone marrow (*BM*) stromal cells expressing SDF-1 or RANTES and MCP-1, respectively (*A*). In the case of BM transplantation, $CD34^+$ cells, known to be latently infected in healthy donors (KONDO and MOCARSKI 1995), might be attracted partially through a pUL33/SDF-1 or a pUS28/MCP-1 interaction (*B*). In the BM, alloreactivity (SODERBERG-NAUCLER et al. 1997) after transplantation could result in the differentiation of transplanted Mo into Mϕ (SINZGER et al. 1997), thereby resulting in full CMV replication in these cells with subsequent infection of stromal cells (*C*). Infected stromal cells (LAGNEAUX et al. 1995) could transmit infection to BM progenitors and assist in the establishment of latency by upregulation of chemokines that inhibit $CD34^+$ proliferation (MIP-1α, MCP-1) (BROXMEYER and KIM 1999; CASHMAN et al. 1990; LAGNEAUX et al. 1996) or by downregulation of necessary stimulatory factors like SCF (LAGNEAUX et al. 1996) (*D*). Latently infected progenitors would carry the CMV genome during their maturation and liberation into the circulation (HAHN et al. 1998) (*E*). Mobilization of matured myeloid cells might be enhanced by pUS27/28 withdrawal of hematopoietic inhibitory factors [MCP-1 (CASHMAN et al. 1990), MIP-1α (BROXMEYER and KIM 1999)] (*F*) and by increased production of IL-8 by infected endothelial cells (CRAIGEN et al. 1997) (*G*). The possible expression of pUS28 on latently infected myeloid cells [Mo, neutrophils (Ne)] in the bloodstream could play a role in their chemoattraction to endothelial cells expressing CX3C (*H*), thereby allowing both transmission of infection to endothelium and transmigration of infected cells into tissues (*I*). CMV transmitted to endothelial cells would become a source for new infection of transmigrating Mo and Ne (GRUNDY et al. 1998; REVELLO et al. 1998) (*J*). Adhesion of uninfected cells might be enhanced by expression of the CMV CXC chemokine, vCXC-1, and/or IL-8 and Groα on infected endothelium and their interaction with CXCR2 on Ne (*K*). Transmigrated Ne and Mo might transmit virus to tissue epithelium, smooth muscle cells, and fibroblasts (SINZGER and JAHN 1996), again via pUS28-facilitated cell fusion (*L*). Differentiation of latently infected Mo into tissue Mϕ at sites of inflammation (*M*) could transmit virus to neighboring tissue components by direct infection with cell-free virus (SINZGER et al. 1996) (*N*). In the early stage of infection, epithelial, endothelial, and smooth muscle cells could attract Mo because of CMV induction of RANTES acting on cellular receptors such as CCR1 and CCR5 (*O*), and later through interaction of vCXC-1 with CXCR2 on Ne (PENFOLD et al. 1999) (*P*). CMV could be transferred from infected smooth muscle cells, fibroblasts, or epithelial cells on interaction of US28 with cell surface-expressed CX3C from surveilling Mϕ or dendritic (*De*) cells (*Q*). Subsequently, CMV could be transported to the lymph nodes for further dissemination. Although the role for lymph nodes in CMV dissemination is unclear, CMV has been localized in these tissues (BORISKIN et al. 1999). Similarly, megakaryocytes (*MK*) and blood platelets (*Pl*) could disseminate CMV (*R*), because it was shown that MK are susceptible to CMV infection (CRAPNELL et al. 2000). Finally, in addition to their function in mediating CMV trafficking, pUL33, pUS27, and pUS28 could establish persistent CMV infection in BM stroma, smooth muscle cells, endothelium (BILLSTROM et al. 1998), fibroblasts (BODAGHI et al. 1998), or epithelial cells (BEISSER et al. 1998). This could be established by autocrine stimulation or constitutive activity of these receptors (CASAROSA et al. 2001; FRAILE-RAMOS et al. 2001). Alternatively, these receptors could act as a chemokine sink, sequestering all extracellular inflammatory chemokines to evade immune surveillance. Both signaling and sequestration might render the local environment favorable for CMV persistence (*S*)

chemokine chimeras under static conditions, they were not immobilized under flow-shear conditions. These results demonstrate that membrane-bound fractalkine is potentially sufficient to immobilize CMV-infected cells in the absence of other adhesion molecules. The US28 gene is transcribed in infected peripheral blood leukocytes from CMV seropositive individuals in vivo (PATTERSON et al. 1998) and in a monocytic cell line, THP-1, in vitro (BEISSER et al. 2001; ZIPETO et al. 1999). These observations indicate that CMV-infected monocytes and possibly also monocytic progenitors may express pUS28. This implies a mechanism for CMV to traffic from monocytes to or through the endothelium either by adhesion and subsequent cell-to-cell transmission of CMV or by transendothelial migration of

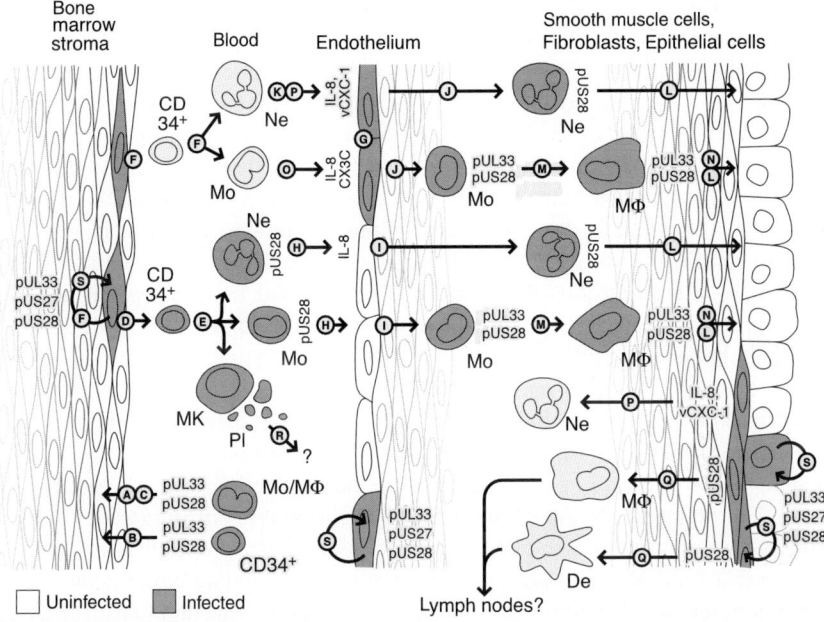

the monocytic cells into underlying tissues (Fig. 6). Smooth muscle cells (SMC) infected with CMV or transfected with US28 display chemokinesis in the presence of MCP-1 and chemotactic properties in a RANTES gradient (STREBLOW et al. 1999). Although this may reflect pUS28-mediated SMC migration in CMV-related vascular disease in vivo, it is less clear what role migrating SMC may have in the dissemination of CMV in healthy individuals.

In addition to the proposed role of pUS28-CX3C interaction in trafficking CMV from monocytes to and across endothelial cells, there exists another possible mode of CMV exchange between cells. Previously, it was shown that macrophages and dendritic cells express CX3C chemokines on their cell surface (BAZAN et al. 1997; IMAI et al. 1997). Because these cell types are involved in immune surveillance, they may encounter CMV-infected cells expressing pUS28. Cells that are fully permissive for CMV infection are likely to express pUS28 after infection in vivo, because the US28 gene was shown to be expressed in fibroblasts, SMC, and endothelial cells in vitro (BILLSTROM et al. 1998; BODAGHI et al. 1999; STREBLOW et al. 1999; VIEIRA et al. 1998). Hence, adhesion between infected cells and antigen-presenting cells (macrophages or dendritic cells) could result in subsequent cell-to-cell transmission from the former to the latter two cell types (Fig. 6)

The UL33 gene product may also play an important role in CMV dissemination. UL33 is transcribed at very early times pi (DAVIS-POYNTER et al. 1995). Consequently, UL33 may be transcribed in latently infected myeloid cells, as are MIE genes and US28. In addition, UL33-like genes were identified and characterized in the genomes of MCMV (M33) and RCMV (R33) CMV (BEISSER et al.

1998; DAVIS-POYNTER et al. 1997). Mutant viruses, from which these UL33 gene homologs were deleted, showed no difference in replication efficiency in vitro compared with wild-type viruses. However, in vivo, these mutant viruses were unable either to enter or to replicate in the salivary gland epithelium of infected mice and rats. Similarly, the UL33 gene, of which both sequence and genome location correspond to those of M33 and R33, may therefore be essential for salivary gland tropism in humans. In Fig. 6, we propose a role for the UL33 gene product as a chemotaxis-driving factor in infected monocytes or macrophages. Similar to what was proposed for pUS28, it is possible that pUL33 mediates CMV trafficking by attracting latently infected cells into solid tissue, in particular, the salivary gland epithelium and possibly other secretory tissues. SDF-1, a CXC chemokine that is constitutively expressed by epithelial cells (PABLOS et al. 1999) is a candidate ligand for UL33, as suggested in Sect. 4.5. Consequently, chemotaxis of infected monocytes toward the epithelial layer could be driven by interaction of pUL33 with SDF-1 (Fig. 6). Alternately, the possibility remains that, to maintain persistent CMV infection in salivary gland epithelial cells, pUL33 may have to be expressed at the surface of these cells. Thus an intracellular activation state could be established by continuous signaling of pUL33 by SDF-1, to establish an environment suitable for CMV persistence. This continuous signaling could occur either through uninterrupted binding of ambient chemokine or through constitutive activity of the receptor (Fig. 6).

5.2 Modulation of Host Cell Chemokine Production in Relation to CMV Dissemination and Persistence

CMV infection rarely causes overt disease in immunocompetent individuals. Even in immuncompromised patients, active viral replication does not necessarily result in end-organ disease. Factors that tilt the balance between active virus replication and CMV disease are not known. It is most likely that CMV utilizes the chemokine network to propel infected cells into an environment conducive for replication, persistence, or latency. Once there, viral modulation of chemokines could assist in avoiding immune detection of the infected cell at that site.

RANTES can induce the release of IFN-γ (APPAY et al. 2000), which not only is an inhibitor of many chemokines (BAGGIOLINI 1998) but also blocks CMV replication after expression of IE proteins (BODAGHI et al. 1999 and references therein). CMV induction of RANTES could thereby indirectly result in a persistent infection.

In the early stages of viral replication, CMV induces production of RANTES. Binding of chemokines to extracellular proteoglycans concentrates them and enhances their activity (DIASBARUFFI et al. 1998; LUSTER et al. 1995; ORAVECZ et al. 1997). However, SCHAARSCHMIDT et al. (1999) reported that CMV infection downregulates proteoglycan transcription. Thus secreted RANTES would be more likely to form a gradient around uninfected, proteoglycan-producing cells, thereby leaving infected cells "sheltered" from attack. In vivo, RANTES production was significantly higher in bronchoalveolar lavage (BAL) fluids during CMV pneumonitis than in lung transplant patients with non-CMV-related allograft rejection or in transplant patients without complications (MONTI et al. 1996). BAL macrophages isolated from patients with CMV pneumonitis spontaneously released more RANTES than those from control patients. This enhanced production returned to baseline with the resolution of infection. Such high production of RANTES could lead to blocking of lymphocyte cytotoxic activity (APPAY et al. 1999).

At a later stage of CMV infection, local inflammatory reactions could be controlled by chemokine downregulation around the infected cells. The pUS28 receptor adsorbs RANTES from the infected cell environment (BILLSTROM et al. 1999; BODAGHI et al. 1998). RANTES, as well as MIP-1α/β and MCP-1 and 3, which also bind pUS28, are chemoattractant for T, dendritic, and natural killer cells (reviewed in LOETSCHER et al. 2000). RANTES adsorption by pUS28 could inhibit establishment of a chemokine gradient and thereby block both lymphocyte attraction and effector mechanism activation (HADIDA et al. 1998).

The majority of CC chemokines inhibit proliferation of hematopoietic progenitors activated by cytokines (reviewed in BROXMEYER and KIM 1999). These include MIP-1α, which is induced by infection of BM stroma (LAGNEAUX et al. 1996). Paradoxically, CMV infection would seem to downregulate some of the inhibitory chemokines. Secretion by BM myofibroblasts of constitutively produced MCP-1, an even more potent inhibitor of progenitor proliferation (CASHMAN et al. 1990), is abolished in CMV-infected stromal myofibroblasts (Michelson and Charbord, unpublished results). This was not seen with ΔUS28 or ΔUS28/27 CMV mutants. Interaction of US28 in progenitors with inhibitory CC chemokines could also play a role in maintaining latency/persistence by inhibiting proliferation of these cells.

In vivo, CMV DNA can be found in circulating $CD34^+$ cells and in BM aspirates of healthy CMV carriers (HAHN et al. 1998; HAHN and MOCARSKI 1996; KONDO et al. 1994; KONDO and MOCARSKI 1995; MENDELSON et al. 1996; MINTON et al. 1994; VONLAER et al. 1995). It was also detected in pretransplant trephine BM biopsies of healthy BM donors and recipients by in situ hybridization and/or immunochemical detection of CMV immediate-early antigens (FEST et al. 1994a,b; PENCHANSKY and KRAUSE 1979). Viral DNA persists within BM progenitors throughout their differentiation and maturation (HAHN et al. 1998), particularly within the myeloid lineage (ZHURAVSKAYA et al. 1996). In vitro CMV infection of BM and cord blood progenitors in the absence of stromal cells causes inhibition of colony formation (reviewed in MICHELSON 1997; see SINDRE et al. 2000). Moreover, CMV has been implicated in pancytopenia after BM transplantation (reviewed in ALMEIDA-PORADA and ASCENSAO 1996). Related to this is the fact that CMV induces IL-8 production (ALMEIDA-PORADA et al. 1997; CRAIGEN and GRUNDY 1996; CRAIGEN et al. 1997; MURAYAMA et al. 1997). This chemokine is a renown mobilizer of $CD34^+$ progenitors into the circulation and could thus play a role in depletion of progenitors from BM. Increased serum levels of IL-8 were found to correlate with CMV infection and antigenemia after BM transplantation (FIETZE et al. 1994; HUMAR et al. 1999). IL-8 plasma levels were also significantly increased, whereas MIP-1α levels decreased, in renal transplant patients who later developed CMV disease (Nordoy et al. 1999). Here, CMV-mediated mobilization of progenitor cells by IL-8 upregulation could play a significant role in the dissemination of latently infected progenitors.

6 Conclusions

From what is known about CMV-encoded chemokines and chemokine receptors, it appears that their participation in immune evasion would be mainly at the level of viral dissemination sheltered from the immune system through (cell-to-cell) passage and movement of receptor bearing infected cells bidirectionally across endothelial barriers. In addition, the ability of pUS28 to withdraw CC chemokines from the environment of infected cells could also confer a measure of immune evasion by blunting effector lymphocyte migration and activation.

So far, there have been reports for up- and downregulation of chemokines and cytokines of the host organism at least at the transcriptional level and by chemokine scavenging via CMV-encoded chemokine receptors. In addition, CMV may contribute to the effects of chemokine/cytokine modulation by expressing viral chemokines. Each of the CMV-encoded chemokine and chemokine receptor genes may exert individual functions in either dissemination or the establishment and maintenance of viral latency in vivo. Several of these putative functions are outlined in this chapter. However, there may also be an intricate interplay between the different cytokines, chemokines, and chemokine receptors of both viral and host origin (Selbie and Hill 1998). For this, we still need to examine especially the kinetics of expression of the CMV-encoded pUL33, pUL78, pUS27, pUS28, and vCXC-1 and the putative chemokine encoded by ORF UL147 in more diverse cellular environments than those that have been studied to date. Special attention should be paid to cytokine–chemokine interactions in CMV-infected cells of the myeloid lineage. Although these cells are in general not permissive for full CMV replication, they are important CMV carriers that are most likely steered by the complex cytokine/chemokine network and probably play an important role in viral dissemination within and between individuals.

In healthy individuals, full, active CMV replication in vivo occurs at limited, confined sites within target organs. Effectively, the CMV genome can be detected in many organs and within many cell types (Hendrix et al. 1997; Myerson et al. 1984; Toorkey and Carrigan 1989), but expression of late antigens (Sinzger et al. 1997) with the development of pathology is rare compared with the incidence of genome-carrying cells detected (Hendrix et al. 1997; Larsson et al. 1998). Infection of innate immune cells by CMV possibly accounts for piloting the virus to sites relevant for replication. Close contact between these infected piloting cells and permissive target cells likely enables passage without CMV ever entering extracellular spaces. Additionally, CMV induces downregulation of HLA molecules (see other chapters in this book), as well as withdrawal of chemokines (BILLSTROM et al. 1999; BODAGHI et al. 1998; VIEIRA et al. 1998). These phenomena acting in concert are likely to result in the typical low-profile distribution of virus in the host to contribute to the immune evasion inherent to CMV infection.

References

Albrecht T, Boldogh I, Fons M, Lee CH, AbuBakar S, Russell JM, Au WW (1989) Cell-activation responses to cytomegalovirus infection relationship to the phasing of CMV replication and to the induction of cellular damage. Sub-Cellular Biochemistry 15:157–202

Almeida-Porada G, Porada CD, Shanley JD, Ascensao JL (1997) Altered production of GM-CSF and IL-8 in cytomegalovirus-infected, IL-1-primed umbilical cord endothelial cells. Exp Hematol 25: 1278–1285

Almeida-Porada GD, Ascensao JL (1996) Cytomegalovirus as a cause of pancytopenia. Leukemia Lymphoma 21:217–223

Appay V, Brown A, Cribbes S, Randle E, Czaplewski LG (1999) Aggregation of RANTES is responsible for its inflammatory properties. J Biol Chem 274:27505–27515

Appay V, Rod-Dunbar P, Cerundolo V, McMichael A, Czaplewski L, Rowland-Jones S (2000) RANTES activates antigen-specific cytotoxic T lymphocytes in a mitogen-like manner through cell surface aggregation. International Immunology 12:1173–1182

Baggiolini M (1998) Chemokines and leukocyte traffic. Nature 392:565–568

Baggiolini M, Dewald B, Moser B (1997) Human chemokines: an update. Annu Rev Immunol 15: 675–705

Baldwin BR, Zhang CO, Keay S (2000) Cloning and epitope mapping of a functional partial fusion receptor for human cytomegalovirus gH. J Gen Virol 81:27–35

Bazan JF, Bacon KB, Hardiman G, Wang W, Soo K, Rossi D, Greaves D R, Zlotnik A, Schall T J (1997) A new class of membrane-bound chemokine with a CX3C motif. Nature 385:640–644

Beisser P, Grauls G, Bruggeman C, Vink C (1999) Deletion of the R78 G protein-coupled receptor gene from rat cytomegalovirus results in an attenuated, syncytium-inducing mutant strain. J Virol 73: 7218–7230

Beisser PS, Laurent L, Virelizier JL, Michelson S (2001) Human cytomegalovirus chemokine receptor gene US28 is transcribed in latently infected THP-1 monocytes. J Virol 75:5949–5957

Beisser PS, Vink C, Vandam JG, Grauls G, Vanherle SJV, Bruggeman CA (1998) The R33 G protein-coupled receptor gene of rat cytomegalovirus plays an essential role in the pathogenesis of viral infection. J Virol 72:2352–2363

Billstrom MA, Johnson GL, Avdi NJ, Worthen GS (1998) Intracellular signaling by the chemokine receptor US28 during human cytomegalovirus infection. J Virol 72:5535–5544

Billstrom MA, Lehman LA, Scott Worthen G (1999) Depletion of extracellular RANTES during human cytomegalovirus infection of endothelial cells. Am J Respir Cell Mol Biol 21:163–167

Bodaghi B, Goureau O, Zipeto D, Laurent L, Virelizier JL, Michelson S (1999) Role of IFN-gamma-induced indoleamine 2,3 dioxygenase and inducible nitric oxide synthase in the replication of human cytomegalovirus in retinal pigment epithelial cells. J Immunol 162:957–964

Bodaghi B, Jones TR, Zipeto D, Vita C, Sun L, Laurent L, Arenzana-Seisdedos F, Virelizier JL, Michelson S (1998) Chemokine sequestration by viral chemoreceptors as a novel viral escape strategy: withdrawal of chemokines from the environment of cytomegalovirus-infected cells. J Exp Med 188:855–866

Bold S, Ohlin M, Garten W, Radsak K (1996) Structural domains involved in human cytomegalovirus glycoprotein B-mediated cell-cell fusion. J Gen Virol 77:2297–2302

Boriskin YS, Moore P, Murday AJ, Booth JC, Butcher PD (1999) Human cytomegalovirus genome sequences in lymph nodes. Microbes Infection 1:279–283

Broxmeyer HE, Kim CH (1999) Regulation of hematopoiesis in a sea of chemokine family members with a plethora of redundant activities. Exp Hematol 27:1113–1123

Bruening W, Giasson B, Mushynski W, Durham HD (1998) Activation of stress-activated MAP protein kinases up-regulates expression of transgenes driven by the cytomegalovirus immediate/early promoter. Nucleic Acids Res 26:486–489

Casarosa P, Bakker R, Verzijl D, Navis M, Timmerman H, Smit M (2001) Constitutive siginally of the human cytomegalovirus-encoded chemokine receptor US28. J Biol Chem 276:1133–1137

Cashman D, Eaves A, Raines E, Ross R, CJ E (1990) Mechanisms that regulate the cell cycle status of very primitive hematopoietic cells in long-term human marrow cultures. I. Stimulatory role of a variety of mesenchymal cell activators and inhibitory role of TGF-beta. Blood 75:96–101

Cha TA, Tom E, Kemble GW, Duke GM, Mocarski ES, Spaete RR (1996) Human cytomegalovirus clinical isolates carry at least 19 genes not found in laboratory strains. J Virol 70:78–83

Chambers J, Angulo A, Amaratunga D, Guo H, Jiang Y, Wan J, Bittner A, Frueh K, Jackson M, Peterson P, Erlander M, Ghazal P (1999) DNA microarrays of the complex human cytomegalovirus genome: Profiling kinetic class with drug sensitivity of viral gene expression. J Virol 73:5757–5766

Chee MS, Bankier AT, Beck S, Bohni R, Brown CM, Cerny R, Horsnell T, Hutchinson CA, Kourisarides T, Martignetti JA, Preddi E, Satchwell SC, Tomlinson P, Weston KM, Barrell BG (1990a) Analysis of the protein-coding content of the sequence of human cytomegalovirus strain AD169. In: McDougall JM (ed) Cytomegaloviruses, Current Topics in Microbiology and Immunology. Springer-Verlag, Berlin, Heidelberg, New York, London, Paris, Tokyo, HongKong, pp 125–169

Chee MS, Satchwell SC, Preddie E, Weston KM, Barrell BG (1990b) Human cytomegalovirus encodes three G protein-coupled receptor homologues. Nature 344:774–777

Choe H, Farzan M, Konkel M, Martin K, Sun Y, Marcon L, Cayabyab M, Berman M, Dorf ME, Gerard N, Gerard C, Sodroski J (1998) The orphan seven-transmembrane receptor apj supports the entry of primary T-cell-line-tropic and dualtropic human immunodeficiency virus type 1. J Virol 72:6113–6118

Craigen JL, Grundy JE (1996) Cytomegalovirus induced up-regulation of LFA-3 (Cd58) and ICAM-1 (Cd54) is a direct viral effect that is not prevented by ganciclovir or foscarnet treatment. Transplantation 62:1102–1108

Craigen JL, Yong KL, Jordan NJ, MacCormac LP, Westwick J, Akbar AN, Grundy JE (1997) Human cytomegalovirus infection up-regulates interleukin-8 gene expression and stimulates neutrophil transendothelial migration. Immunology 92:138–145

Crapnell K, Zanjani ED, Chaudhuri A, Ascensao H, St Jeor S, Maciejewski JP (2000) In vitro infection of megakaryocytes and their precursors by human cytomegalovirus. Blood 95:487–493

Davis-Poynter NJ, Lynch DM, Vally H, Shellam GR, Rawlinson WD, Barrell BG, Farrell HE (1997) Identification and characterization of a G protein-coupled receptor homolog encoded by murine cytomegalovirus. J Virol 71:1521–1529

Davis-Poynter NJ, Rawlinson WD, Barrell BG, Shellam GR, Farrell HE (1995) Identification and characterisation of a G-protein coupled receptor homologue encoded by murine cytomegalovirus. The 20th International Herpesvirus Workshop, Groningen, University of Groningen, Program and Abstracts: 88 (Abstract)

Diasbaruffi M, Pereiradasilva G, Jamur MC, Roquebarreira MC (1998) Heparin potentiates in vivo neutrophil migration induced by Il-8. Glycoconjugate Journal 15:523–526

Faure M, Voyno-Yasenetskaya TA, Bourne HR (1994) cAMP and beta subunits of heterotrimeric G proteins stimulate the mitogen-activated protein kinase pathway in COS-7 cells

Fest T, Angonin R, Mougin C, Deschaseaux M, Lab M, Cahn JY, Herve P (1994a) Detection of cytomegalovirus-infected cells in bone marrow biopsy specimens obtained before allogeneic bone marrow transplantation from donors and recipients. Transplantation 57:1681–1683

Fest T, Deschaseaux M, Mougin C, Cahn JY, Dupond JL, Herve P (1994b) In situ hybridization for the detection of cytomegalovirus in blood or bone marrow leucocytes after allogeneic bone marrow transplantation. Br J Haematol 86:619–623

Fietze E, Prösch S, Reinke P, et al. (1994) Cytomegalovirus infection in transplant recipients. The role of tumor necrosis factor. Transplantation 58:675–680

Fraile-Ramos A, Kledal TN, Pelchen-Matthews A, Bowers K, Schwartz TW, Marsh M (2001) The human cytomegalovirus us28 protein is located in endocytic vesicles and undergoes constitutive endocytosis and recycling. Mol Biol Cell 12:1737–1749

Gao JL, Kuhns DB, Tiffany HL, McDermott D, Li X, Francke U, Murphy PM (1993) Structure and functional expression of the human macrophage inflammatory protein 1 alpha/RANTES receptor. J Exp Med 177:1421–1427

Gao JL, Murphy PM (1994) Human cytomegalovirus open reading frame US28 encodes a functional beta chemokine receptor. J Biol Chem 269:28539–28542

Garnett HM (1979) The early effects of human cytomegalovirus infection on macromolecular synthesis in human embryonic fibroblasts. Arch Virol 60:147–151

Gerna G, Percivalle E, Baldanti F, Sozzani S, Lanzarini P, Genini E, Lilleri D, Revello MG (2000) Human cytomegalovirus replicates abortively in polymorphonuclear leukocytes after transfer from infected endothelial cells via transient microfusion events. J Virol 74:5629–5638

Goh C-S, Bogan AA, Joachimiak M, Walther D, Cohen FE (2000) Co-evolution of proteins with their interaction partners. J Mol Biol 299:283–293

Gompels UA, Nicholas J, Lawrence G, Jones M, Thomson BJ, Martin ME, Efstathiou S, Craxton M, Macaulay HA (1995) The DNA sequence of human herpesvirus-6: structure, coding content, and genome evolution. Virology 209:29–51

Grundy JE, Lawson KM, MacCormac LP, Fletcher JM, Yong KL (1998) Cytomegalovirus-infected endothelial cells recruit neutrophils by the secretion of C-X-C chemokines and transmit virus by direct neutrophil-endothelial cell contact and during neutrophil transendothelial migration. J Infect Dis 177:1465–1474

Gutkind JS (1998) The pathways connecting G protein-coupled receptors to the nucleus through divergent mitogen-activated protein kinase cascades. J Biol Chem 273:1839–1842

Hadida F, Vieillard V, Autran B, Lewis-Clark I, Baggiolini M, Debré P (1998) HIV-specific T cell cytotoxicity mediated by RANTES via the chemokine receptor CCR3. J Exp Med 188:609–614

Hahn G, Jores R, Mocarski ES (1998) Cytomegalovirus remains latent in a common precursor of dendritc and myeloid cells. Proc. Natl. Acad. Sci (USA) 95:3937–3942

Hahn G, Mocarski E (1996) Human cytomegalovirus latency and latently-infected cell types in hematopoeitic progenitors and peripheral blood (Abs. No 303). In: 21st Herpesvirus Workshop. Northern Illinois University, DeKalb, IL, USA

Hamm HE (1998) The many faces of G-protein signaling. J Biol Chem 273:669–672

Haskell CA, Cleary MD, Charo IF (2000) Unique role of the chemokine domain of fractalkine in cell capture: Kinetics of receptor dissociation correlate with cell adhesion. J Biol Chem 275:34183–34189

Hendrix RM, Wagenaar M, Slobbe RL, Bruggeman CA (1997) Widespread presence of cytomegalovirus DNA in tissues of healthy trauma victims. J Clin Pathol 50:59–63

Hirsch AJ, Shenk T (1999) Human cytomegalovirus inhibits transcription of the CC chemokine MCP-1 gene. J Virol 73:404–410

Humar A, St Louis P, Mazzulli T, McGeer A, Lipton J, Messner H, MacDonald KS (1999) Elevated serum cytokines are associated with cytomegalovirus infection and disease in bone marrow transplant recipients. J Infect Dis 179:484–488

Imai T, Hieshima K, Haskell C, Baba M, Nagira M, Nishimura M, Kakizaki M, Takagi S, Nomiyama H, Schall TJ, Yoshie O (1997) Identification and molecular characterization of fractalkine receptor CX(3)CR1, which mediates both leukocyte migration and adhesion. Cell 91:521–530

Isegawa Y, Ping Z, Nakano K, Sugimoto N, Yamanishi K (1998) Human herpesvirus 6 open reading frame U12 encodes a functional beta-chemokine receptor. J Virol 72:6104–6112

Kledal TN, Rosenkilde MM, Schwartz TW (1998) Selective recognition of the membrane-bound CX3C chemokine, fractalkine, by the human cytomegalovirus-encoded broad-spectrum receptor US28. FEBS Lett 441:209–214

Kondo K, Kaneshima H, Mocarski ES (1994) Human cytomegalovirus latent infection of granulocyte-macrophage progenitors. Proc Natl Acad Sci USA 91:11879–11883

Kondo K, Mocarski ES (1995) Cytomegalovirus latency and latency-specific transcription in hematopoietic progenitors. Scand J Infect Dis pp 63–67

Kuhn DE, Beall CJ, Kolattukudy PE (1995) The cytomegalovirus US28 protein binds multiple CC chemokines with high affinity. Biochem Biophys Res Commun 211:325–330

Lagneaux L, Delforge A, Bron D, Bosmans E, Stryckmans P (1995) Comparative analysis of cytokines released by bone marrow stromal cells from normal donors and B-cell chronic lymphocytic leukemic patients. Leuk Lymphoma 17:127–133

Lagneaux L, Delforge A, Snoek R, Bosmans E, Schols D, De Clercq E, Stryckmans P, Bron D (1996) Imbalance in production of cytokines by bone marrow stromal cells following cytomegalovirus infection. J Infect Dis 174:913–919

Larsson S, Soderberg-Nauclér C, Wang FZ, Moller E (1998) Cytomegalovirus DNA can be detected in peripheral blood mononuclear cells from all seropositive and most seronegative healthy blood donors over time. Transfusion 38:271–278

Locati M, Lamorte G, Luini W, Introna M, Bernasconi S, Mantovani A, Sozzani S (1996) Inhibition of monocyte chemotaxis to C-C chemokines by antisense oligonucleotide for cytosolic phospholipase A(2). J Biol Chem 271:6010–6016

Loetscher P, Moser B, Baggiolini M (2000) Chemokines and their receptors in lymphocyte traffic and HIV infection. Adv Immunol 74:127–180

Luster AD, Greenberg SM, Leder P (1995) The IP-10 chemokine binds to a specific cell surface heparan sulfate site shared with platelet factor 4 and inhibits endothelial cell proliferation. J Exp Med 182: 219–231

Margulies BJ, Browne H, Gibson W (1996) Identification of the human cytomegalovirus G protein-coupled receptor homologue encoded by UL33 in infected cells and enveloped virus particles. Virology 225:111–125

Mendelson M, Monard S, Sissons P, Sinclair J (1996) Detection of endogenous human cytomegalovirus in CD34(+) bone marrow progenitors. J Gen Virol 77:3099–3102

Menotti L, Mirandola P, Locati M, Campadelli-Fiume G (1999) Trafficking to the plasma membrane of the seven-transmembrane protein encoded by human herpesvirus 6 U51 gene involves a cell-specific function present in T lymphocytes. J Virol 73:325–333

Michelson S (1997) Interaction of human cytomegalovirus with monocytes/macrophages: a love-hate relationship. Pathol Biol (Paris) 45:146–158

Michelson S, Dal Monte P, Zipeto D, Bodaghi B, Laurent L, Oberlin E, Arenzana-Seisdedos F, Virelizier JL, Landini MP (1997) Modulation of RANTES production by human cytomegalovirus infection of fibroblasts. J Virol 71:6495–6500

Milne RBS, Mattick C, Nicholson L, Alcami A, Gompels UA (2000) RANTES binding and down-regulation by a novel human herpesvirus-6 chemokine receptor. J Immunol 164:2396–2404

Minton EJ, Tysoe C, Sinclair JH, Sissons JG (1994) Human cytomegalovirus infection of the monocyte/macrophage lineage in bone marrow. J Virol 68:4017–4021

Mizoue LS, Sullivan SK, King DS, Kledal TN, Schwartz TW, Bacon KB, Handel TM (2001) Molecular determinants of receptor binding and signaling by the CX3C chemokine fractalkine. J Biol Chem 29:29

Monti G, Magnan A, Fattal M, Rain B, Humbert M, Mege JL, Noirclerc M, Dartevelle P, Cerrina J, Simonneau G, Galanaud P, Emilie D (1996) Intrapulmonary production of RANTES during rejection and CMV pneumonitis after lung transplantation. Transplantation 61:1757–1762

Murayama T, Ohara Y, Obuchi M, Khabar KS, Higashi H, Mukaida N, Matsushima K (1997) Human cytomegalovirus induces interleukin-8 production by a human monocytic cell line, THP-1, through acting concurrently on AP-1- and NF-κB-binding sites of the interleukin-8 gene. J Virol 71:5692–5695

Murphy PM (2000) Viral antichemokines: From pathogenesis to drug discovery. J Clin Invest 105: 1515–1517

Murphy PM, Baggiolini M, Charo IF, Herbert CA, Horuk R, Matsushima K, Miller LH, Oppenheim JJ, Power CA (2000) International Union of Pharmacology. XXII. Nomenclature for chemokine receptors. Pharmacological Reviews 52:145–176

Myerson D, Hackman RC, Nelson JA, Ward DC, McDougall JK (1984) Widespread presence of histologically occult cytomegalovirus. Hum Pathol 15:430–439

Navarro D, Paz P, Tugizov S, Topp K, La Vail J, Pereira L (1993) Glycoprotein B of human cytomegalovirus promotes virion penetration into cells, transmission of infection from cell to cell, and fusion of infected cells. Virology 197:143–158

Neote K, DiGregorio D, Mak JY, Horuk R, Schall TJ (1993) Molecular cloning, functional expression, and signaling characteristics of a C-C chemokine receptor. Cell 72:415–425

Nordoy I, Muller F, Nordal KP, Rollag H, Lien E, Aukrust P, Froland SS (1999) Immunologic parameters as predictive factors of cytomegalovirus disease in renal allograft recipients. J Infect Dis 180:195–198

Ohagen A, Li L, Rosenzweig A, Gabuzda D (2000) Cell-dependent mechanisms restrict the HIV type 1 coreceptor activity of US28, a chemokine receptor homolog encoded by human cytomegalovirus. AIDS Res Hum Retroviruses 16:27–35

Oliveira SA, Shenk TE (2001) Murine cytomegalovirus M78 protein, a G protein-coupled receptor homologue, is a constituent of the virion and facilitates accumulation of immediate-early viral mRNA. Proc Natl Acad Sci USA 98:3237–3242

Oravecz T, Pall M, Wang J, Roderiquez G, Ditto M, Norcross MA (1997) Regulation of anti-HIV-1 activity of RANTES by heparan sulfate proteoglycans. J Immunol 159:4587–4592

Pablos JL, Amara A, Bouloc A, Santiago B, Caruz A, Galindo M, Delaunay T, Virelizier JL, Arenzana-Seisdedos F (1999) Stromal-cell derived factor is expressed by dendritic cells and endothelium in human skin. Am J Pathol 155:1577–1586

Patterson BK, Landay A, Andersson J, Brown C, Behbahani H, Jiyamapa D, Burki Z, Stanislawski D, Czerniewski MA, Garcia P (1998) Repertoire of chemokine receptor expression in the female genital tract: implications for human immunodeficiency virus transmission. Am J Pathol 153:481–490

Penchansky L, Krause JR (1979) Identification of cytomegalovirus in bone marrow biopsy. Southern Medical Journal 72:500–501

Penfold ME, Dairaghi DJ, Duke GM, Saederup N, Mocarski ES, Kemble GW, Schall TJ (1999) Cytomegalovirus encodes a potent alpha chemokine. Proc Natl Acad Sci USA 96:9839–9844

Pleskoff O, Treboute C, Alizon M (1998) The cytomegalovirus-encoded chemokine receptor US28 can enhance cell-cell fusion mediated by different viral proteins. J Virol 72:6389–6397

Pleskoff O, Treboute C, Brelot A, Heveker N, Seman M, Alizon M (1997) Identification of a chemokine receptor encoded by human cytomegalovirus as a cofactor for HIV-1 entry. Science 276:1874–1878

Randolph-Habecker J, Beall CJ, Kolattukudy PE, Sedmak DD (1997) Monocyte chemoattractant protein-1 binding by cytomegalovirus-infected endothelial cells. Transplant Proc 29:807–808

Reschke M, Revello MG, Percivalle E, Radsak K, Landini MP (1999) Constitutive expression of human cytomegalovirus (HCMV) glycoprotein gpUL75 (gH) in astrocytoma cells: a study of the specific humoral immune response. Viral Immunol 12:249–262

Revello MG, Percivalle E, Arbustini E, Pardi R, Sozzani S, Gerna G (1998) In vitro generation of human cytomegalovirus pp65 antigenemia, viremia, and leukodnaemia. J Clin Invest 101:2686–2692

Rodems SM, Spector DH (1998) Extracellular signal-regulated kinase activity is sustained early during human cytomegalovirus infection. J Virol 72:9173–9180

Rucker J, Edinger AL, Sharron M, Samson M, Lee B, Berson JF, Yi Y, Margulies B, Collman RG, Doranz BJ, Parmentier M, Doms RW (1997) Utilization of chemokine receptors, orphan receptors, and herpesvirus-encoded receptors by diverse human and simian immunodeficiency viruses. J Virol 71:8999–9007

Saederup N, Lin YC, Dairaghi DJ, Schall TJ, Mocarski ES (1999) Cytomegalovirus-encoded beta chemokine promotes monocyte-associated viremia in the host. Proc Natl Acad Sci USA 96: 10881–10886

Schaarschmidt P, Reinhardt B, Michel D, Vaida B, Mayr K, Luske A, Baur R, Gschwend J, Kleinschmidt K, Kountidis M, Wenderoth U, Voisard R, Mertens T (1999) Altered expression of extracellular matrix in human-cytomegalovirus-infected cells and a human artery organ culture model to study its biological relevance. Intervirology 42:365–372

Selbie LA, Hill SJ (1998) G protein-coupled-receptor cross-talk: the fine-tuning of multiple receptor-signalling pathways. Trends Pharmacol Sci 19:87–93

Shibutani T, Johnson TM, Yu ZX, Ferrans VJ, Moss J, Epstein SE (1997) Pertussis toxin-sensitive G proteins as mediators of the signal transduction pathways activated by cytomegalovirus infection of smooth muscle cells. J Clin Invest 100:2054–2061

Signoret N, Marsh M (2000) Analysis of chemokine receptor endocytosis and recycling. In: "Chemokine Protocols". Humana Press, Inc., Totowa, New Jersey, USA

Sindre H, Rollag H, Degre M, Hestdal K (2000) Human cytomegalovirus induced inhibition of hematopoietic cell line growth is initiated by events taking place before translation of viral gene products. Arch Virol 145:99–111

Sinzger C, Jahn G (1996) Human cytomegalovirus cell tropism and pathogenesis. Intervirology 39:302–319

Sinzger C, Knapp J, Plachter B, Schmidt K, Jahn G (1997) Quantification of replication of clinical cytomegalovirus isolates in cultured endothelial cells and fibroblasts by a focus expansion assay. J Virol Methods 63:103–112

Sinzger C, Plachter B, Grefte A, The TH, Jahn G (1996) Tissue macrophages are infected by human cytomegalovirus in vivo. J Infect Dis 173:240–245

Soderberg-Naucler C, Fish KN, Nelson JA (1997) Reactivation of latent human cytomegalovirus by allogeneic stimulation of blood cells from healthy donors. Cell 91:119–126

Speir E, Shibutani T, Yu ZX, Ferrans V, Epstein SE (1996) Role of reactive oxygen intermediates in cytomegalovirus gene expression and in the response of human Smooth muscle cells to viral infection. Circ Res 79:1143–1152

Speir E, Yu ZX, Ferrans VJ, Huang ES, Epstein SE (1998) Aspirin attenuates cytomegalovirus infectivity and gene expression mediated by cyclooxygenase-2 in coronary artery smooth muscle cells. Circ Res 83:210–216

Streblow DN, Soderberg-Naucler C, Vieira J, Smith P, Wakabayashi E, Ruchti F, Mattison K, Altschuler Y, Nelson JA (1999) The human cytomegalovirus chemokine receptor US28 mediates vascular smooth muscle cell migration. Cell 99:511–520

Toorkey CB, Carrigan DR (1989) Immunohistochemical detection of an immediate early antigen of human cytomegalovirus in normal tissues. J Infect Dis 160:741–751

Vieira J, Schall TJ, Corey L, Geballe AP (1998) Functional analysis of the human cytomegalovirus US28 gene by insertion mutagenesis with the green fluorescent protein gene. J Virol 72:8158–8165

Vonlaer D, Meyerkoenig U, Serr A, Finke J, Kanz L, Fauser AA, Neumannhaefelin D, Brugger W, Hufert FT (1995) Detection of cytomegalovirus DNA in CD34(+) cells from blood and bone marrow. Blood 86:4086–4090

Welch AR, McGregor LM, Gibson W (1991) Cytomegalovirus homologs of cellular G protein-coupled receptor genes are transcribed. J Virol 65:3915–3918

Yurochko AD, Huang ES (1999) Human cytomegalovirus binding to human monocytes induces immunoregulatory gene expression. J Immunol 162:4806–4816

Zhu H, Cong JP, Shenk T (1997) Use of differential display analysis to assess the effect of human cytomegalovirus infection on the accumulation of cellular RNAs -induction of interferon-responsive RNAs. Proc Natl Acad Sci USA 94:13985–13990

Zhuravskaya T, Maciejewski J, Massey R, St. Jeor S. (1996) Human cytomegalovirus (HCMV) infection of hematopoietic progenitor cells (Abs No. 301). In: 21st Herpesvirus Workshop, Northern Illinois University, DeKalb, IL, USA

Zipeto D, Bodaghi B, Laurent L, Virelizier JL, Michelson S (1999) Kinetics of transcription of human cytomegalovirus chemokine receptor US28 in different cell types. J Gen Virol 80:543–547

Fatal Attraction:
Cytomegalovirus-Encoded Chemokine Homologs

N. Saederup and E.S. Mocarski Jr

Members of the cytomegalovirus (CMV) subfamily of betaherpesviruses infecting primates and rodents encode divergent proteins with sequence characteristics and activities of chemokines, a class of small, secreted proteins that control leukocyte migration and trafficking behavior. Human CMV genes UL146 and UL147 encode proteins with sequence characteristics of CXC chemokines, whereas, murine CMV encodes a CC chemokine homolog (MCK-2). Human CMV UL146 encodes a neutrophil-attracting chemokine denoted viral CXC chemokine-1 (vCXCL1) that is as potent as host IL-8 and functions via the CXCR2 receptor, one of two human IL-8 receptors. Murine CMV MCK-2 is composed of a chemokine domain derived from open reading frame (ORF) m131 (and denoted MCK-1) as well as a domain derived from m129 that does not have sequence similarity to any known class of proteins. A synthetic version of murine CMV m131 (MCK-1) protein carries out many of the activities of a positive-acting chemokine, including transient release of intracellular calcium stores and cell adhesion of peritoneal macrophage populations. In the context of the viral genome and infection of the mouse host, the m131-m129 (MCK-2) gene product confers increased inflammation, higher levels of viremia, and higher titers of virus in salivary glands, consistent with a role in promoting dissemination by attracting an important mononuclear leukocyte population. Other characterized primate CMVs, but not other primate betaherpesviruses, encode gene products similar to human UL146 and UL147. Other characterized rodent CMVs encode a gene product similar to the murine CMV chemokine homolog, although not as a spliced gene product. Thus chemokines, like viral proteins that downmodulate MHC class I expression or have sequence homology to host MHC class I proteins, have evolved in primate and rodent CMVs to carry out an analogous set of immunomodulatory functions during infection of the host even though they arise from distinct origins.

1	Introduction	236
2	Chemokine Biology	237
3	The Murine CMV Chemokine Homolog, MCK-2	238
3.1	Features of CMV Infection Relevant to Chemokine Expression	238
3.2	Structure and Expression of the *mck* Gene	240
3.3	Chemokine-Like Properties of MCK-1, the Chemokine Domain of *mck*	241
3.4	Investigations of *mck*-Mutant Viruses	243
3.5	Concluding Remarks on *mck* Function In Vivo	245
4	The Human CMV Chemokine Homologs vCXCL-1 and vCXCL-2	246
4.1	Expression and Structure of vCXCL-1 and vCXCL-2	248
4.2	vCXCL-1 Chemokine Activities	248
4.3	Concluding Remarks on Human CMV Chemokines	249
5	Conclusion	250
	References	251

Department of Microbiology and Immunology, Stanford University School of Medicine, Stanford, CA 94305-5124, USA

1 Introduction

Animal viruses have evolved elaborate mechanisms to promote survival in their hosts. The genomes of all large DNA viruses contain accessory genes that are dispensable for replication in tissue culture but that modulate the virus–host interaction in vivo during the course of viral infection. These roles include facilitating immune evasion, ensuring efficient dissemination, and providing determinants of tissue tropism affecting acute and latent infection. Many viral accessory genes are homologs of host genes that were almost certainly pirated from the host via recombination events. Multiple genes encoding chemokine and chemokine receptor homologs are found in herpesviruses such as human cytomegalovirus (CMV) and Kaposi's sarcoma herpesvirus (KSHV), leading to speculation on their roles in viral infection (DAIRAGHI et al. 1998; ALCAMI and KOSZINOWSKI 2000). The chemokine network normally regulates trafficking behavior of leukocyte subsets in the host, so the existence of viral homologs suggests a role in modulating the host immune response or altering leukocyte trafficking to benefit virus infection. In addition to encoding components of the chemokine network, viruses such as Epstein-Barr induce expression of host chemokines and chemokine receptors during productive or latent infection (MCCOLL et al. 1997; YOSHIE et al. 1997). Beyond speculation from behavior of these gene products in vitro, any understanding of how virus-encoded or virus-induced chemokines or chemokine receptors contribute to viral pathogenesis requires an animal model. Unfortunately, all of the medically significant DNA viruses that carry or induce this class of gene function are highly host restricted. Viruses such as human CMV, KSHV, and Epstein-Barr virus do not infect any laboratory animals. Surrogate systems employing relatives of these viruses that naturally infect murine and simian hosts are among the only avenues to understanding of the role that accessory functions play viral pathogenesis or latency.

The cytomegaloviruses constitute a distinct family within the betaherpesvirus subgroup of herpesviruses. Human CMV infects a majority of the population by early adulthood, and, although generally asymptomatic in immunocompetent individuals, CMV infection remains the most important congenital disease as well as a major problem in immunocompromised patients. CMV is also associated with exacerbation of vascular disease, particularly after solid organ transplantation. In the immunocompetent host, CMV establishes a lifelong latent infection with periodic episodes of reactivation and shedding (PASS 2001). The immune system plays a paradoxical role in both the acute and latent phases of CMV infection (Ho 1991; MOCARSKI and COURCELLE 2001). On the one hand, cytomegaloviruses such as murine CMV rely on host mononuclear leukocytes to serve as vehicles for dissemination from a portal of entry during acute infection (BALE and O'NEIL 1989; STODDART et al. 1994). On the other hand, these viruses face the consequences of the immune response where leukocytes control acute infection through both innate and adaptive clearance mechanisms (KOSZINOWSKI et al. 1990; HANSON et al. 1999; JONJIC et al. 1994; SCALZO et al. 1999; PASS 2001). During latency, mononuclear leukocytes and their progenitors are likely to play key roles in harboring latent virus (KOFFRON et al. 1998; POLLOCK et al. 1997) as well as playing key roles in suppressing reactivation. Similar interactions are suspected to occur with human CMV; however, studies on this virus are complicated by having to follow natural infection in humans. Human CMV associates with neutrophils and monocytes during acute infection (REVELLO et al. 1998; ZANGHELLINI et al. 1999) and enters latency in a monocyte progenitor (HAHN et al. 1998; KONDO et al. 1994; SLOBEDMAN and MOCARSKI 1999; SODERBERG-NAUCLER et al. 1997). Immune clearance relies on the cellular arm of the adaptive immune response (RIDDELL and GREENBERG 1997). Given such a complex relationship, it is not surprising that cytomegaloviruses have evolved a sophisticated armament of gene products that influence leukocyte behavior.

This review will focus on the biological properties and activities of chemokine homologs encoded by human and murine CMV. Section 2 will provide an overview of host chemokines as a backdrop for understanding the possible roles of viral chemokine homologs in viral pathogenesis. Section 3 will discuss the activities of the murine CMV chemokine homolog MCK-2 in mouse CMV pathogenesis. Section 4 will summarize the properties of the human CMV chemokine vCXC-1 and the chemokine homolog vCXC-2. The Conclusion (Sect. 5) will explore the implications of this type of function in the biology of CMV and other large DNA viruses.

2 Chemokine Biology

The chemokines are a family of small (8–12kDa) secreted chemoattractant cytokines that share a common structure dictated by the pattern of disulfide bonds that form between conserved cysteine (C) residues. The number and spacing of cysteines in the amino-terminus defines four chemokine classes, CXC, CC, XC and CX_3C, where X represents any amino acid (aa) except C (BAGGIOLINI and LOETSCHER 2000). A majority of chemokines have two sets of disulfide bonds formed by the pairing the first two cysteines (C_1 and C_2) with two others in a C_1-C_3 and C_2-C_4 pattern. These secreted proteins bind to G protein-coupled heptahelical receptors found on leukocytes that are designated by the class of chemokine bound (CXCRs, CCRs, XCRs, and CX_3CR, respectively) (ROSSI and ZLOTNIK 2000).

Some chemokines are expressed constitutively, in secondary lymphoid tissues such as lymph nodes and spleen as well as in tissue such as the gastrointestinal tract, where they control trafficking and homing behavior of leukocytes (CAMPBELL and BUTCHER 2000). Expression of other chemokines is induced as part of an inflammatory response after tissue damage or infection (BAGGIOLINI and LOETSCHER 2000). Once made, chemokines diffuse and attach to proteoglycans in the extracellular matrix, forming concentration gradients that attract specific leukocytes to sites of production (KUSCHERT et al. 1999). Leukocytes circulating in peripheral blood or resident in tissues that carry the appropriate receptors become activated through this process (TANAKA et al. 1993). Activation via a chemokine receptor induces cell signaling cascades, including the mitogen-activated protein kinase and phospholipase C pathways (KNALL et al. 1996; LI et al. 2000). One signaling event common to all chemokines is the transient release (mobilization) of intracellular stores of calcium from the endoplasmic reticulum (BAGGIOLINI et al. 1997). This release stimulates the rapid phosphorylation of integrins that increases cellular adhesion (CAMPBELL et al. 1998). Through these events, leukocytes become stimulated to bind to the endothelial cells lining blood vessels near sites of injury or infection. The adherent leukocytes extravasate into the surrounding tissues and migrate by chemotaxis in response to local chemokine gradients (CAMPBELL and BUTCHER 2000; SPRINGER 1994). Chemokine receptors are expressed in distinct patterns on different leukocyte subpopulations in ways that dictate the timing and intensity of an inflammatory response (SALLUSTO et al. 2000). In addition to controlling leukocyte trafficking, chemokines are important regulators of effector functions such as granule release from neutrophils or mast cells and cytokine expression in lymphocytes or macrophages (LOETSCHER et al. 1996; SALLUSTO et al. 2000; TAUB et al. 1996).

3 The Murine CMV Chemokine Homolog, MCK-2

3.1 Features of CMV Infection Relevant to Chemokine Expression

The human and murine CMV genomes exhibit a colinear organization in which approximately 80 open reading frames (ORFs) are conserved (RAWLINSON et al. 1996). These include many viral gene products involved in gene regulation, DNA synthesis, and virion maturation. Other human and murine CMV-encoded genes have diverged markedly at the aa sequence level but are known to carry out similar functions. A striking level of sequence divergence has occurred in the viral genes that modulate interactions with the host (MOCARSKI and COURCELLE 2001). Examples include those functions that downmodulate major histocompatibility complex (MHC) class I gene expression to inhibit cytotoxic T lymphocyte (CTL) effector function (AHN et al. 1996; HENGEL et al. 1998, 1999; JONES and SUN 1997; KRMPOTIC et al. 1999; WIERTZ et al. 1997; ZIEGLER et al. 2000; ALCAMI and KOSZINOWSKI 2000). Others may modulate the natural killer (NK) cell response (FARRELL et al. 1997; LEONG et al. 1998; TOMASEC et al. 2000). These functions retain neither predicted aa sequence identity nor relative genome position in the two viruses despite having similar impact on pathogenesis. Included in this set of human and murine CMV accessory functions are chemokine homologs MCK-2 in murine CMV and vCXC-1 and vCXC-2 in human CMV (MACDONALD et al. 1997, 1999; PENFOLD et al. 1999). These represent different types of chemokines that have apparently been pirated independently during the course of evolution in separate hosts. MCK-2 is retained in both strains of murine CMV that are in widespread use, Smith (distributed by the ATCC) and K181, isolated as a more virulent derivative of Smith (OSBORN 1982) but actually a different strain altogether (MOCARSKI et al. 1990).

Dissemination to salivary glands, where virus replication leads to shedding in saliva, is independent of inoculation route. Peak titers of 10^7–10^9 plaque-forming units (PFU)/g of tissue at 14 days p.i. are several orders of magnitude higher than detected in any other tissue (CARDIN et al. 1993; STODDART et al. 1994). Depending on the strain of mouse used and on the inoculum dose, virus may persist in the salivary glands and be shed in saliva for several months. BALB/c mice exhibit a complete and rapid clearance of infectious virus within 5 or 6 weeks p.i. (JONJIC et al. 1989; LUCIN et al. 1992; PAVIC et al. 1993; CARDIN et al. 1993), which has contributed to the value of this strain for experimental work. Serous acinar cells, which are epithelial in origin and are responsible for producing components of saliva, are the principal cell type supporting replication in salivary glands (HENSON and STRANO 1972; STODDART et al. 1994). Virus shed in saliva fosters transmission to new hosts, although this aspect of viral pathogenesis has not been subjected to detailed study. Host-to-host transmission is most readily observed in mice through activities such as biting rather than casual contact. Salivary glands are also major sites of CMV replication in humans, where transmission occurs via more casual direct contact with saliva (Ho 1991; PASS 2001).

One central step in CMV pathogenesis is dissemination from a portal of entry via the bloodstream to organs in the host. Cell-free infectious CMV is not readily detected in the blood or plasma. When human CMV DNA is detected by polymerase chain reaction in plasma from viremic immunocompromised patients, infectious virus and viral antigens are present in cells in the bloodstream (WOLF and SPECTOR 1993). Much less is known about natural infection of immunocompetent individuals. Limited studies suggest that virus remains leukocyte associated in the blood stream (REVELLO et al. 1998; ZANGHELLINI et al. 1999). Murine CMV-infected mononuclear leukocytes are responsible for viral dissemination from sites of inoculation (BALE and O'NEIL 1989; STODDART et al. 1994), and this appears to occur in two distinct phases (COLLINS et al. 1994). Primary viremia occurs within 2 days after inoculation (COLLINS et al. 1994) and seeds sites such as the spleen, liver, adrenal gland, lungs, and brown fat tissue (STODDART et al. 1994). Peak levels of virus-positive leukocytes can be detected between days 4 and 6 p.i. After replication at these sites (COLLINS et al. 1994) salivary glands become seeded (STODDART et al. 1994). The cell type carrying infectious murine CMV to salivary glands is not well defined but appears to be a macrophage, monocyte, or dendritic cell based on morphological appearance and phagocytic activity (STODDART et al. 1994).

During the peak of infection, inflammatory foci surrounding sites of murine CMV replication at the inoculation site as well as in the liver and spleen include macrophages, other mononuclear cells, and polymorphonuclear leukocytes. Macrophages likely contribute to clearance during the innate response to this virus (ANDREWS et al. 2001; SALAZAR-MATHER et al. 1998). Depletion of macrophages by silica or liposomal intoxication increases virus titers in both liver and spleen (HANSON et al. 1999). These methods are known to have broad effects on the host beyond depletion of phagocytic leukocytes, so the results may not be unambiguous. A dual role of macrophages in clearance as well as dissemination or latency illustrates how one cell type can have two roles in this virus–host interaction.

3.2 Structure and Expression of the *mck* Gene

Murine CMV ORF m131, although not initially noted to be similar to any host gene (RAWLINSON et al. 1996), encodes an 81-aa protein, designated MCK-1 (MACDONALD et al. 1997), that is similar to host chemokines. MCK-1 contains four cysteines, including an amino-terminal CC motif, and a putative amino-terminal signal peptide as well as weak similarity to other chemokines (MACDONALD et al. 1997). An mRNA splicing event causes the major product of the *mck* gene to consist of MCK-1 as an amino-terminal component fused in-frame to 199 additional codons including ORF m129 (Fig. 1; FLEMING et al. 1999; MACDONALD et al. 1999). All 81 codons of MCK-1 are included in this larger gene product as the mRNA splice donor is adjacent to the m131 termination codon. The product of the spliced m131-m129 fusion has been denoted MCK-2. This protein is encoded by a true late (γ2) transcript in infected fibroblasts (FLEMING et al. 1999; MACDONALD et al. 1999; VIEIRA et al. 1994) and has a predicted size of 31kDa with five potential

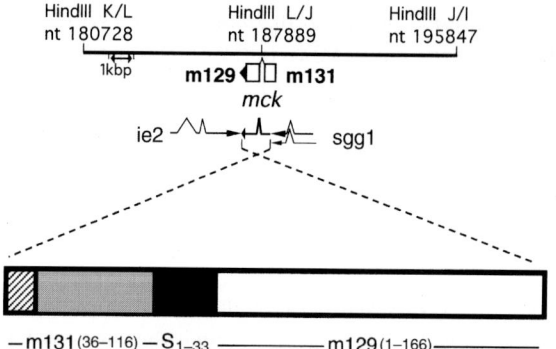

Fig. 1. Schematic representation of the structure of the *mck* gene and gene product. *Top line*: Restriction map of the *Hin*dIII L and J DNA fragments in murine CMV strain K181[+] corresponding to nt 180728 to 195847 of the Smith strain genome sequence (accession no. U68299). Restriction sites for *Hin*dIII are indicated *above* and a 1-kbp scale marker is indicated by a *double-headed arrow* below the line. *Open boxes with arrowheads* below the line depict the position of viral m131 and m129 ORFs containing coding sequences for MCK-2 (MACDONALD et al. 1999). *Arrows* depict transcripts for the *mck* gene and genes (*ie2*, *sgg1*) that flank *mck* in wild-type virus. *Bottom*: Schematic representation of MCK-2 protein based on cDNA sequence (accession no. AF124602). The *hatched* and *stippled boxes* correspond to the predicted 18-aa signal peptide and the 63-aa chemokine domain of the mature protein, respectively. The *solid box* represents the 33-aa sequence encoded within a spacer (*S*) region between m131 and m129, and m129 is represented by the *open box*. The *numbers in parentheses* indicate the amino acids derived from m131 or m129 present in each portion of MCK-2

N-linked glycosylation sites. Immunoblot analysis was used to show that several forms of MCK-2 ranging from 31 to 45kDa are present in infected cells. PNGase F treatment of larger species reduces all species to an apparent size of 31kDa, indicating that some of the N-linked glycosylation sites are used. Glycosylated MCK-2 is secreted into the medium of murine CMV infected cells (MACDONALD et al. 1999).

The structure of MCK-2 is unusual compared with that of other CC chemokines. The amino-terminal domain, derived from ORF m131, has a sequence and size typical of CC chemokines. The 199-aa carboxyl-terminal portion shares no apparent sequence homology with any genes in the public databases (BLASTP; 3/3/01), nor does it contain any known protein structures or motifs. MCK-2 may be compared to the mammalian CX_3C chemokine fractalkine (CX_3CL1), which is composed of an amino-terminal chemokine domain or "head" attached to a long carboxyl-terminal extension (BAZAN et al. 1997) with similarity to a mucin stalk. Most fractalkine is embedded plasma membranes via a carboxyl-terminal transmembrane segment; however, proteolytic cleavage near the base of the mucin stalk releases a secreted version of fractalkine (BAZAN et al. 1997). Directed mutagenesis experiments have shown that the classic chemokine properties (receptor binding, signaling) of the chemokine domain of fractalkine are independent of the carboxyl-terminal mucin-like stalk (BAZAN et al. 1997; HASKELL et al. 2000; IMAI et al. 1997). Other chemokines have been fused to the fractalkine stalk as well as to other carboxyl-terminal fusion partners, while retaining both binding and signaling properties of the original chemokine stalk (BAZAN et al. 1997; HASKELL et al. 2000; IMAI et al. 1997). MCK-2 contains neither a putative carboxyl-terminal transmembrane region nor a GPI-anchor motif (FLEMING et al. 1999) and thus would be expected to be produced only as a secreted protein. On the basis of the data that have been generated on fractalkine and chemokines that have been carboxyl-terminally fused to other proteins, we would predict that the chemokine activity of MCK-2 would be represented by the amino-terminal MCK-1 chemokine domain.

3.3 Chemokine-Like Properties of MCK-1, the Chemokine Domain of *mck*

To investigate the chemokine properties of the *mck* gene product, the chemokine domain (MCK-1) was chemically synthesized starting from a predicted signal sequence cleavage site between aa 18 and 19. This MCK-1 polypeptide was tested for ability to induce calcium mobilization in different primary leukocyte populations from murine CMV-infected mice and uninfected controls with fluorescent video microscopy (SAEDERUP et al. 1999). MCK-1-induced mobilization in a small percentage (2%–5%) of resident peritoneal exudate cells (PECs) isolated from uninfected mice by glass adherence. Interestingly, a much higher percentage of cells responded to MCK-1 when PECs were isolated 48h after intraperitoneal inoculation of either *mck*-expressing or *mck*-mutant virus. Only a small proportion of PECs became infected with virus and these exhibited a morphology typical of macrophages. Thus host mononuclear cells that are found in the peritoneum are highly responsive to MCK-1 with a percentage of responsive cells increasing dramatically after viral infection. This process occurs whether or not the virus itself expresses *mck*. Synthetic MCK-1 failed to induce calcium flux in any other population of cells, including nonadherent PECs and either adherent or nonadherent populations of peripheral blood leukocytes from either virus- or mock-infected animals. The release of intracellular calcium in PECs occurred with the same time course in cells from either infected or uninfected mice, with maximal calcium release at approximately 12 s after exposure to MCK-1 that returned to prestimulation levels within 2 min. This pattern is typical of mammalian chemokines. Also characteristic of other chemically synthesized chemokines (BOSHOFF et al. 1997; VAN DAMME et al. 1994), activity was found to be independent of side chain modifications such as N-linked glycosylation present on the natural protein. These observations indicated that MCK-1 normally targets activated macrophages, particularly those responding to viral infection. The positive impact of MCK-1 suggests a natural role of the protein in either attracting or activating this cell type (SAEDERUP et al. 1999).

The adherence behavior of individual MCK-1 responsive PECs from murine CMV-infected mice was also evaluated. A significant proportion of adherent cells would slowly roll across the glass surface as a result of turbulence associated with chemokine injection into the assay chamber. Rolling cells stopped moving within seconds after fluxing and remained tightly anchored to the glass surface for the duration of the experiment. This is consistent with cell surface changes that increase adherence and suggests that leukocyte behavior may be altered by exposure to this viral gene product during the course of CMV infection (SAEDERUP et al. 1999).

These observations are consistent with the suggestion that MCK-1 functions as a positive-acting chemokine. When a number of cell lines expressing individual chemokine receptors were assayed for responsiveness to MCK-1, only human CCR3, a natural receptor for CCL11 (eotaxin), CCL5 (RANTES), and CCL7 (MCP-3), was found to mediate an MCK-1-dependent response (SAEDERUP et al. 1999). Cross-ligand desensitization on CCR3 cells showed that MCK-1-induced signaling via human CCR3 was specific, because prior exposure to human CCL11 (eotaxin) completely abrogated any MCK-1 response. Synthetic MCK-1 also induced a response on the THP-1 human monocyte cell line. Cells bearing other chemokine receptors, including human CCR1, CCR2, and CCR5, murine CCR3, or human CMV US28, neither responded to MCK-1 nor exhibited altered response to a subsequent positive-control chemokine (SAEDERUP et al. 1999). MCK-1 does not exhibit antagonist activity under any circumstances and is neither an agonist nor antagonist on cells bearing murine CCR3, suggesting that this is not a natural receptor for this chemokine. It seems likely that MCK-1 engages a murine chemokine receptor with structural properties related to human CCR3, although the identity of this receptor remains to be determined. It is interesting to note that human THP-1 cells, which respond to MCK-1, express CCR1, CCR2, CCR5, CCR8, and CX3CR but not CCR3, CCR4, CCR6, or CCR7 (Gosling and Schall, personal communication). Given the THP-1 chemokine receptor profile, CCR8 represents a potential MCK-1 receptor. CCR8 is expressed in murine monocyte-macrophage lineage cells (GOYA et al. 1998; LUO and DORF 1996).

Altogether, these observations suggest that MCK-1, the chemokine domain of MCK-2, retains agonist activity that targets murine monocytes or macrophages at sites of murine CMV infection in the host. Similar activity would be expected from the MCK-2 protein; however, this protein has not been evaluated directly. Recent work (CHENG et al. 2001; INNGJERDINGEN et al. 2001; SALLUSTO et al. 2000) has highlighted the importance of both activation and differentiation on the profile of chemokine receptors displayed on NK cells, dendritic cells, lymphocytes, and monocytes. MCK-2 activity in vivo is predicted to be directed at monocytes and/or macrophages based on the behavior of synthetic MCK-1, although other leukocyte subsets may be influenced by this function depending on expression of an appropriate receptor.

3.4 Investigations of *mck*-Mutant Viruses

The first studies that employed an *mck*-mutant virus were published before the recognition that this gene encoded a chemokine homolog (CARDIN et al. 1993; STODDART et al. 1994). This mutant virus, designated RM461, contained a *lacZ* reporter gene inserted in the region between m131 and m129, resulting in disruption of the *mck* transcript. RM461 replicated as well as wild-type virus strain K181$^+$ in NIH 3T3 fibroblasts, allowing the conclusion that this region was not required for growth in tissue culture cells (CARDIN et al. 1993; STODDART et al. 1994; CARDIN et al. 1995). This conclusion was supported by further work after *mck* was predicted to encode a chemokine homolog (MACDONALD et al. 1997). After inoculation into BALB/c mice, RM461 grew as well as wild-type virus in most organs tested, including spleen, liver, adrenal gland, and lungs, but replication in the salivary glands, where titers peak between days 14 and 21 p.i., was reduced by two to three orders of magnitude (CARDIN et al. 1993; STODDART et al. 1994; FLEMING et al. 1999; SAEDERUP et al. 1999). This same phenotype was observed after inoculation of BALB/c (*nu/nu*) or C.B17 (*scid/scid*) mice, indicating that the phenotype is independent of B- or T lymphocyte function (CARDIN et al. 1993; STODDART et al. 1994). A different pattern was reported by FLEMING et al. (1999) in which *mck* mutants were cleared more rapidly than wild-type virus from spleen and liver. Differences in virus titers in these organs at 5 days p.i. was further evaluated in animals depleted with anti-asialoGM1 or anti-CD4 plus anti-CD8 antibodies. Either of these treatments partially restored *mck*-mutant virus growth in the spleen but did not increase titers in the salivary glands at later times. Depletion studies using anti-CD4 plus anti CD8 antibodies suggested an impact of *mck* on the adaptive immune response; however, more recent evaluation comparing *mck*-mutant viruses to control viruses did not find any consistent indication that clearance of mutant was more rapid (SAEDERUP et al. 2001). The reduced levels of *mck*-mutant virus in salivary glands can first be detected by 7 days p.i., soon after initial seeding of this tissue and soon after the peak viremia that is responsible for disseminating the virus to this tissue. The impact of *mck* on virus dissemination to or replication in salivary glands occurred well in advance of immune clearance of infectious virus from this tissue, which starts at about 3 weeks p.i. in BALB/c mice. When the levels of infectious virus in salivary glands was evaluated over a 6-week time course, mutant virus remained at significantly lower levels than wild type but exhibited a similar pattern of clearance (CARDIN et al. 1993). These data, together with the behavior of *mck*-mutant virus in immunodeficient strains of mice, suggested that *mck* does not alter host clearance mechanisms. Thus, in balance, experiments comparing *mck* mutants to rescued viruses made in the same virus strain (K181) and evaluated in BALB/c mice (FLEMING et al. 1999; SAEDERUP et al. 2001) have consistently found that *mck* expression is important to have maximal dissemination to or replication in salivary glands. Mouse age remains a potentially important difference in the studies that have been reported, and this characteristic is known to impact the course of infection (OSBORN 1982). Finally, all *mck* mutants appear to be as capable of entering latency as wild-type viruses based on experiments after spleen explant reactivation in tissue culture or reactivation in vivo (CARDIN et al. 1993; STODDART et al. 1994; Lin, unpublished data).

Murine CMV recombinants carrying the *lac*Z reporter insert allow tracking of infection in the host (STODDART et al. 1994). The viral *ie2* gene is completely dispensable for replication in tissue culture cells as well as for replication, dissemination, transmission, and latency in the mouse (CARDIN et al. 1993, 1995). Viruses with marker genes at this site have been used to study the tissue distribution of *mck*-mutant and control viruses. Tissue sections made from salivary glands infected with *lac*Z-tagged *mck*-expressing virus were compared with tissues from *mck*-mutant virus-infected mice by staining for expression of β-galactosidase. Mutant virus foci of infection were similar in size to control virus but were present in reduced numbers (Lin, Lagenaur and Mocarski, unpublished data), suggesting that *mck*-mutant viruses were deficient in their ability to seed salivary glands. When levels of peripheral blood leukocyte-associated viremia were examined by cocultivation with permissive NIH 3T3 cell monolayers, *mck*-mutant virus exhibited 20- to 50-fold lower peak than control virus (SAEDERUP et al. 1999, 2001). The inability of *mck*-mutant viruses to reach levels of viremia as high as control *mck*-expressing viruses correlated with the reduced numbers of foci and titers in the salivary glands. When mice were coinfected with wild-type and *mck*-mutant viruses together, both reached the high peak viremia levels of wild type, consistent with a role for MCK-2 as a secreted factor capable of complementing mutant virus behavior in *trans* (SAEDERUP et al. 1999). In recent work (SAEDERUP et al. 2001), coinfection with wild-type and *mck*-mutant viruses was shown to complement the mutant phenotype in salivary glands, further supporting a link between levels of viremia and efficiency of seeding salivary glands. Combined with the observations on MCK-1 activity on macrophages and monocytes, this work suggests that the function of *mck* is to alter the behavior of monocyte/macrophages in a manner that enhances viremia. One prediction from all of the previous work is that *mck* would display a proinflammatory activity in vivo.

Recent work (SAEDERUP et al. 2001) has shown that expression of *mck* leads to increased inflammation in footpads 2 days after inoculation. Levels of replication by these viruses at the site of inoculation and in the draining lymph nodes are equivalent, suggesting that the difference in inflammation does not result from varying amounts of virus or viral antigen. When administered to mouse footpads, *mck*-mutant virus exhibits less than 50% of the swelling and markedly fewer infiltrating leukocytes (SAEDERUP et al. 2001) at day 2 or 3 p.i. Histological evaluation reveals less cellular infiltration, reduced necrosis, and a lower level of edema in mutant virus-infected tissues. Interestingly, the work of FLEMING et al. (1999) included the observation that *mck*-mutant virus induced fewer and smaller inflammatory foci relative to wild-type virus in livers at times when mutant and wild-type viral titers in this organ were equivalent. All of these effects appear to be consistent with a proinflammatory, chemokine-like activity of the *mck* gene product. A recombinant virus with a mutated *mck* CC chemokine motif exhibits the same poor inflammation, viremia, and dissemination phenotypes as other *mck* mutants, highlighting the connection between phenotype and likely chemokine activity (SAEDERUP et al. 2001). The levels of viral replication in other tissues remained the same as control viruses as in previous work (CARDIN et al. 1993; SAEDERUP et al. 1999; STODDART et al. 1994). All of these data support a model in which the *mck* gene product increases local inflammation to enhance the reservoir of mononuclear cells that can taxi the virus via the blood stream to the salivary glands.

3.5 Concluding Remarks on *mck* Function In Vivo

It appears that murine CMV has evolved a strategy of provoking an marked inflammatory response via the expression of *mck*. This is an unusual viral strategy; the vast majority of characterized viral immunomodulatory functions are believed to decrease immune responses, not to provoke them (ALCAMI and KOSZINOWSKI 2000). *mck*-Enhanced inflammation appears to be important to achieve a peak viremia for dissemination to the salivary glands, where virus replication leads to shedding and facilitates transmission to new hosts. The ability of the MCK-1 polypeptide to activate peritoneal macrophages in vitro, together with the observed impact of *mck* on the course of viral infection in vivo (SAEDERUP et al. 2001; SAEDERUP et al. 1999), suggest a link between *mck*-induced local inflammation and efficient dissemination to the salivary glands. The major *mck* gene product, MCK-2, may function to (1) recruit a specific subset of inflammatory cells, (2) enhance overall levels of inflammation, (3) alter the permissiveness of leukocytes for infection, or (4) alter the trafficking behavior of infected leukocytes (Fig. 2). Any or all of these effects would serve to enhance levels of viremia. The ability of wild-type virus to complement the reduced titers of virus in both the peripheral blood and salivary glands suggests that the two phenomena are linked.

The variable observations regarding the impact of *mck* on immune clearance in spleen and liver remain an issue that needs to be resolved. Adaptive and NK cell-

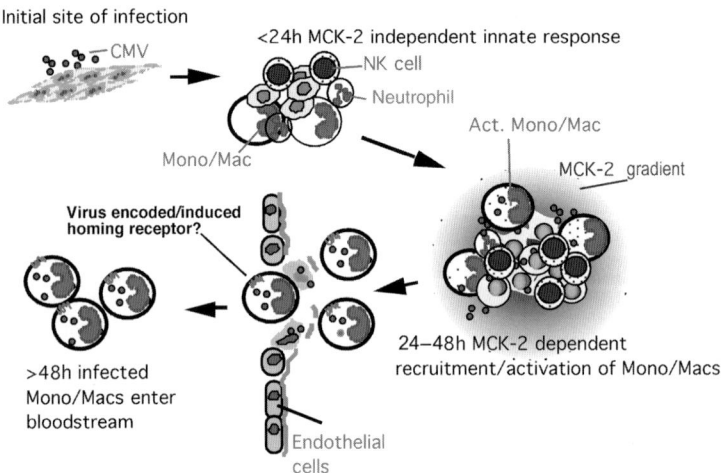

Fig. 2. Model for impact of MCK-2 at sites of viral infection. After inoculation of murine CMV, virus gene expression and DNA replication ensues within the first 24h and is accompanied by an innate inflammatory response. As virus-infected cells begin producing progeny virus, MCK-2 is orchestrating the recruitment of a permissive monocyte/macrophage (*Mono/Mac*) population. This attraction confers in higher levels of viremia and more efficient dissemination on wild-type viruses compared with *mck* mutants. This population is shown to be directed efficiently into the bloodstream by an as yet undefined processs

mediated immune clearance mechanisms are not responsible for the *mck* impact on dissemination to salivary glands (FLEMING et al. 1999; STODDART et al. 1994). This, combined with the initial manifestation of the phenotype in salivary glands by 7 days p.i (CARDIN et al. 1993), strongly suggests that the role of *mck* is in ensuring levels of viremia that can sustain maximal dissemination. Further elucidation of the mechanism(s) of *mck* function will require additional characterization of the chemokine receptor(s) engaged by MCK-2, as well as the identification of cell types recruited in an MCK-2-dependent fashion to sites of infection that appear in the peripheral blood during peak viremia.

4 The Human CMV Chemokine Homologs vCXCL-1 and vCXCL-2

Early attempts to derive human CMV strains for vaccine purposes produced two high-passage laboratory strains that are attenuated for humans and are popular for laboratory investigations (ELEK and STERN 1974; HO 1991; PLOTKIN et al. 1989). Comparisons of the genomes of the highly passaged AD169 and Towne strains with the Toledo strain, which exhibits definite virulence characteristics not observed with attenuated strains (PLOTKIN et al. 1989) led to the discovery of a cluster of CMV genes that had been lost or modified during tissue culture passage. Two of the genes that were present in Toledo but completely lost from AD169 were the adjacent ORFs UL146 and UL147 (CHA et al. 1996). Both ORFs exhibited C spacing that was reminiscent of CXC chemokines (PENFOLD et al. 1999). The adjacent ORFs may have originated from a duplication event during evolution; however, UL146 and UL147 retain very limited homology to each other and remarkably little similarity to known mammalian CXC chemokines. The absence of UL146 and UL147 from the laboratory-adapted strain AD169 genome does not affect viral replication levels in fibroblasts. An impact on infection in the natural host is likely but difficult to assess experimentally because of the species specificity of this virus.

Human CMV replication in the natural host is not well understood. Viral pathogenesis, tropism for epithelial and endothelial cells, and immune clearance mechanisms are generally believed to be similar to murine CMV, but there are important distinctions as well (KOSZINOWSKI et al. 1990; PASS 2001). In natural infections with human CMV, viremia persists for months to years after initial infection, accompanied by persistent replication and asymptomatic shedding of virus in both saliva and urine (REVELLO et al. 1998). Eventually, viral replication is believed to be brought under control by the development of an adaptive immune response where the contribution of a virus-specific CTL response to clearance is well documented in patient populations (RIDDELL and GREENBERG 1997). Virus remains latent in hematopoietic progenitors, sporadically reactivates, and is shed in saliva.

Human CMV viremia also appears to differ from murine CMV in that neutrophils, in addition to monocytes and endothelial cells, have been ascribed a role as carriers of infectious virus (PASS 2001). Human CMV interacts and intersects with neutrophils in a number of interesting, although unconventional ways. Whereas human CMV replicates in subsets of endothelial cells and monocytes, neutrophils do not support productive replication but can be passive carriers of infectious virus (GRUNDY et al. 1998). Although best characterized in immunocompromised individuals who have high levels of viremia, the mechanisms of this interaction and the nature of the infectivity of virus-positive neutrophils are still unclear. As many as 0.1% of peripheral blood neutrophils can be CMV positive in immunocompromised patients (SALTZMAN et al. 1988). Neutrophil-associated pathologies such as retinitis and pneumonitis occur in these patients (DUGGAN et al. 1986; PEPOSE et al. 1985). These diseases are mediated by active CMV replication, as they can be treated with drugs that target the viral DNA polymerase (PASS 2001). An association has also been found between CMV and vascular disease, particularly arteriosclerosis after cardiac transplantation, but also possibly restenosis after coronary angioplasty (CHENG and RIVERA 1998; PASS 2001). A common element of all these diseases is the inflammatory insult leading to proliferation and/or activation of immune effector cells. One proposed mechanism suggests that CMV, carried by endothelial cells or monocytes, reactivates during the inflammatory processes characteristic of vascular injury (SODERBERG-NAUCLER et al. 1997; GUETTA et al. 1997; STREBLOW et al. 1999), resulting in an exacerbation of inflammation. Both human and murine CMV encode genes that regulate interaction with endothelial cells and macrophages (BRUNE et al. 2001; HANSON et al.

1999; KAHL et al. 2000; SAEDERUP et al. 1999; SINZGER et al. 1999, 2000; TOMAZIN et al. 1999). These genes may be expected to influence pathogenesis of CMV disease (HANSON et al. 1999; SAEDERUP et al. 1999). Human CMV encodes at least one chemokine that has the potential to influence all of the cell types associated with viremia.

4.1 Expression and Structure of vCXCL-1 and vCXCL-2

UL146 and UL147 have conserved chemokine motifs including signal peptide sequences, four appropriately spaced cysteines, and size (predicted sizes of 117 aa and 159 aa, respectively). The predicted product of UL146, designated vCXCL-1, contains an "ELR" (glutamic acid-leucine-arginine) motif adjacent to the amino-terminal CXC motif and characteristic of neutrophil-attracting chemokines (BAGGIOLINI et al. 1997). The product of UL147, designated vCXCL-2, does not contain an ELR motif but does have four cysteines reminiscent of non-ELR CXC chemokines. This section will focus on the known properties of vCXCL-1, the product of UL146 that behaves as a positive-acting chemokine (PENFOLD et al. 1999).

UL146 mRNA is expressed as a true late (γ2) gene whose product is secreted from cultures of infected fibroblasts and influences the neutrophil-attracting chemokine activity of conditioned medium (PENFOLD et al. 1999). The protein encoded by UL146 has three predicted N-linked glycosylation sites. Investigation of recombinant CMV expressing FLAG epitope-tagged UL146 protein by immunoblot analysis with or without PNGase F digestion revealed that the predominant gene product is glycosylated at two sites and exhibits a size ($\sim$14kDa) consistent with the removal of a signal-sequence after translation.

4.2 vCXCL-1 Chemokine Activities

Supernatant from cells infected with strain Toledo, which has a complete UL146 gene, induced a fourfold greater level of neutrophil chemotaxis than did mock-infected cell supernatants (PENFOLD et al. 1999). Supernatants from cells infected with recombinant Toledo strains carrying mutations affecting vCXCL-1 displayed a reduced ability to recruit neutrophils. These results suggest that UL146 expression contributed to the induction of neutrophil chemotaxis. To investigate the chemokine properties of this protein more directly, two recombinant proteins, starting at one of two predicted signal sequence cleavage sites, Y_{18} and T_{23}, were prepared by expression cloning into *E. coli*. These proteins were assayed for binding, calcium mobilization, and chemotaxis activity on human peripheral blood neutrophils and on a panel of cell lines expressing known CC, CXC, XC, or CX_3C receptors, including human CMV US28. The T_{23} form of recombinant UL146 protein strongly competes for IL-8 binding ($K_I = 3$nM) and signals specifically through CXCR2, one of two IL-8 receptors. UL146 protein does not interact with the other IL8 receptor, CXCR1, and so is similar to GROα in using only CXCR2. On the basis of the behavior of the T_{23} form of UL146 protein, this gene was denoted vCXCL-1 as the first representative of a CXC chemokine to be characterized in a virus. vCXCL-1 is as potent as IL-8 for signaling on CXCR2 and fails to compete with the natural ligands for any of 16 other human chemokine receptors. The T_{23} form of vCXCL-1 induced neutrophil chemotaxis as potently as the ELRCXC chemokines IL-8 and GROα. In contrast, the Y_{18} form of vCXCL-1 has not shown activity in a variety of chemotaxis and binding assays seeking to reveal its role as an agonist or antagonist. These studies clearly show that vCXCL-1 is a proinflammatory chemokine that influences the migration of neutrophils and would be expected to impact trafficking behavior of any CXCR2-bearing lineage of leukocyte (PENFOLD et al. 1999).

4.3 Concluding Remarks on Human CMV Chemokines

As expected from studies on mammalian ELRCXC chemokines, vCXCL-1 can activate neutrophils via binding to CXCR2, a major ELRCXC chemokine receptor (MURPHY et al. 2000). According to recent reports, activated monocytes and endothelial cells can be induced to express CXCR2 (BOISVERT et al. 2000; BONECCHI et al. 2000; MURDOCH et al. 1999). These observations are intriguing because vCXCL-1 may attract and/or activate cell types other than neutrophils, possibly those that play a more direct role as hosts for CMV replication. Human CMV may use vCXCL-1 to recruit leukocyte to sites of infection to enhance viral dissemination, analogous to the role of *mck* in murine CMV, or to attract hematopoietic cells that become latently infected. Future research will determine whether the activities of vCXCL-1 extend beyond attracting neutrophils and include the ability to modulate the behavior of monocytes, macrophages, or endothelial cells. Unfortunately, the murine receptor that recognizes human IL-8 does not bind vCXCL-1 (Sparer and Gossling, unpublished data), complicating direct evaluation of vCXCL-1 function in the mouse. The aa sequence of this chemokine shows substantial divergence from human ELRCXC chemokines, which likely contributes to its failure to bind any receptors in mouse leukocytes.

Interestingly, UL146 genes from different CMV isolates show a remarkable degree of strain-to-strain aa sequence variability. All vCXC-1 isolates preserve the ELRCXC motif and fall into identifiable clusters but display as much as 50% variation in aa sequence. This gene is therefore one of the most variable CMV genes (Penfold, personal communication). The biological relevance of this diversity will not be revealed until UL146 from a number of strains are functionally characterized, although it is already clear that, given the poor conservation of aa sequence to human ELRCXC chemokines whose function it mimics, vCXCL-1 primary sequence says very little about conserved function. The chemokine homolog adjacent to UL146 on the viral genome, UL147 (vCXCL-2), exhibits a higher level of sequence conservation and is more than 90% identical in strains that exhibit striking divergence in UL146 sequence. How these two human CMV chemokine homologs contribute to pathogenesis of this virus remains an intriguing question.

Although clearly a potent chemokine, a role for vCXCL-1 or the more poorly characterized vCXCL-2 in viral pathogenesis, virulence, or latency remains difficult to assess without an animal model. Human CMV-associated diseases in immunocompromised hosts, including retinitis, gastroenteritis, and posttransplant arteriosclerosis, certainly display a prominent involvement of neutrophils, endothelial cells, and monocytes. The extent to which elaboration of vCXCL-1 contributes to increasing inflammation that has pathological consequences in disease states remains to be determined. It may be possible to study the impact of vCXCL-1-like chemokine homologs in animal species that are more closely related to humans. Rhesus macaque or African green monkey cytomegalovirus may be predicted to contain at least two genes, one possibly related to human CMV UL146 and the other related to UL147, based on analysis of published nucleotide sequence (Saederup, unpublished data) of a CMV of likely simian origin (MARTIN 1999). The presence of other immunomodulatory genes such as US28 and IL-10 homologs in cytomegaloviruses of simian origin (LOCKRIDGE et al. 2000; MARTIN 1999, 2000) suggests that their modulatory functions are closer to human CMV than are the rodent viruses.

5 Conclusion

Research on the functional activities of cytomegalovirus chemokine homologs has just begun, yet results have already provided a paradigm shift in how virus-encoded immunomodulatory functions control interaction with the host. Rather than having a primary role dedicated to downmodulating innate or adaptive components of the host immune response, human and murine CMV both encode proinflammatory chemokine homologs that attract host cells otherwise known for their ability to eliminate invading pathogens. How either virus balances the attraction of cells that should constitute a threat with the use of these cells to disseminate will require considerably more understanding of responsive cell types. In the case of murine CMV, the viral CC chemokine homolog MCK-2 seems to attract cells that help disseminate the virus without increasing clearance of infectious virus or virus-infected cells at initial sites of infection or in a variety of organs other than the salivary glands. This activity optimizes the cellular response, facilitating viremia that promotes dissemination to the salivary glands. Whether the cells that are attracted carry virus from sites of infection via draining lymph or by entering the blood stream is still unclear. No matter the course of dissemination, replication in the salivary glands produces the progeny that allows more efficient spread to additional host animals in the population. The murine CMV chemokine homolog has properties consistent with a chemokine-like role in natural infection. Human CMV vCXCL-1 exhibits all of the known properties of a bona fide chemokine but cannot be evaluated during infection in the natural host. The more we learn about the cellular targets and activities of these CMV-encoded chemokines, the better we will appreciate the manner in which CMV controls the interaction with its host.

With our current knowledge, it is reasonable to speculate that the human CMV chemokines may influence viremia and dissemination in much the same way as *mck* affects dissemination, via the recruitment or activation of neutrophils as well as perhaps monocytes and other cells. The chemokines of murine and human CMV appear to have been acquired independently rather than to have evolved from a common ancestor. Other positive-acting chemokines have been acquired by gammaherpesviruses such as KSHV, where three CC chemokine homologs unrelated to MCK-2 display diverse activities ranging from broad-spectrum antagonism (vMIP-II) (KLEDAL et al. 1997) to receptor-specific CCR8 (vMIP-I) and CCR4 (vMIP-III) signaling (DAIRAGHI et al. 1999; ENDRES et al. 1999; STINE et al. 2000). Other betaherpesviruses such as HHV-6, and even alphaherpesviruses such as Marek's disease virus encode chemokine homologs (LIU et al. 1999; ZOU et al. 1999). Clarifying the roles of any of these chemokine homologs in pathogenesis will help provide clues to common properties in the many herpesviruses that encode this class of function.

The investigation of viral genes that modulate the immune system has provided insights into viral pathogenesis and leads to possible novel therapeutic agents (ALCAMI and KOSZINOWSKI 2000; DEBRUYNE et al. 2000). The propensity for herpesviruses and poxviruses to acquire chemokine and chemokine receptor homologs underscores the central position that the chemokine network holds in the host defense against viruses and microbes. It also gives evidence of the ability of viruses to adapt their hosts' defenses to their own advantage, as in the case of the murine CMV MCK-2 functioning to promote a "Trojan horse" strategy of leukocyte-associated dissemination. Clearly, this strategy would only work for a virus that comes equipped to modulate host immune clearance at many levels. The extent to which chemokine homologs work together with other viral immunomodulatory functions to orchestrate the successful infection of a new host remains among the most intriguing problems in viral pathogenesis. Can the evolutionary change possible in the ancient relationship that herpesviruses enjoy with vertebrate hosts mean that these viruses outwit the adaptability of the mammalian immune system? It may all rest in the balance.

References

Ahn K, Angulo A, Ghazal P, Peterson PA, Yang Y, Fruh K (1996) Human cytomegalovirus inhibits antigen presentation by a sequential multistep process. Proc Natl Acad Sci USA 93:10990–10995
Alcami A, Koszinowski UH (2000) Viral mechanisms of immune evasion. Immunol Today 21:447–455
Andrews DM, Farrell HE, Densley EH, Scalzo AA, Shellam GR, Degli-Esposti MA (2001) NK1.1(+) Cells and murine cytomegalovirus infection: What happens in situ? J Immunol 166:1796–1802

Baggiolini M, Dewald B, Moser B (1997) Human chemokines: an update. Annu Rev Immunol 15: 675–705
Baggiolini M, Loetscher P (2000) Chemokines in inflammation and immunity. Immunol Today 21: 418–420
Bale JF, Jr., O'Neil ME (1989) Detection of murine cytomegalovirus DNA in circulating leukocytes harvested during acute infection of mice. J Virol 63:2667–2673
Bazan JF, Bacon KB, Hardiman G, Wang W, Soo K, Rossi D, Greaves DR, Zlotnik A, Schall TJ (1997) A new class of membrane-bound chemokine with a CX3C motif. Nature 385:640–644
Boisvert WA, Curtiss LK, Terkeltaub RA (2000) Interleukin-8 and its receptor CXCR2 in atherosclerosis. Immunol Res 21:129–137
Bonecchi R, Facchetti F, Dusi S, Luini W, Lissandrini D, Simmelink M, Locati M, Bernasconi S, Allavena P, Brandt E, Rossi F, Mantovani A, Sozzani S (2000) Induction of functional IL-8 receptors by IL-4 and IL-13 in human monocytes. J Immunol 164:3862–3869
Boshoff C, Endo Y, Collins PD, Takeuchi Y, Reeves JD, Schweickart VL, Siani MA, Sasaki T, Williams TJ, Gray PW, Moore PS, Chang Y, Weiss RA (1997) Angiogenic and HIV-inhibitory functions of KSHV-encoded chemokines. Science 278:290–294
Brune W, Menard C, Heesemann J, Koszinowski UH (2001) A ribonucleotide reductase homolog of cytomegalovirus and endothelial cell tropism. Science 291:303–305
Campbell JJ, Butcher EC (2000) Chemokines in tissue-specific and microenvironment-specific lymphocyte homing. Curr Opin Immunol 12:336–341
Campbell JJ, Hedrick J, Zlotnik A, Siani MA, Thompson DA, Butcher EC (1998) Chemokines and the arrest of lymphocytes rolling under flow conditions. Science 279:381–384
Cardin RD, Abenes GB, Stoddart CA, Mocarski ES (1995) Murine cytomegalovirus IE2, an activator of gene expression, is dispensable for growth and latency in mice. Virology 209:236–241
Cardin RD, Boname JM, Abenes GB, Jennings SA, Mocarski ES (1993) Reactivation of murine cytomegalovirus from latency. In: Plotkin S, Michelson S (eds) Multidisciplinary Approaches to Understanding Cytomegalovirus Disease. Elsevier, Amsterdam, pp 101–110
Cha TA, Tom E, Kemble GW, Duke GM, Mocarski ES, Spaete RR (1996) Human cytomegalovirus clinical isolates carry at least 19 genes not found in laboratory strains. J Virol 70:78–83
Cheng JW, Rivera NG (1998) Infection and atherosclerosis–focus on cytomegalovirus and *Chlamydia pneumoniae*. Ann Pharmacother 32:1310–1316
Cheng SS, Lai JJ, Lukacs NW, Kunkel SL (2001) Granulocyte-macrophage colony stimulating factor up-regulates CCR1 in human neutrophils. J Immunol 166:1178–1184
Collins TM, Quirk MR, Jordan MC (1994) Biphasic viremia and viral gene expression in leukocytes during acute cytomegalovirus infection of mice. J Virol 68:6305–6311
Dairaghi DJ, Fan RA, McMaster BE, Hanley MR, Schall TJ (1999) HHV8-encoded vMIP-I selectively engages chemokine receptor CCR8. Agonist and antagonist profiles of viral chemokines. J Biol Chem 274:21569–21574
Dairaghi DJ, Greaves DR, Schall TJ (1998) Abduction of chemokine elements by herpesviruses. Sem Virol 8:377–385
DeBruyne LA, Li K, Bishop DK, Bromberg JS (2000) Gene transfer of virally encoded chemokine antagonists vMIP-II and MC148 prolongs cardiac allograft survival and inhibits donor-specific immunity. Gene Ther 7:575–582
Duggan MA, Pomponi C, Robboy SJ (1986) Pulmonary cytology of the acquired immune deficiency syndrome: an analysis of 36 cases. Diagn Cytopathol 2:181–186
Elek SD, Stern H (1974) Development of a vaccine against mental retardation caused by cytomegalovirus infection in utero. Lancet 1:1–5
Endres MJ, Garlisi CG, Xiao H, Shan L, Hedrick JA (1999) The Kaposi's sarcoma-related herpesvirus (KSHV)-encoded chemokine vMIP- I is a specific agonist for the CC chemokine receptor (CCR)8. J Exp Med 189:1993–1998
Farrell HE, Vally H, Lynch DM, Fleming P, Shellam GR, Scalzo AA, Davis-Poynter NJ (1997) Inhibition of natural killer cells by a cytomegalovirus MHC class I homologue in vivo. Nature 386: 510–514
Fleming P, Davis-Poynter N, Degli-Esposti M, Densley E, Papadimitriou J, Shellam G, Farrell H (1999) The murine cytomegalovirus chemokine homolog, m131/129, is a determinant of viral pathogenicity. J Virol 73:6800–6809
Goya I, Gutierrez J, Varona R, Kremer L, Zaballos A, Marquez G (1998) Identification of CCR8 as the specific receptor for the human beta- chemokine I-309: cloning and molecular characterization of murine CCR8 as the receptor for TCA-3. J Immunol 160:1975–1981

Grundy JE, Lawson KM, MacCormac LP, Fletcher JM, Yong KL (1998) Cytomegalovirus-infected endothelial cells recruit neutrophils by the secretion of C-X-C chemokines and transmit virus by direct neutrophil-endothelial cell contact and during neutrophil transendothelial migration. J Infect Dis 177:1465–1474

Guetta E, Guetta V, Shibutani T, Epstein SE (1997) Monocytes harboring cytomegalovirus: interactions with endothelial cells, smooth muscle cells, and oxidized low-density lipoprotein. Possible mechanisms for activating virus delivered by monocytes to sites of vascular injury. Circ Res 81:8–16

Hahn G, Jores R, Mocarski ES (1998) Cytomegalovirus remains latent in a common precursor of dendritic and myeloid cells. Proc Natl Acad Sci USA 95:3937–3942

Hanson LK, Slater JS, Karabekian Z, Virgin HWt, Biron CA, Ruzek MC, van Rooijen N, Ciavarra RP, Stenberg RM, Campbell AE (1999) Replication of murine cytomegalovirus in differentiated macrophages as a determinant of viral pathogenesis. J Virol 73:5970–5980

Haskell CA, Cleary MD, Charo IF (2000) Unique role of the chemokine domain of fractalkine in cell capture. J Biol Chem 275:34183–34189

Hengel H, Brune W, Koszinowski UH (1998) Immune evasion by cytomegalovirus – survival strategies of a highly adapted opportunist. Trends Microbiol 6:190–197

Hengel H, Reusch U, Gutermann A, Ziegler H, Jonjic S, Lucin P, Koszinowski UH (1999) Cytomegaloviral control of MHC class I function in the mouse. Immunol Rev 168:167–176

Henson D, Strano AJ (1972) Mouse cytomegalovirus. Clin J Path 68:183–195

Ho M (1991) Cytomegalovirus: Biology and Infection, 2nd Edition. Plenum Publishing Corp, New York

Imai T, Hieshima K, Haskell C, Baba M, Nagira M, Nishimura M, Kakizaki M, Takagi S, Nomiyama H, Schall TJ, Yoshie O (1997) Identification and molecular characterization of fractalkine receptor CX3CR1, which mediates both leukocyte migration and adhesion. Cell 91:521–530

Inngjerdingen M, Damaj B, Maghazachi AA (2001) Expression and regulation of chemokine receptors in human natural killer cells. Blood 97:367–375

Jones TR, Sun L (1997) Human cytomegalovirus US2 destabilizes major histocompatibility complex class I heavy chains. J Virol 71:2970–9

Jonjic S, Mutter W, Weiland F, Reddehase MJ, Koszinowski UH (1989) Site-restricted persistent cytomegalovirus infection after selective long-term depletion of CD4+ T lymphocytes. J Exp Med 169:1199–1212

Jonjic S, Pavic I, Polic B, Crnkovic I, Lucin P, Koszinowski UH (1994) Antibodies are not essential for the resolution of primary cytomegalovirus infection but limit dissemination of recurrent virus. J Exp Med 179:1713–1717

Kahl M, Siegel-Axel D, Stenglein S, Jahn G, Sinzger C (2000) Efficient lytic infection of human arterial endothelial cells by human cytomegalovirus strains. J Virol 74:7628–7635

Kledal TN, Rosenkilde MM, Coulin F, Simmons G, Johnsen AH, Alouani S, Power CA, Luttichau HR, Gerstoft J, Clapham PR, Clark-Lewis I, Wells TNC, Schwartz TW (1997) A broad-spectrum chemokine antagonist encoded by Kaposi's sarcoma-associated herpesvirus. Science 277:1656–1659

Knall C, Young S, Nick JA, Buhl AM, Worthen GS, Johnson GL (1996) Interleukin-8 regulation of the Ras/Raf/mitogen-activated protein kinase pathway in human neutrophils. J Biol Chem 271:2832–2838

Koffron AJ, Hummel M, Patterson BK, Yan S, Kaufman DB, Fryer JP, Stuart FP, Abecassis MI (1998) Cellular localization of latent murine cytomegalovirus. J Virol 72:95–103

Kondo K, Kaneshima H, Mocarski ES (1994) Human cytomegalovirus latent infection of granulocyte-macrophage progenitors. Proc Natl Acad Sci USA 91:11879–11883

Koszinowski UH, del Val M, Reddehase MJ (1990) Cellular and molecular basis of the protective immune response to cytomegalovirus infection. Curr Top Microbiol Immunol 154:189–220

Krmpotic A, Messerle M, Crnkovic-Mertens I, Polic B, Jonjic S, Koszinowski UH (1999) The immunoevasive function encoded by the mouse cytomegalovirus gene m152 protects the virus against T cell control in vivo. J Exp Med 190:1285–1296

Kuschert GS, Coulin F, Power CA, Proudfoot AE, Hubbard RE, Hoogewerf AJ, Wells TN (1999) Glycosaminoglycans interact selectively with chemokines and modulate receptor binding and cellular responses. Biochemistry 38:12959–12968

Leong CC, Chapman TL, Bjorkman PJ, Formankova D, Mocarski ES, Phillips JH, Lanier LL (1998) Modulation of natural killer cell cytotoxicity in human cytomegalovirus infection: the role of endogenous class I major histocompatibility complex and a viral class I homolog. J Exp Med 187:1681–1687

Li Z, Jiang H, Xie W, Zhang Z, Smrcka AV, Wu D (2000) Roles of PLC-beta2 and -beta3 and PI3K-gamma in chemoattractant-mediated signal transduction. Science 287:1046–1049

Liu JL, Lin SF, Xia L, Brunovskis P, Li D, Davidson I, Lee LF, Kung HJ (1999) MEQ and V-IL8: cellular genes in disguise? Acta Virol 43:94–101

Lockridge KM, Zhou SS, Kravitz RH, Johnson JL, Sawai ET, Blewett EL, Barry PA (2000) Primate cytomegaloviruses encode and express an IL-10-like protein. Virology 268:272–280

Loetscher P, Seitz M, Clark-Lewis I, Baggiolini M, Moser B (1996) Activation of NK cells by CC chemokines. Chemotaxis, Ca2+ mobilization, and enzyme release. J Immunol 156:322–327

Lucin P, Pavic I, Polic B, Jonjic S, Koszinowski UH (1992) Gamma interferon-dependent clearance of cytomegalovirus infection in salivary glands. J Virol 66:1977–1984

Luo Y, Dorf ME (1996) Beta-chemokine TCA3 binds to mesangial cells and induces adhesion, chemotaxis, and proliferation. J Immunol 156:742–748

MacDonald MR, Burney MW, Resnick SB, Virgin HI (1999) Spliced mRNA encoding the murine cytomegalovirus chemokine homolog predicts a beta chemokine of novel structure. J Virol 73:3682–3691

MacDonald MR, Li XY, Virgin HWt (1997) Late expression of a beta chemokine homolog by murine cytomegalovirus. J Virol 71:1671–1678

Martin WJ (1999) Melanoma growth stimulatory activity (MGSA/GRO-alpha) chemokine genes incorporated into an African green monkey simian cytomegalovirus-derived stealth virus. Exp Mol Pathol 66:15–18

Martin WJ (2000) Chemokine receptor-related genetic sequences in an african green monkey simian cytomegalovirus-derived stealth virus. Exp Mol Pathol 69:10–16

McColl SR, Roberge CJ, Larochelle B, Gosselin J (1997) EBV induces the production and release of IL-8 and macrophage inflammatory protein-1 alpha in human neutrophils. J Immunol 159:6164–6168

Mocarski ES, Courcelle CT (2001) Cytomegaloviruses and their replication. In: Knipe DM, Howley PM (eds) Fields Virology. 4th edn. Lippincott Williams and Wilkins Publishers, Philadelphia, pp 2629–2673

Mocarski ES, Abenes GB, Manning WC, Sambucetti LC, Cherrington JM (1990) Molecular genetic analysis of cytomegalovirus gene regulation in growth, persistence and latency. Curr Top Microbiol Immunol 154:47–74

Murdoch C, Monk PN, Finn A (1999) CxC chemokine receptor expression on human endothelial cells. Cytokine 11:704–712

Murphy PM, Baggiolini M, Charo IF, Hebert CA, Horuk R, Matsushima K, Miller LH, Oppenheim JJ, Power CA (2000) International union of pharmacology. XXII. Nomenclature for chemokine receptors. Pharmacol Rev 52:145–176

Osborn JE (1982) Cytomegaloviruses and other herpesviruses. In: Foster HL, Small JD, Fox JG (eds) The Mouse in Biomedical Research. Academic Press, New York, pp 267–293

Pass RF (2001) Cytomegalovirus. In: Fields BN, Knipe DM, Howley PM (eds) Fields Virology. 3rd edn. Lippincott Williams and Wilkins Publishers, Philadelphia, pp 2675–2705

Pavic I, Polic B, Crnkovic I, Lucin P, Jonjic S, Koszinowski UH (1993) Participation of endogenous tumour necrosis factor alpha in host resistance to cytomegalovirus infection. J Gen Virol 74:2215–23

Penfold ME, Dairaghi DJ, Duke GM, Saederup N, Mocarski ES, Kemble GW, Schall TJ (1999) Cytomegalovirus encodes a potent alpha chemokine. Proc Natl Acad Sci USA 96:9839–9844

Pepose JS, Holland GN, Nestor MS, Cochran AJ, Foos RY (1985) Acquired immune deficiency syndrome: Mechanisms of ocular disease. J Ophthalmol 92:472–484

Plotkin SA, Starr SE, Friedman HM, Gonczol E, Weibel RE (1989) Protective effects of Towne cytomegalovirus vaccine against low-passage cytomegalovirus administered as a challenge. J Infect Dis 159:860–865

Pollock JL, Presti RM, Paetzold S, Virgin HWT (1997) Latent murine cytomegalovirus infection in macrophages. Virology 227:168–179

Rawlinson WD, Farrell HE, Barrell BG (1996) Analysis of the complete DNA sequence of murine cytomegalovirus. J Virol 70:8833–8849

Revello MG, Zavattoni M, Sarasini A, Percivalle E, Simoncini L, Gerna G (1998) Human cytomegalovirus in blood of immunocompetent persons during primary infection: prognostic implications for pregnancy. J Infect Dis 177:1170–5117

Riddell SR, Greenberg PD (1997) T cell therapy of human CMV and EBV infection in immunocompromised hosts. Rev Med Virol 7:181–192

Rossi D, Zlotnik A (2000) The biology of chemokines and their receptors. Annu Rev Immunol 18:217–242

Saederup N, Aguirre SA, Sparer TE, Boulez DM, Mocarski ES (2001) Murine cytomegalovirus CC chemokine MCK-2 is a determinant of dissemination that increases inflammation at initial sites of infection. J Virol 75:9966–9976

Saederup N, Lin YC, Dairaghi DJ, Schall TJ, Mocarski ES (1999) Cytomegalovirus-encoded beta chemokine promotes monocyte-associated viremia in the host. Proc Natl Acad Sci USA 96: 10881–10886

Salazar-Mather TP, Orange JS, Biron CA (1998) Early murine cytomegalovirus (MCMV) infection induces liver natural killer (NK) cell inflammation and protection through macrophage inflammatory protein 1alpha (MIP-1alpha)-dependent pathways. J Exp Med 187:1–14

Sallusto F, Mackay CR, Lanzavecchia A (2000) The role of chemokine receptors in primary, effector, and memory immune responses. Annu Rev Immunol 18:593–620

Saltzman RL, Quirk MR, Jordan MC (1988) Disseminated cytomegalovirus infection. Molecular analysis of virus and leukocyte interactions in viremia. J Clin Invest 81:75–81

Sinzger C, Kahl M, Laib K, Klingel K, Rieger P, Plachter B, Jahn G (2000) Tropism of human cytomegalovirus for endothelial cells is determined by a post-entry step dependent on efficient translocation to the nucleus. J Gen Virol 81:3021–3025

Sinzger C, Schmidt K, Knapp J, Kahl M, Beck R, Waldman J, Hebart H, Einsele H, Jahn G (1999) Modification of human cytomegalovirus tropism through propagation in vitro is associated with changes in the viral genome. J Gen Virol 80:2867–2877

Slobedman B, Mocarski ES (1999) Quantitative analysis of latent human cytomegalovirus. J Virol 73:4806–4812

Soderberg-Naucler C, Fish KN, Nelson JA (1997) Reactivation of latent cytomegalovirus by allogeneic stimulation of blood cells from healthy donors. Cell 91:119–126

Springer TA (1994) Traffic signals for lymphocyte recirculation and leukocyte emigration: the multistep paradigm. Cell 76:301–314

Stine JT, Wood C, Hill M, Epp A, Raport CJ, Schweickart VL, Endo Y, Sasaki T, Simmons G, Boshoff C, Clapham P, Chang Y, Moore P, Gray PW, Chantry D (2000) KSHV-encoded CC chemokine vMIP-III is a CCR4 agonist, stimulates angiogenesis, and selectively chemoattracts TH2 cells. Blood 95:1151–1157

Stoddart CA, Cardin RD, Boname JM, Manning WC, Abenes GB, Mocarski ES (1994) Peripheral blood mononuclear phagocytes mediate dissemination of murine cytomegalovirus. J Virol 68:6243–6253

Tanaka Y, Adams DH, Hubscher S, Hirano H, Siebenlist U, Shaw S (1993) T-cell adhesion induced by proteoglycan-immobilized cytokine MIP-1 beta. Nature 361:79–82

Taub DD, Ortaldo JR, Turcovski-Corrales SM, Key ML, Longo DL, Murphy WJ (1996) Beta chemokines costimulate lymphocyte cytolysis, proliferation, and lymphokine production. J Leukoc Biol 59:81–89

Tomasec P, Braud VM, Rickards C, Powell MB, McSharry BP, Gadola S, Cerundolo V, Borysiewicz LK, McMichael AJ, Wilkinson GW (2000) Surface expression of HLA-E, an inhibitor of natural killer cells, enhanced by human cytomegalovirus gpUL40. Science 287:1031–1033

Tomazin R, Boname J, Hegde NR, Lewinsohn DM, Altschuler Y, Jones TR, Cresswell P, Nelson JA, Riddell SR, Johnson DC (1999) Cytomegalovirus US2 destroys two components of the MHC class II pathway, preventing recognition by CD4+ T cells. Nat Med 5:1039–1043

van Bruggen I, Price P, Robertson TA, Papadimitriou JM (1989) Morphological and functional changes during cytomegalovirus replication in murine macrophages. J Leukoc Biol 46:508–520

Van Damme J, Proost P, Put W, Arens S, Lenaerts JP, Conings R, Opdenakker G, Heremans H, Billiau A (1994) Induction of monocyte chemotactic proteins MCP-1 and MCP-2 in human fibroblasts and leukocytes by cytokines and cytokine inducers. Chemical synthesis of MCP-2 and development of a specific RIA. J Immunol 152:5495–502

Vieira J, Farrell HE, Rawlinson WD, Mocarski ES (1994) Genes in the HindIII J fragment of the murine cytomegalovirus genome are dispensable for growth in cultured cells: insertion mutagenesis with a lacZ/gpt cassette. J Virol 68:4837–4846

Wiertz E, Hill A, Tortorella D, Ploegh H (1997) Cytomegaloviruses use multiple mechanisms to elude the host immune response. Immunol Lett 57:213–216

Wolf DG, Spector SA (1993) Early diagnosis of human cytomegalovirus disease in transplant recipients by DNA amplification in plasma. Transplantation 56:330–334

Yoshie O, Imai T, Nomiyama H (1997) Novel lymphocyte-specific CC chemokines and their receptors. J Leukoc Biol 62:634–644

Zanghellini F, Boppana SB, Emery VC, Griffiths PD, Pass RF (1999) Asymptomatic primary cytomegalovirus infection: virologic and immunologic features. J Infect Dis 180:702–707

Ziegler H, Muranyi W, Burgert HG, Kremmer E, Koszinowski UH (2000) The luminal part of the murine cytomegalovirus glycoprotein gp40 catalyzes the retention of MHC class I molecules. EMBO J 19:870–881
Zou P, Isegawa Y, Nakano K, Haque M, Horiguchi Y, Yamanishi K (1999) Human herpesvirus 6 open reading frame U83 encodes a functional chemokine. J Virol 73:5926–5933

Herpesviral Proteins Regulating Apoptosis

T. DERFUSS and E. MEINL*

The induction of apoptosis of virus-infected cells is an important defense mechanism of the host. Apoptosis of an infected cell can be induced cell autonomously as a consequence of viral replication or can be mediated by CTLs attacking the infected cells. Herpesviruses have developed different strategies to interfere with cell-autonomous apoptosis and to block CTL-induced apoptosis mediated by death receptors such as Fas and TRAIL. Herpesviruses, which establish a lifelong persistence in the infected host, can be found principally in two different conditions, episomal persistence with a limited number of genes expressed and lytic replication with expression of almost all genes. To meet the need of the virus to enhance survival of the infected cell, herpesviruses have evolved different strategies that function during both episomal persistence and lytic replication. Herpesviruses, which encode 70 to more than 200 genes have incorporated cell homologous antiapoptotic genes, they code for multifunctional genes that can also regulate apoptosis, and, finally, they modulate the expression of cellular apoptosis-regulating genes to favor survival of the infected cells. Viral interference with host cell apoptosis enhances viral replication, facilitates virus spread and persistence, and may promote the development of virus-induced cancer.

1	Introduction	258
1.1	Herpesviruses	259
1.2	Virus-Induced and -Inhibited Apoptosis	259
2	Mechanisms of Viral Inhibition of Apoptosis	260
2.1	α-Herpesviruses	260
2.1.1	γ1-34.5 Protein	260
2.1.2	Latency-Associated Transcript	261
2.1.3	Infected Cell Polypeptide 4, US3, and US5	261
2.2	β-Herpesviruses	261
2.3	γ-Herpesviruses	262
2.3.1	Viral FLICE-Inhibitory Proteins (FLIPs)	262
2.3.2	Viral Bcl-2 Homologs	264
2.3.3	Modulation of Cellular Apoptotic Pathways	266
3	Potential Role of Viral Antiapoptotic Mechanisms in Cancer and Viral Persistence	267
References		268

Abbreviations

Apaf-1   apoptotic protease activating factor 1
BH   Bcl-2 homology domain

Department of Neuroimmunology, Max Planck Institute of Neurobiology, Martinsried, Germany
* Institute for Clinical Neuroimmunology, Ludwig Maximilians University, Munich, Germany

CTL	cytotoxic T lymphocyte
DED	death effector domain
DD	death domain
EBV	Epstein-Barr virus
eIF-2alpha	alpha subunit of the translation initiation factor 2
FADD	Fas-associated death domain
FLICE	FADD-like interleukin-converting enzyme
FLIP	FLICE inhibitory protein
HCMV	human cytomegalovirus
HHV 8	human herpesvirus type 8
H. saimiri	*Herpesvirus saimiri*
HSV	herpes simplex virus
ICP	infected cell polypeptide
LAT	latency-associated transcript
LMP-1	latent membrane protein 1
NGF	nerve growth factor
PKR	double-stranded RNA-dependent protein kinase R
TNF	tumor necrosis factor
TRADD	TNF receptor-associated death domain
TRAIL	TNF-related apoptosis-inducing ligand
US	unique short
VDAC	voltage-dependent anion channel
vICA	viral inhibitor of caspase-8-induced apoptosis
vIRF	viral interferon regulatory factor
vMIA	viral mitochondrial inhibitor of apoptosis
VZV	varicella-zoster virus

1 Introduction

The induction of apoptosis in virus-infected cells is an important defense mechanism of the host. Apoptosis of a virus-infected cell occurs either as a direct response to viral infection or on recognition of infection by the host immune system. Viruses have evolved different strategies to evade host immune responses including inhibition of antibody- and complement-mediated effects, interference with interferons, inhibition and modulation of cytokines and chemokines, blockade of antigen presentation, and also inhibition of apoptosis (SPRIGGS 1996; ALCAMI and KOSZINOWSKI 2000).

Herpesviruses, which establish a lifelong persistence in the infected host, can be found principally in two different conditions, episomal persistence with a limited number of gene expression and lytic replication with expression of almost all genes. To meet the need of the virus to enhance survival of the infected cell, herpesviruses have developed different strategies that function during both episomal persistence and lytic replication. Herpesviruses, which encode 70 to more than 200 genes have incorporated cell homologous anti-apoptotic genes, they code for multifunctional genes that can also regulate apoptosis, and, finally, they modulate the expression of cellular apoptosis-regulating genes to favor survival of the infected cells. Viral interference with host cell apoptosis enhances viral replication, facilitates virus spread and persistence, and may promote the development of virus-induced cancer.

1.1 Herpesviruses

On the basis of sequence homology and biological properties the family of herpesviruses has been subdivided in three subfamilies α-, β-, and γ-herpesviruses. α-herpesviruses are neurotropic with a rapid replication cycle and a broad host and cell range. HSV-1, HSV-2, and VZV belong to the subfamily of α-herpesviruses. After oral or genital infection HSV migrates retrograde to the cranial nerve ganglia or spinal ganglia, where it can enter a latent state. In the latent phase of infection only a limited group of viral genes is expressed, yet productive infection can be readily induced to provide a continuing reservoir of infectious virus. Reactivation can be triggered by certain stimuli like heat, stress, or immunosuppression. Members of the subfamilies of β- and γ-herpesviruses replicate more slowly and in a much more restricted range of cells of glandular and/or lymphatic origin. HCMV, which belongs to the subfamily of β-herpesviruses, can cause retinitis and encephalitis mainly in immunocompromised patients and approximately one-half of the adult population is latently infected with this virus. EBV, HHV-8, and *H. saimiri* are examples of γ-herpesviruses. EBV infects B cells and is associated with Burkitt's lymphoma, and HHV-8 is associated with Kaposi's sarcoma. *H. saimiri* is a lymphotropic and oncogenic virus. It persists in its natural host, the squirrel monkey, without inducing disease, whereas it can cause leukemia and lymphoma in other New World primates. Human T cells can be transformed to stable growth in vitro by this virus.

1.2 Virus-Induced and -Inhibited Apoptosis

We face the seeming paradox that, on the one hand, infection with herpesviruses can induce apoptosis during replication and, on the other hand, herpesviruses interfere with host cell apoptosis. HSV induces apoptosis in activated T cells (TROPEA et al. 1995; ITO et al. 1997) and VZV induces apoptosis in Vero cells (SADZOT-DELVAUX et al. 1995). Why has the viral host relationship evolved such that virus-infected cells are killed by apoptosis? One may speculate that an apoptotic death of the infected cell might be beneficial for the virus, because cells dying by apoptosis do not trigger an inflammation as opposed to necrotic cells. This review focuses on different strategies evolved by herpesviruses to prolong the life of the infected cell to enhance viral replication.

Apoptosis of an infected cell can either be caused by a signal from outside the cell, e.g., a virus-specific cytotoxic T cell, or caused by a signal from inside the cell and may occur cell autonomously in the absence of an immune response toward the virus. Herpesviruses have evolved a number of strategies to interfere with both cell-autonomous and CTL-induced apoptosis.

2 Mechanisms of Viral Inhibition of Apoptosis

2.1 α-Herpesviruses

α-Herpesviruses infect a variety of different cells and persist in neurons. Human members of this subfamily are HSV-1, HSV-2, and VZV. A number of different effectors encoded by HSV-1 have been linked to inhibition of apoptosis, namely the immediate-early proteins ICP27 (AUBERT and BLAHO 1999) and ICP4 (LEOPARDI and ROIZMAN 1996), the protein kinase US3 (LEOPARDI et al. 1997; MUNGER et al. 2001), US5 (JEROME et al. 1999), the γ1-34.5 protein (CHOU and ROIZMAN 1992), and the latency-associated transcript (LAT) (PERNG et al. 2000). In addition to coding for antiapoptotic genes, HSV also uses cellular regulators of apoptosis to prolong the life of the infected cells. HSV-2 upregulates the expression of the Fas-L (SIEG et al. 1996). Remarkably, differences in the regulation of apoptosis by HSV-1 and HSV-2 were observed (JEROME et al. 1999). HSV-1 infection of activated T cells resulted in fratricide of antiviral T cells (RAFTERY et al. 1999). It was proposed that the induction of fratricide is an important immune evasion mechanism of HSV-1, helping the virus to persist in the host organism.

2.1.1 γ1-34.5 Protein

During the course of infection with a number of viruses including herpesviruses, double-stranded RNA occurs (JACQUEMONT and ROIZMAN 1975). Thereby, a double-stranded RNA-dependent protein kinase R (PKR) is activated. Subsequently, the α-subunit of the translation initiation factor 2 (eIF-2α) is phosphorylated and the total protein synthesis is shut off. Viruses have evolved a number of strategies to block this shutoff of protein synthesis. HSV-1 and HSV-2 are especially vulnerable to the shutoff of protein synthesis (CASSADY et al. 1998b). In HSV-infected cells, viral γ1-34.5 protein blocks the shutoff of protein synthesis by directing the protein phosphatase 1α to dephosphorylate the α-subunit of eukaryotic translation initiation factor 2 (eIF-2α) (HE et al. 1997). The amino acid sequence of the γ1-34.5 protein that interacts with the phosphatase shows a high homology to a domain of the eukaryotic protein GADD34 (CHOU and ROIZMAN 1992, 1994). By this means, the γ1-34.5 protein of HSV prevents the total shutoff of protein synthesis and premature cell death of HSV-infected neuronal cells (CHOU and ROIZMAN 1992, 1994). In addition, HSV has a second strategy to prevent the phosphorylation of eIF-2α: US11 can under appropriate conditions substitute for the γ1-34.5 protein (CASSADY et al. 1998a).

2.1.2 Latency-Associated Transcript

Whereas the γ1-34.5 protein functions during replication, HSV has also evolved strategies to favor the survival of latently infected neurons. During latency in neurons one characteristic transcript, termed latency-associated transcript (LAT), is found. LAT is required for establishment and maintenance of latency of HSV and plays a role in the reactivation process (LEIB et al. 1989). The survival of the latently infected neurons is obviously essential for the life cycle of HSV. Recently it was reported that LAT blocks neuronal apoptosis (PERNG et al. 2000). LAT inhibited apoptosis induced by different mechanisms, including ceramide-induced apoptosis and apoptosis induced by DNA damage via etoposide. This suggested that LAT interferes with a downstream effector of apoptosis that is common to many apoptotic pathways (PERNG et al. 2000). This report, however, was disputed by another group (THOMPSON 2000).

2.1.3 Infected Cell Polypeptide 4, US3, and US5

Infected cell polypeptide (ICP)4, which is a repressor and transactivator for viral gene expression, not only plays a role in reactivation of the virus from latency (HALFORD et al. 2001) but can also modulate apoptosis. It has been shown that apoptosis induced by hyperthermia and by the virus itself can be inhibited by ICP4 (LEOPARDI and ROIZMAN 1996). This effect could be related to the ability of HSV-1 to maintain bcl-2 RNA and protein levels during infection. ICP4 and ICP27 prevent the p38 mitogen-activated protein kinase-dependent destabilization of Bcl-2 and protect from a cisplatin-induced decrease of bcl-2 RNA (ZACHOS et al. 2001). By this means the virus could also achieve the observed inhibition of cytochrome c release from mitochondria (ZHOU et al. 2000). The protein kinases US3 and US5 encoded by HSV-1 are principal viral proteins required to block apoptosis. US3 blocks virus-induced apoptosis at a premitochondrial level because cytochrome c release and activation of caspase 3 are inhibited (LEOPARDI et al. 1997; MUNGER et al. 2001). CD95- and UV-induced apoptosis are also inhibited by US3 and US5 in HSV-1 infected cells but not in HSV-2 infected cells (JEROME et al. 1999).

2.2 β-Herpesviruses

HCMV is the human herpesvirus with the largest genome, about 230kb. Two HCMV-encoded proteins, IE-1 and IE-2, have been described to inhibit apoptosis mediated by tumor necrosis factor (TNF)-α, although the mechanism underlying this process has not yet been elucidated (ZHU et al. 1995). It has also been shown that overexpression of IE-2, but not IE-1, can protect smooth muscle cells from apoptosis induced by overexpression of p53 or by doxorubicin (TANAKA et al. 1999).

Another report provides data that HCMV infection could contribute to the reduced sensitivity of tumor cells to chemotherapy by induction of the cellular Bcl-2. HCMV infection of a human neuroblastoma cell line was shown to render the cells less sensitive to cytotoxic agents. These effects were associated with increased levels of Bcl-2 and decreased ability of the cells to undergo apoptosis. After antiviral treatment of the cells with ganciclovir the cells became again sensitive to cytotoxic agents, the Bcl-2 levels normalized, and apoptosis was inducible to the same degree as in the uninfected parental cell line (CINATL et al. 1998).

Recently, based on a screening of a genomic HCMV DNA library in an expression vector, an antiapoptotic gene of HCMV could be identified. The product of *UL37*, designated viral mitochondrial inhibitor of apoptosis (vMIA) blocked CD95- and doxorubicin mediated apoptosis in HeLa cells. vMIA functioned downstream of the activation of caspase-8 and Bid cleavage, resided in mitochondria, and associated with adenine nucleotide translocator (GOLDMACHER et al. 1999). vMIA acted upstream of the release of cytochrome c and resembled thereby the *H. saimiri*-encoded Bcl-2 homolog (DERFUSS et al. 1998) but did not show sequence homology to Bcl-2 family members.

Using the same approach of screening a genomic HCMV library the Goldmacher group found another apoptosis-modulating protein called viral inhibitor of caspase-8-induced apoptosis (vICA). vICA also blocks apoptosis induced by CD95. It binds to the prodomain of caspase-8 and thereby inhibits its activation. Surprisingly, it has no significant sequence homology to other viral or cellular inhibitors of caspase-8 (FLIPs) and therefore represents a new class of apoptosis-inhibitory proteins (SKALEISKAYA et al. 2001).

2.3 γ-Herpesviruses

The subfamily of γ-herpesviruses is further divided into two subgroups, γ-1 herpesviruses, such as EBV, and γ-2 herpesviruses, such as *H. saimiri* and HHV-8. It is a peculiar feature of γ-herpesviruses, in particular γ-2 herpesviruses, to contain a number of genes with homology to cellular genes (NEIPEL et al. 1997; MEINL et al. 1998).

γ-1 (MESEDA et al. 2000; HENDERSON et al. 1993) and γ-2 (ENSSER et al. 1997; DERFUSS et al. 1998; SARID et al. 1997) herpesviruses encode a homolog of Bcl-2, whereas FLIP is found in γ-2, but not γ-1 herpesviruses (MEINL et al. 1998).

2.3.1 Viral FLICE Inhibitory Proteins (FLIPs)

Apoptosis can be induced from outside the cell by involvement of specialized death receptors. Death receptors are members of the TNF/nerve growth factor (NGF) family receptors (NAGATA 1997). To date, six human transmembrane death receptors have been identified, namely Fas (CD95, APO-1), TNFRI, TRAMP (also named DR3, WSL-1, Apo-3, or LARD) and two receptors for TRAIL, TRAIL-R1 (DR4) and TRAIL-R2 (DR5) and DR6 (ASHKENAZI and DIXIT 1999). These receptors share a cytoplasmic domain that is required for transmission of cell death and is therefore called death domain (DD). Cytotoxic T cells use the Fas-L and the directed secretion of perforin/granzymes as effector mechanisms.

Apoptotic signalling of Fas has been most extensively elaborated, but there is evidence that the other death receptors use similar, though distinct, death signaling pathways. Binding of Fas to its ligand leads to the recruitment of a death-inducing signaling complex (NAGATA 1997). This complex comprises adapter proteins that contain protein-protein interaction motifs (Fig. 1). The adapter molecule FADD is recruited to Fas by interactions of their respective DD. FADD then recruits FLICE (caspase-8, MACH, Mch-5) via interactions of their death effector domains (DEDs), leading to FLICE activation by autocleavage. Activated FLICE subsequently cleaves caspase-3 and initiates a proteolytic cascade that results in apoptosis. TNFRI and TRAMP use an additional adapter molecule, named TRADD, to recruit FADD and then FLICE (Fig. 1).

An understanding of the adapter molecules involved in death signaling provided the basis for identifying a novel class of viral antiapoptotic effectors (THOME

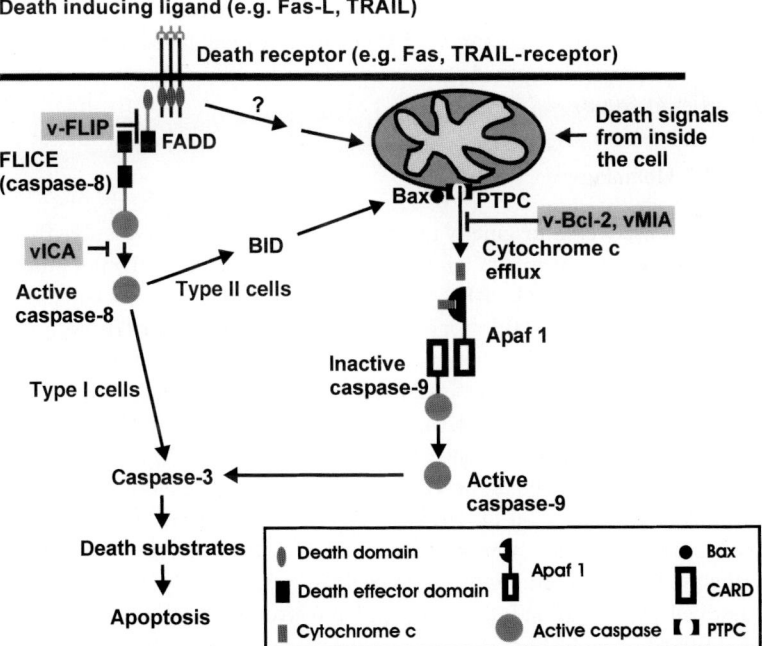

Fig. 1. Modulation of apoptosis by v-FLIP and v-Bcl-2. v-FLIPs specifically inhibit apoptosis mediated by death receptors. v-ICA specifically targets caspase-8 and inhibits its activation. v-Bcl-2 and vMIA inhibit those apoptotic pathways that are signaled through mitochondrial release of cytochrome c. *FADD*, Fas-associated death domain; *FLICE*, FADD-like interleukin-converting enzyme; *CARD*, caspase-recruiting domain; *PTPC*, permeability transition pore complex; *FLIP*, FLICE-inhibitory protein; *vICA*, viral inhibitor of caspase 8-induced-apoptosis; *vMIA*, viral mitochondrial inhibitor of apoptosis; *Apaf-1*, apoptotic protease-activating factor 1

et al. 1997; BERTIN et al. 1997; HU et al. 1997). Searches for homology to DEDs yielded viral proteins with two DEDs. These viral antiapoptotic molecules block the activation of FLICE (caspase-8) and were therefore named FLIPs (FLICE-inhibitory proteins) (THOME et al. 1997) (Fig. 1). Shortly thereafter, cellular FLIPs (IRMLER et al. 1997) were discovered by different groups (reviewed in WALLACH 1997). The homology between viral FLIPs and their cellular counterpart is restricted to key amino acids of the DEDs (approximately 25% sequence identity). In addition, there is a splice variant of cellular FLIP that contains a C-terminal caspase-like domain that, however, lacks the active cysteine site and is therefore inactive. The identified viral proteins have two DEDs and bind to cellular DED-containing proteins like FADD and FLICE and thereby block the formation of the death-inducing signaling complex (THOME et al. 1997). Viral FLIPs block cell death induced by the several death receptors (THOME et al. 1997; BERTIN et al. 1997; HU et al. 1997).

vFLIPs have been detected in several γ-2 herpesviruses, including HHV-8 and *H. saimiri*, but also in the poxvirus molluscum contagiosum virus (THOME et al. 1997). The antiapoptotic activity of γ-2 herpesviruses-encoded FLIPs was further elaborated: Cleavage of procaspase-8 into its catalytic subunits is reduced after expression of HHV-8-FLIP (BELANGER et al. 2001). Permissive cells infected with a deletion mutant of *H. saimiri* lacking v-FLIP showed an enhanced sensitivity to CD95-mediated apoptosis compared with cells infected with wild-type *H. saimiri* (GLYKOFRYDES et al. 2000).

2.3.2 Viral Bcl-2 Homologs

Cellular Bcl-2 was discovered originally as an oncogenic protein in follicular B-cell lymphoma. Homology to members of this family is restricted to distinct Bcl-2 homology (BH) domains: BH1, BH2, BH3, and BH4. Cellular Bcl-2 family members comprise apoptosis-inhibitory molecules like Bcl-2, or Bcl-x_L, and apoptosis-inducing effectors, like Bax or Bak (KROEMER 1997; REED 1997). All viral Bcl-2 family members studied so far blocked the development of apoptosis. Cellular Bcl-2 and *H. saimiri* encoded-Bcl-2 inhibit apoptosis mediated by irradiation, oxygen radicals, or DNA damage (DERFUSS et al. 1998), but inhibition of Fas-mediated apoptosis by cellular or viral Bcl-2 proteins is cell type dependent (SCAFFIDI et al. 1998; DERFUSS et al. 1998).

Interestingly, evidence has been obtained that Fas may induce apoptosis via different pathways that are used almost exclusively in different cells (SCAFFIDI et al. 1998). In some cells, Fas ligation induced a rapid and strong activation of caspase-8 and caspase-3 (see above and Fig. 1) and subsequent apoptosis was not blocked by Bcl-2. In other cells, Fas-ligation resulted in a considerably weaker activation of FLICE (caspase-8), which cleaves Bid, a proapoptotic member of the Bcl-2 family (LUO et al. 1998; LI et al. 1998). Cleaved Bid subsequently translocates to mitochondria, where it promotes, either alone or together with Bax, cytochrome *c* efflux (ESKES et al. 2000). Subsequently, cytochrome *c* can interact with the cytosolic apoptotic protease-activating factor 1 (Apaf-1), which in turn activates caspase 9 and thereby initiates the caspase cascade (LI et al. 1997). Although Bcl-2 could not block caspase-8 activation and Bid cleavage in these cells, the effects of cleaved Bid and Bax on mitochondria were prevented by Bcl-2 (MURPHY et al. 2000). Another apoptotic pathway involving the Fas binding molecule Daxx and Jun kinases was reported to be sensitive to Bcl-2 (YANG et al. 1997).

Cellular Bcl-2 and Bcl-x_L have a hydrophobic membrane-interacting domain at their C-terminus and are found mainly in mitochondria and in the endoplasmic reticulum. Bcl-2 and Bcl-x_L prevent the release of mitochondrial cytochrome c and subsequent activation of caspases (KLUCK et al. 1997), presumably by interacting with mitochondrial membrane channels like VDAC and adenine nucleotide translocator (SHIMIZU et al. 1999). Very similar to its cellular homologs, the *H. saimiri*-encoded Bcl-2 stabilizes mitochondria and inhibits the activation of caspase-3 (DERFUSS et al. 1998).

Compared with cellular Bcl-2, viral Bcl-2 family members are typically shorter and lack a strong homology to the BH3 and BH4 domains. Cellular Bcl-2 can be converted to a death promotor on cleavage by caspases, and this death-promoting activity is dependent on BH3 (CHENG et al. 1997). Interestingly, the herpesvirus-encoded Bcl-2 homologs escape caspase-mediated conversion to a proapoptotic molecule (BELLOWS et al. 2000). These differences between cellular and viral Bcl-2 homologs might be explained as follows. The conversion of cellular Bcl-2 to a proapoptotic molecule in response to caspase activation serves as a kind of positive feedback to ensure execution of a started cell death program. Conversely, it is useful for the virus to prolong the life of the infected cell by its own Bcl-2, but conversion of the viral Bcl-2 to a death-promoting agent might lead to a premature cell death.

Although the biological importance of Fas, the prototype of death receptors, is well appreciated, recent reports point to an involvement of the TRAIL system in overcoming certain malignancies (WALCZAK et al. 1999) and also viral infections (SEDGER et al. 1999). TRAIL induces apoptosis in various tumor cells but generally not in normal cells (WILEY et al. 1995). HCMV infection upregulates the expression of TRAIL and TRAIL-Rs on virus-infected fibroblasts, and IFN-γ potentiates these effects (SEDGER et al. 1999).

The biological importance of the TRAIL system becomes evident when considering that in the cellular genome four different receptors for the cytotoxic ligand TRAIL are encoded (ASHKENAZI and DIXIT 1999) that form a distinct subgroup within the TNF-R superfamily. Two of the TRAIL receptors 1–4 function as a decoy. TRAIL binds also to a more distantly related TNFR homolog, osteoprotegerin, which is a secreted, soluble receptor for the TNF homolog TRANCE/RANKL/OPGL.

Using Jurkat cells stably transfected with the *H. saimiri*-encoded Bcl-2 homolog, we found that this viral protein interfered with TRAIL mediated apoptosis (unpublished data). When testing different cell lines other groups observed no protection against TRAIL-induced apoptosis by the cellular Bcl-2 (WALCZAK et al. 2000; KEOGH et al. 2000; GAZITT et al. 1999). Perhaps these contradicting results are an indication for a similar dichotomy in TRAIL-mediated apoptosis as has been described for Fas-mediated apoptosis (SCAFFIDI et al. 1998). TRAIL receptors 1, 2, and 4 mediate activation of NF-κB, which has antiapoptotic activity. It is unclear whether v-Bcl-2 or cellular Bcl-2 also interfere with TRAIL-mediated activation of NF-κB.

Comparing anti-apoptotic activities of v-Bcl-2 and v-FLIP, the picture is emerging that these two antiapoptotic proteins are complementary. The antiapoptotic activity of v-FLIP is restricted to apoptosis mediated via death receptors and is thereby different from and partially overlapping of the antiapoptotic activity of v-Bcl-2. v-Bcl-2 inhibits apoptosis induced by a variety of stimuli including irradiation, oxygen radicals, and dexamethasone, whereas the inhibition of death receptor-mediated apoptosis is cell type dependent (Fig. 1).

Recently, a second Bcl-2 homolog present in the genome of EBV was identified (MARSHALL et al. 1999). This viral protein has several unique features. In the BH1 domain, glycine at position 149 (numbering according to the sequence of BALF1) is replaced by a serine. Surprisingly, mutation of this glycine in cellular Bcl-2 abolishes the antiapoptotic activity. This viral protein also shows some homology in the BH4 domain, but it lacks the DXXD motif suggested to be important in the regulation of Bcl-2 by caspases (MARSHALL et al. 1999). This novel Bcl-2 homolog inhibited apoptosis in HeLa cells induced via Fas and via camptothecin.

2.3.3 Modulation of Cellular Apoptotic Pathways

In addition to coding for specialized antiapoptotic proteins, many viruses also upregulate cellular modulators of apoptosis to prolong the life of infected cells. This phenomenon has been particularly elaborated for EBV. The EBV-encoded oncoprotein LMP-1 induces the expression of the antiapoptotic molecule A20 (FRIES et al. 1996; SPENDER et al. 1999), of Bcl-2 (HENDERSON et al. 1991), and also of the recently discovered Bcl-2 family member *bfl-1* (D'SOUZA et al. 2000). BARF1, known for inducing malignant transformation in B-cell lines and fibroblasts, is another gene of EBV that can induce expression of cellular Bcl-2 (SHENG et al. 2001). EBV transforms B lymphocytes into lymphoblastoid cell lines usurping the Notch and TNF pathways to effect transcription including NF-κB activation. The activation of NF-κB has been found to be critical for the survival of EBV-transformed B lymphocytes (CAHIR-MCFARLAND et al. 2000), and also HHV-8 seems to use induction of NF-κB to prevent apoptosis in primary effusion lymphoma (KELLER et al. 2000).

Another way of modulating apoptosis of the host cell is the expression of viral IFN-regulatory factors (vIRF). HHV-8 codes for different vIRFs that can interfere with the intracellular signal transduction of type-I interferons. vIRF-2 has been shown to interact with PKR, thereby inhibiting its autophosphorylation and blocking the phosphorylation of PKR substrates like eIF-2α (BURYSEK and PITHA 2001). The γ1-34.5 protein of HSV-1 described above uses a similar approach to block the shutoff of protein synthesis. The mechanism by which vIRF-1 of HHV-8 can prevent apoptosis is different. vIRF-1 binds to p53 with its DNA binding region. This interaction inhibits the transcriptional activation of p53 and thereby prevents apoptosis and facilitates uncontrolled cell proliferation (NAKAMURA et al. 2001; SEO et al. 2001).

3 Potential Role of Viral Antiapoptotic Mechanisms in Cancer and Viral Persistence

The lymphotropic herpesviruses, which encode a v-Bcl-2 and/or a v-FLIP, are oncogenic. HHV-8, the most recently discovered human tumor virus, is involved in the pathogenesis of Kaposi's sarcoma, primary effusion lymphoma, and the plasma cell variant of multicentric Castleman's disease. HHV-8 is not pathogenic in most otherwise healthy individuals but is highly oncogenic in HIV-1-infected and iatrogenically immunosuppressed individuals. It establishes a latent infection in most KS spindle (endothelial tumor) cells and in the neoplastic B cells of primary effusion lymphomas and has – controversially discussed – been linked with plasmocytoma (SCHULZ 2000). EBV has been linked to different human malignancies (FARRELL 1995). *H. saimiri* induces fulminant leukemia and lymphoma in a number of susceptible primate species and transforms human T cells in vitro (MEINL et al. 1995). The issue of whether the viral antiapoptotic genes are involved in oncogenesis must be answered for each virus and each tumor separately, because tumor development may be associated with viral replication (HHV-8 in Kaposi's sarcoma) or with limited viral gene expression (EBV in lymphoblastoid B cell lymphoma, Hodgkin's disease, or Burkitt's lymphoma).

The EBV-encoded Bcl-2 homologs are expressed during lytic replication but usually not under latent conditions. Also, an EBV deletion mutant lacking the Bcl-2 homolog BHRF-1 was able to transform B cells in vitro (MARCHINI et al. 1991). However, the survival of EBV-transformed B cells is critically dependent on the antiapoptotic activity of NF-κB (CAHIR-MCFARLAND et al. 2000). EBV seems to use antiapoptotic genes mainly during lytic replication but utilizes cellular antiapoptotic pathways to ensure survival of the transformed lymphoblastoid cells.

The *H. saimiri*-encoded FLIP and Bcl-2 are expressed late during lytic replication in permissive owl monkey kidney cells (KRAFT et al. 1998). At this time, the infected cultures are partially protected from Fas-mediated cell death (THOME 1997). The expression of both *H. saimiri*-encoded antiapoptotic genes was found to be restricted to cultures with viral replication, and both genes were not expressed during episomal persistence in human growth-transformed T cells that do not produce infectious virus (KRAFT et al. 1998). These in vitro experiments suggest that the main function of the *H. saimiri*-encoded FLIP and Bcl-2 is to enhance viral replication by prolonging the life of the productively infected cell. By contrast, expression of the Bcl-2 homolog of murine γ-herpesvirus 68 was detected not only during lytic replication but also during viral persistance in the absence of replication (ROY et al. 2000).

Different sets of experiments strongly suggest that the HHV-8-encoded FLIP is directly linked to development of malignancies. The HHV-8-encoded FLIP may be transcribed during episomal persistence, as seen in a tumor cell line derived from a body cavity-based B cell lymphoma (SARID et al. 1998). The HHV-8-FLIP has also been found to be transcribed in Kaposi's sarcoma lesions, especially in older lesions (STURZL et al. 1999). HHV-8-FLIP could act as a tumor progression factor by promoting tumor establishment and growth in vivo. When injected into immunocompetent recipient mouse strains, murine B lymphoma cells (A20) transduced with this viral gene rapidly induced aggressive tumors showing a high rate of survival and growth. The tumor-progressive activity of HHV-8-FLIP was mediated by prevention of death receptor-induced apoptosis triggered by conventional T cells (DJERBI et al. 1999). Expression of HHV-8-Bcl-2 was also detected in Kaposi's sarcoma tissue and in B cell lines that were derived from body cavity-based B cell lymphoma (SARID et al. 1997). Another study did not detect constitutive expression of HHV-8-Bcl-2 (CHENG et al. 1997). Both groups, however, described a strong expression after activation of these B cell lines with phorbol ester acetate, indicating that HHV-8-Bcl-2 is mainly expressed during lytic replication.

In addition to oncogenesis viral antiapoptotic effects have been linked to viral persistence. The γ1-34.5 gene of HSV is essential for neurovirulence but not essential for growth in non-neuronal cells in culture. This gene precludes neuronal cells from triggering total host shutoff of protein synthesis characteristic of programmed cell death (CHOU and ROIZMAN 1992). Viruses lacking this gene showed a lower incidence of establishment of latency or reactivation from latency (WHITLEY et al. 1993).

Acknowledgements. This work was supported by the DFG (SFB 466 and 571) and by the Wilhelm Sander-Stiftung (97.081.1). The Institute for Clinical Neuroimmunology is supported by the Hermann and Lilly Schilling Foundation.

References

Alcami A, Koszinowski UH (2000) Viral mechanisms of immune evasion Immunol. Today 21:447–455
Ashkenazi A, Dixit VM (1999) Apoptosis control by death and decoy receptors. Curr Opin Cell Biol 11:255–260
Aubert M, Blaho JA (1999) The herpes simplex virus type 1 regulatory protein ICP27 is required for the prevention of apoptosis in infected human cells. J Virol 73:2803–2813
Belanger C, Gravel A, Tomoiu A, Janelle ME, Gosselin J, Tremblay MJ, Flamand L (2001) Human herpesvirus 8 viral FLICE-inhibitory protein inhibits Fas-mediated apoptosis through binding and prevention of procaspase-8 maturation. J Hum Virol 4:62–73
Bellows DS, Chau BN, Lee P, Lazebnik Y, Burns WH, Hardwick JM (2000) Antiapoptotic herpesvirus Bcl-2 homologs escape caspase-mediated conversion to proapoptotic proteins. J Virol 74:5024–5031
Bertin J, Armstrong RC, Ottilie S, Martin DA, Wang Y, Banks S, Wang GH, Senkevich TG, Alnemri ES, Moss B, Lenardo MJ, Tomaselli KJ, Cohen JI (1997) Death effector domain-containing herpesvirus and poxvirus proteins inhibit both Fas- and TNFR1-induced apoptosis. Proc Natl Acad Sci USA 94:1172–1176

Burysek L, Pitha PM (2001) Latently expressed human herpesvirus 8-encoded interferon regulatory factor 2 inhibits double-stranded RNA-activated protein kinase. J Virol 75:2345–2352

Cahir-McFarland ED, Davidson DM, Schauer SL, Duong J, Kieff E (2000) NF-kappa B inhibition causes spontaneous apoptosis in Epstein-Barr virus-transformed lymphoblastoid cells. Proc Natl Acad Sci USA 97:6055–6060

Cassady KA, Gross M, Roizman B (1998a) The herpes simplex virus US11 protein effectively compensates for the gamma1(34.5) gene if present before activation of protein kinase R by precluding its phosphorylation and that of the alpha subunit of eukaryotic translation initiation factor 2. J Virol 72:8620–8626

Cassady KA, Gross M, Roizman B (1998b) The second-site mutation in the herpes simplex virus recombinants lacking the gamma134.5 genes precludes shutoff of protein synthesis by blocking the phosphorylation of eIF-2alpha. J Virol 72:7005–7011

Cheng EHY, Kirsch DG, Clem RJ, Ravi R, Kastan MB, Bedi A, Ueno K, Hardwick JM (1997) Conversion of Bcl-2 to a Bax-like death effector by caspases. Science 278:1966–1968

Cheng EH, Nicholas J, Bellows DS, Hayward GS, Guo HG, Reitz MS, Hardwick JM (1997) A Bcl-2 homolog encoded by Kaposi sarcoma-associated virus, human herpesvirus 8, inhibits apoptosis but does not heterodimerize with Bax or Bak. Proc Natl Acad Sci USA 94:690–694

Chou J, Roizman B (1992) The gamma 1(34.5) gene of herpes simplex virus 1 precludes neuroblastoma cells from triggering total shutoff of protein synthesis characteristic of programmed cell death in neuronal cells. Proc Natl Acad Sci USA 89:3266–3270

Chou J, Roizman B (1994) Herpes simplex virus 1 gamma(1)34.5 gene function, which blocks the host response to infection, maps in the homologous domain of the genes expressed during growth arrest and DNA damage. Proc Natl Acad Sci USA 91:5247–5251

Cinatl JJ, Cinatl J, Vogel JU, Kotchetkov R, Driever PH, Kabickova H, Kornhuber B, Schwabe D, Doerr HW (1998) Persistent human cytomegalovirus infection induces drug resistance and alteration of programmed cell death in human neuroblastoma cells. Cancer Res 58:367–372

Derfuss T, Fickenscher H, Kraft MS, Henning G, Lengenfelder D, Fleckenstein B, Meinl E (1998) Anti-apoptotic activity of the Herpesvirus saimiri encoded Bcl-2 homolog: stabilization of mitochondria and inhibition of caspase-3 like activity. J Virol 72:5897–5904

Djerbi M, Screpanti V, Catrina AI, Bogen B, Biberfeld P, Grandien A (1999) The inhibitor of death receptor signaling, FLICE-inhibitory protein defines a new class of tumor progression factors [see comments]. J Exp Med 190:1025–1032

D'Souza B, Rowe M, Walls D (2000) The bfl-1 gene is transcriptionally upregulated by the Epstein-Barr virus LMP1, and its expression promotes the survival of a Burkitt's lymphoma cell line. J Virol 74:6652–6658

Ensser A, Pflanz R, Fleckenstein B (1997) Primary structure of the alcelaphine herpesvirus 1 genome. J Virol 71:6517–6525

Eskes R, Desagher S, Antonsson B, Martinou JC (2000) Bid induces the oligomerization and insertion of Bax into the outer mitochondrial membrane. Mol Cell Biol 20:929–935

Farrell PJ (1995) Epstein-Barr virus immortalizing genes. Trends Microbiol 3:105–109

Fries KL, Miller WE, Raab Traub N (1996) Epstein-Barr virus latent membrane protein 1 blocks p53-mediated apoptosis through the induction of the A20 gene. J Virol 70:8653–8659

Gazitt Y, Shaughnessy P, Montgomery W (1999) Apoptosis-induced by TRAIL and TNF-alpha in human multiple myeloma cells is not blocked by BCL-2 Cytokine. 11:1010–1019

Glykofrydes D, Niphuis H, Kuhn EM, Rosenwirth B, Heeney JL, Bruder J, Niedobitek G, Muller-Fleckenstein I, Fleckenstein B, Ensser A (2000) Herpesvirus saimiri vFLIP provides an antiapoptotic function but is not essential for viral replication, transformation, or pathogenicity. J Virol 74:11919–11927

Goldmacher VS, Bartle LM, Skaletskaya A, Dionne CA, Kedersha NL, Vater CA, Han J, Lutz RJ, Watanabe S, McFarland ED, Kieff ED, Mocarski ES, Chittenden T (1999) A cytomegalovirus-encoded mitochondria-localized inhibitor of apoptosis structurally unrelated to Bcl-2. Proc Natl Acad Sci USA 96:12536–12541

Halford WP, Kemp CD, Isler JA, Davido DJ, Schaffer PA (2001) ICP0, ICP4, or VP16 expressed from adenovirus vectors induces reactivation of latent herpes simplex virus type 1 in primary cultures of latently infected trigeminal ganglion cells. J Virol 75:6143–6153

He B, Gross M, Roizman B (1997) The gamma(1)34.5 protein of herpes simplex virus 1 complexes with protein phosphatase 1alpha to dephosphorylate the alpha subunit of the eukaryotic translation initiation factor 2 and preclude the shutoff of protein synthesis by double-stranded RNA-activated protein kinase. Proc Natl Acad Sci USA 94:843–848

Henderson S, Huen D, Rowe M, Dawson C, Johnson G, Rickinson A (1993) Epstein-Barr virus-coded BHRF1 protein, a viral homologue of Bcl- 2, protects human B cells from programmed cell death. Proc Natl Acad Sci USA 90:8479–8483

Henderson S, Rowe M, Gregory C, Croom Carter D, Wang F, Longnecker R, Kieff E, Rickinson A (1991) Induction of bcl-2 expression by Epstein-Barr virus latent membrane protein 1 protects infected B cells from programmed cell death. Cell 65:1107–1115

Hu S, Vincenz C, Buller M, Dixit VM (1997) A novel family of viral death effector domain-containing molecules that inhibit both CD-95- and tumor necrosis factor-1-induced apoptosis. J Biol Chem 272:9621–9624

Irmler M, Thome M, Hahne M, Schneider P, Hofmann K, Steiner V, Bodmer J-L, Schröter M, Burns K, Mattmann C, Rimoldi D, French LE, Tschopp J (1997) Identification of death receptor signals by cellular FLIP. Nature 388:190–195

Ito M, Watanabe M, Kamiya H, Sakurai M (1997) Herpes simplex virus type 1 induces apoptosis in peripheral blood T lymphocytes. J Infect Dis 175:1220–1224

Jacquemont B, Roizman B (1975) RNA synthesis in cells infected with herpes simplex virus. X. Properties of viral symmetric transcripts and of double-stranded RNA prepared from them J Virol 15:707–713

Jerome KR, Fox R, Chen Z, Sears AE, Lee H, Corey L (1999) Herpes simplex virus inhibits apoptosis through the action of two genes, Us5 and Us3. J Virol 73:8950–8957

Keller SA, Schattner EJ, Cesarman E (2000) Inhibition of NF-kappaB induces apoptosis of KSHV-infected primary effusion lymphoma cells. Blood 96:2537–2542

Keogh SA, Walczak H, Bouchier-Hayes L, Martin SJ (2000) Failure of Bcl-2 to block cytochrome c redistribution during TRAIL-induced apoptosis FEBS. Lett 471:93–98

Kluck RM, Bossy Wetzel E, Green DR, Newmeyer DD (1997) The release of cytochrome c from mitochondria: a primary site for Bcl-2 regulation of apoptosis. Science 275:1132–1136

Kraft MS, Henning G, Fickenscher H, Lengenfelder D, Tschopp J, Fleckenstein B, Meinl E (1998) Herpesvirus saimiri transforms human T cells to stable growth without inducing resistance to apoptosis. J Virol 72:3138–3145

Kroemer G (1997) The proto-oncogene Bcl-2 and its role in regulating apoptosis. Nat Med 3:614–620

Leib DA, Bogard CL, Kosz-Vnenchak M, Hicks KA, Coen DM, Knipe DM, Schaffer PA (1989) A deletion mutant of the latency-associated transcript of herpes simplex virus type 1 reactivates from the latent state with reduced frequency. J Virol 63:2893–2900

Leopardi R, Roizman B (1996) The herpes simplex virus major regulatory protein ICP4 blocks apoptosis induced by the virus or by hyperthermia. Proc Natl Acad Sci USA 93:9583–9587

Leopardi R, Van Sant C, Roizman B (1997) The herpes simplex virus 1 protein kinase US3 is required for protection from apoptosis induced by the virus. Proc Natl Acad Sci USA 94:7891–7896

Li H, Zhu H, Xu C-J, Yuan J (1998) Cleavage of BID by caspase 8 mediates the mitochondrial damage in the Fas pathway of apoptosis. Cell 94:491–500

Li P, Nijhawan D, Budihardjo I, Srinivasula SM, Ahmad M, Alnemri ES, Wang X (1997) Cytochrome c and dATP-dependent formation of Apaf-1/caspase-9 complex initiates an apoptotic protease cascade. Cell 91:479–489

Luo X, Budihardjo I, Zou H, Slaughter C, Wang X (1998) Bid, a Bcl2 interacting protein, mediates cytochrome c release from mitochondria in response to activation of cell surface death receptors. Cell 94:481–490

Marchini A, Tomkinson B, Cohen JI, Kieff E (1991) BHRF1, the Epstein-Barr virus gene with homology to Bcl2, is dispensable for B-lymphocyte transformation and virus replication. J Virol 65:5991–6000

Marshall WL, Yim C, Gustafson E, Graf T, Sage DR, Hanify K, Williams L, Fingeroth J, Finberg RW (1999) Epstein-Barr virus encodes a novel homolog of the bcl-2 oncogene that inhibits apoptosis and associates with Bax and Bak. J Virol 73:5181–5185

Meinl E, Fickenscher H, Thome M, Tschopp J, Fleckenstein B (1998) Anti-apoptotic strategies of lymphotropic viruses. Immunol Today 19:474–479

Meinl E, Hohlfeld R, Wekerle H, Fleckenstein B (1995) Immortalization of human T cells by herpesvirus saimiri. Immunol Today 16:55–58

Meseda CA, Arrand JR, Mackett M (2000) Herpesvirus papio encodes a functional homologue of the Epstein-Barr virus apoptosis suppressor, BHRF1. J Gen Virol 81 Pt 7:1801–5:1801–1805

Munger J, Chee AV, Roizman B (2001) The U(S)3 protein kinase blocks apoptosis induced by the d120 mutant of herpes simplex virus 1 at a premitochondrial stage. J Virol 2001 75:5491–5497

Murphy KM, Streips UN, Lock RB (2000) Bcl-2 inhibits a Fas-induced conformational change in the Bax N terminus and Bax mitochondrial translocation. J Biol Chem 275:17225–17228

Nagata S (1997) Apoptosis by death factor. Cell 88:355–365

Nakamura H, Li M, Zarycki J, Jung JU (2001) Inhibition of p53 tumor suppressor by viral interferon regulatory factor. J Virol 75:7572–7582

Neipel F, Albrecht J-C, Fleckenstein B (1997) Cell-homologous genes in the Kaposi's sarcoma-associated rhadinovirus human herpesvirus 8: determinants of its pathogenicity. J Virol 71:4187–4192

Perng GC, Jones C, Ciacci-Zanella J, Stone M, Henderson G, Yukht A, Slanina SM, Hofman FM, Ghiasi H, Nesburn AB, Wechsler SL (2000) Virus-induced neuronal apoptosis blocked by the herpes simplex virus latency-associated transcript. Science 287:1500–1503

Raftery MJ, Behrens CK, Muller A, Krammer PH, Walczak H, Schonrich G (1999) Herpes simplex virus type 1 infection of activated cytotoxic T cells: Induction of fratricide as a mechanism of viral immune evasion. J Exp Med 190:1103–1114

Reed JC (1997) Double identity for proteins of the Bcl-2 family. Nature 387:773–776

Roy DJ, Ebrahimi BC, Dutia BM, Nash AA, Stewart JP (2000) Murine gammaherpesvirus M11 gene product inhibits apoptosis and is expressed during virus persistence. Arch Virol 145:2411–2420

Sadzot-Delvaux C, Thonard P, Schoonbroodt S, Piette J, Rentier B (1995) Varicella-zoster virus induces apoptosis in cell culture. J Gen Virol 76:2875–2879

Sarid R, Flore O, Bohenzky RA, Chang Y, Moore PS (1998) Transcription mapping of the Kaposi's sarcoma-associated herpesvirus (human herpesvirus 8) genome in a body cavity-based lymphoma cell line (BC-1). J Virol 72:1005–1012

Sarid R, Sato T, Bohenzky RA, Russo JJ, Chang Y (1997) Kaposi's sarcoma-associated herpesvirus encodes a functional bcl- 2 homologue. Nat Med 3:293–298

Scaffidi C, Fulda S, Srinivasan A, Friesen C, Li F, Tomaselli KJ, Debatin KM, Krammer PH, Peter ME (1998) Two CD95 (Apo-1/Fas) signaling pathways. EMBO J 17:1675–1687

Schulz TF (2000) Kaposi's sarcoma-associated herpesvirus (human herpesvirus 8): epidemiology and pathogenesis. J Antimicrob Chemother 2000 45 Suppl T3:15–27:15–27

Sedger LM, Shows DM, Blanton RA, Peschon JJ, Goodwin RG, Cosman D, Wiley SR (1999) IFN-gamma mediates a novel antiviral activity through dynamic modulation of TRAIL and TRAIL receptor expression. J Immunol 163:920–926

Seo T, Park J, Lee D, Hwang SG, Choe J (2001) Viral interferon regulatory factor 1 of Kaposi's sarcoma-associated herpesvirus binds to p53 and represses p53-dependent transcription and apoptosis. J Virol 75:6193–6198

Sheng W, Decaussin G, Sumner S, Ooka T (2001) N-terminal domain of BARF1 gene encoded by Epstein-Barr virus is essential for malignant transformation of rodent fibroblasts and activation of BCL-2. Oncogene 20:1176–1185

Shimizu S, Narita M, Tsujimoto Y (1999) Bcl-2 family proteins regulate the release of apoptogenic cytochrome c by the mitochondrial channel VDAC [see comments]. Nature 399:483–487

Sieg S, Yildirim Z, Smith D, Kayagaki N, Yagita H, Huang Y, Kaplan D (1996) Herpes simplex virus type 2 inhibition of Fas ligand expression. J Virol 70:8747–8751

Skaletskaya A, Bartle LM, Chittenden T, McCormick AL, Mocarski ES, Goldmacher VS (2001) A cytomegalovirus-encoded inhibitor of apoptosis that suppresses caspase-8 activation. Proc Natl Acad Sci USA 98:7829–7834

Spender LC, Cannell EJ, Hollyoake M, Wensing B, Gawn JM, Brimmell M, Packham G, Farrell PJ (1999) Control of cell cycle entry and apoptosis in B lymphocytes infected by Epstein-Barr virus. J Virol 73:4678–4688

Spriggs MK (1996) One step ahead of the game: viral immunomodulatory molecules. Annu Rev Immunol 14:101–130

Sturzl M, Hohenadl C, Zietz C, Castanos-Velez E, Wunderlich A, Ascherl G, Biberfeld P, Monini P, Browning PJ, Ensoli B (1999) Expression of K13/v-FLIP gene of human herpesvirus 8 and apoptosis in Kaposi's sarcoma spindle cells. J Natl Cancer Inst 91:1725–1733

Tanaka K, Zou JP, Takeda K, Ferrans VJ, Sandford GR, Johnson TM, Finkel T, Epstein SE (1999) Effects of human cytomegalovirus immediate-early proteins on p53-mediated apoptosis in coronary artery smooth muscle cells. Circulation 99:1656–1659

Thome M, Schneider P, Hofmann K, Fickenscher H, Meinl E, Neipel F, Mattmann C, Burns K, Bodmer J-L, Schröter M, Scaffidi C, Krammer PH, Peter ME, Tschopp J (1997) Viral FLICE-inhibitory proteins (FLIPs) prevent apoptosis induced by death receptors. Nature 386:517–521

Thompson RLSNM (2000) HSV latency-associated transcript and neuronal apoptosis. Science 289:1651a–1651a

Tropea F, Troiano L, Monti D, Lovato E, Malorni W, Rainaldi G, Mattana P, Viscomi G, Ingletti MC, Portolani M (1995) Sendai virus and herpes virus type 1 induce apoptosis in human peripheral blood mononuclear cells. Exp Cell Res 218:63–70

Walczak H, Bouchon A, Stahl H, Krammer PH (2000) Tumor necrosis factor-related apoptosis-inducing ligand retains its apoptosis-inducing capacity on Bcl-2- or Bcl-xL-overexpressing chemotherapy-resistant tumor cells. Cancer Res 60:3051–3057

Walczak H, Miller RE, Ariail K, Gliniak B, Griffith TS, Kubin M, Chin W, Jones J, Woodward A, Le T, Smith C, Smolak P, Goodwin RG, Rauch CT, Schuh JC, Lynch DH (1999) Tumoricidal activity of tumor necrosis factor-related apoptosis-inducing ligand in vivo [see comments] Nat.Med. 5:157–163

Wallach D (1997) Placing death under control. Nature 388:123–126

Whitley RJ, Kern ER, Chatterjee S, Chou J, Roizman B (1993) Replication, establishment of latency, and induced reactivation of herpes simplex virus gamma 1 34.5 deletion mutants in rodent models. J Clin Invest 91:2837–2843

Wiley SR, Schooley K, Smolak PJ, Din WS, Huang CP, Nicholl JK, Sutherland GR, Smith TD, Rauch C, Smith CA, et al (1995) Identification and characterization of a new member of the TNF family that induces apoptosis. Immunity 3:673–682

Yang X, Khosravi-Far R, Chang HY, Baltimore D (1997) Daxx, a novel Fas-binding protein that activates JNK and apoptosis. Cell 89:1067–1076

Zachos G, Koffa M, Preston CM, Clements JB, Conner J (2001) Herpes simplex virus type 1 blocks the apoptotic host cell defense mechanisms that target Bcl-2 and manipulates activation of p38 mitogen-activated protein kinase to improve viral replication. J Virol 75:2710–2728

Zhou G, Galvan V, Campadelli-Fiume G, Roizman B (2000) Glycoprotein D or J delivered in trans blocks apoptosis in SK-N-SH cells induced by a herpes simplex virus 1 mutant lacking intact genes expressing both glycoproteins. J Virol 74:11782–11791

Zhu H, Shen Y, Shenk T (1995) Human cytomegalovirus IE1 and IE2 proteins block apoptosis. J Virol 69:7960–7970

Subversion of Host Defense Mechanisms by Adenoviruses

H.-G. Burgert*, Z. Ruzsics, S. Obermeier, A. Hilgendorf, M. Windheim, and A. Elsing

Adenoviruses (Ads) cause acute and persistent infections. Alike the much more complex herpesviruses, Ads encode numerous immunomodulatory functions. About a third of the viral genome is devoted to counteract both the innate and the adaptive antiviral immune response. Immediately upon infection, E1A blocks interferon-induced gene expression and the VA-RNA inhibits interferon-induced PKR activity. At the same time, E1A reprograms the cell for DNA synthesis and induces the intrinsic cellular apoptosis program that is interrupted by E1B/19K and E1B/55K proteins, the latter inhibits p53-mediated apoptosis. Most other viral stealth functions are encoded by a separate transcription units, E3. Several E3 products prevent death receptor-mediated apoptosis. E3/14.7K seems to interfere with the cytolytic and pro-inflammatory activities of TNF while E3/10.4K and 14.5K proteins remove Fas and TRAIL receptors from the cell surface by inducing their degradation in lysosomes. These and other functions that may afect granule-mediated cell death might drastically limit lysis by NK cells and cytotoxic T cells (CTL). Moreover, Ads interfere with recognition of infected cell by CTL. The paradigmatic E3/19K protein subverts antigen presentation by MHC class I molecules by inhibiting their transport to the cell surface. In concert, these viral countermeasures ensure prolonged survival in the infected host and, as a consequence, facilitate transmission. Elucidating the molecular mechanisms of Ad-mediated immune evasion has stimulated corresponding research on other viruses. This knowledge will also be instrumental for designing better vectors for gene therapy and vaccination, and may lead to a more rational treatment of life-threatening Ad infections, e.g. in transplantation patients.

1	Introduction	274
2	Adenovirus Classification and Induced Diseases	275
3	The Adenovirus Replication Cycle	276
4	Inhibition of the IFN Response	279
4.1	The IFN Response	279
4.2	Inhibition of IFN-Induced Gene Expression by E1A	279
4.3	Inhibition of IFN-Induced PKR Activity by VA-RNA	280
5	Inhibition of the E1A-Induced Intrinsic Apoptosis Program by E1B Proteins	281
5.1	Apoptosis Induction by E1A	281
5.2	Inhibition of p53-Mediated Apoptosis by E1B/55K	282
5.3	Interference of E1B/19K with E1A-Induced Apoptotic Pathways	283
6	Blocking Death Receptor-Mediated Apoptosis	284
6.1	Apoptosis Induced by the Immune System	284
6.2	Modulation of TNF Activities	285
6.2.1	Cytolytic and Proinflammatory Activities of TNF	285
6.2.2	Induction of TNF Susceptibility by E1A	285

Max von Pettenkofer-Institut, Lehrstuhl Virologie, Genzentrum der Ludwig-Maximilians-Universität, Feodor-Lynen-Str. 25, 81377 München, Germany
* Present address: The University of Warwick, Department of Biological Sciences, Coventry, CV4 7AL, UK

6.2.3 Inhibition of TNF Activities by E3/14.7K 286
6.2.4 Inhibition of TNF Activities by E3/10.4K–14.5K 289
6.2.5 Influence of E1B/19K on TNF-Mediated Cytolysis 289
6.3 TNF-Induced Upregulation of E3 Protein Expression:
  Keeping Up with Host Cytokines 290
6.4 Inhibition of Fas-Mediated Apoptosis 291
6.4.1 Downregulation of Fas Surface Expression by E3/10.4K–14.5K 291
6.4.2 Influence of E3/14.7K on Fas-Mediated Signaling 296
6.5 Downregulation of TRAIL Receptors by E3 Proteins 296

7 Influence of Ad on NK Cell Activity 298
7.1 Role of NK Cells in Antiviral Immunity 298
7.2 Inhibition of E1A-Induced NK Sensitivity During Infection 299

8 Inhibition of CTL Recognition and Lysis 300
8.1 Antigen Presentation by MHC Class I Molecules 300
8.2 Subversion of Antigen Presentation by the E3/19K Protein 301
8.3 Transcriptional Repression of Antigen Presentation Functions by Ad12 E1A 305
8.4 Inhibition of CTL Lysis 306

9 Summary of Countermeasures:
  The Chess Game Between Virus and Host; Facts and Fiction 306

10 Perspectives 308

References 310

1 Introduction

The coexistence of human adenoviruses (Ads) with their natural host implies that coevolution resulted in a balance in which host defense mechanisms control the virus infection but are unable to eradicate Ads from the human population. Obviously, the host response is insufficient to prevent spread of the virus among individuals. As a result, more than 90% of adults have antibodies against Ads in their blood (STRAUS 1984; WADELL 1990). Successful spread is facilitated by the propensity of Ads to establish persistent infections, during which the virus seems to be continuously produced and can be recovered intermittently from stool, urine, or other tissues for variable lengths of time. It is commonly believed that this apparent balance between virus and host is maintained by countermeasures of the virus, resulting in a partial neutralization of antiviral immune responses. Therefore, diseases associated with adenovirus infections are mostly mild and self-limiting, unless the host defense, and in particular the cellular immune defense, is suppressed, e.g., during AIDS or allogeneic bone marrow transplantations (HORWITZ 1996).

Little is known about the in vivo status of human adenovirus during persistence. Because of the strict host range, no suitable and relevant animal model is available to study Ad persistence. In many virus systems it became clear that studies on the molecular interactions of animal viruses with their natural host (e.g., murine cytomegalovirus) provide only limited clues as to the situation of the related human viruses (e.g., human cytomegalovirus) with their host (HENGEL et al. 1998).

However, in the past decade a wealth of molecular interactions between viral proteins and cellular proteins involved in the host response have been described that are thought to represent the basis for persistence and spread of the virus.

In this review, we will focus on the molecular mechanisms of immune evasion by Ads of subgenus C, because essentially all functional data reported until today have been obtained for proteins of this subgenus. After a brief description of the known diseases and the adenovirus replication cycle, the response of the virus to individual antiviral host mechanisms will be discussed, following primarily their presumed timely appearance. Each section starts with an introduction to the antiviral immune response, often referring to other viruses, which is followed by a description of the Ad countermeasures. A whole group of immunomodulatory functions is encoded in a special transcription unit, called early region 3 (E3). E3 proteins seem to be primarily dedicated to limit the detection and destruction of the infected cell by the immune system. We provide new sequence information for the Ad4 E3 proteins and discuss structure/function relationships. A number of excellent reviews have been published that focus on certain aspects of Ad immune evasion. These will be mentioned throughout the text. In general, we will concentrate on more recent data (mostly from the last 5 years) that are relevant for the Ad infection and the virus life cycle. By studying the mechanisms by which Ads counteract host immune defenses, we gain a better understanding of the molecular basis for Ad persistence and, sometimes solely based on the existence of a viral stealth function, we uncover novel host responses directed against Ads. A better knowledge about these host functions will be instrumental for improving the success of Ad-mediated gene therapy.

2 Adenovirus Classification and Induced Diseases

Ads are nonenveloped viruses of icosahedral structure. The virion is composed of at least 10 different structural proteins, the major component being the so-called hexon protein. From the 12 vertices of the virion, antenna-like projections are formed by the homotrimeric fiber protein (SHENK 1996). So far, 51 human Ad serotypes have been distinguished based on their serum neutralization and their DNA restriction pattern (DE JONG et al. 1999). They are classified into six subgenera, A–F, according to their hemagglutination pattern and the polypeptide composition of the virions (WADELL 1990; SHENK 1996). The DNA homology, estimated from extensive restriction cuts, is higher than 50% within one subgroup and lower than 20% between subgenera. The capsid encloses a double-stranded DNA genome, varying in size from 34.1kb in Ad12 (subgenus A) to 35.9kb in Ad2/Ad5 of subgenus C (CHROBOCZEK et al. 1992; SPRENGEL et al. 1994). Although each Ad serotype can infect a wide variety of cells and tissues, distinct, yet partially overlapping, disease patterns have emerged for Ads of the different subgenera: Ads of subgenus A and F cause gastrointestinal infections. Ads of subgenus B and C are mainly associated with infections of the respiratory tract. Clinical symptoms range from a mild pharyngitis to severe acute respiratory diseases, such as pertussis-like syndrome or pneumonia (WADELL 1990; HORWITZ 1996). Ads may account for about 10% of pneumonia in childhood. Ad4, the only member of subgenus E, commonly causes acute respiratory disease in military recruits. Subgenus D contains 31 Ad serotypes and is by far the largest subgroup. Ads of this subgroup have the propensity to cause ocular diseases. A rather severe eye disease, epidemic keratoconjunctivitis (EKC), involving both the conjunctiva and the cornea, is predominantly associated with Ad8, Ad19a, and Ad37 (KEMP et al. 1983).

As mentioned above, after acute Ad disease, a significant percentage of people develop persistent infections that may last months or even years (Fox et al. 1977; HORWITZ 1996). This ability of Ads to persist was already demonstrated by the first isolation of Ads in 1953 from explanted tonsillar tissue (adenoids) of children (ROWE et al. 1953): 33 of 53 adenoids contained an adenoid-degenerating agent (AD), later designated adenoviruses. Subsequent epidemiological studies confirmed this property (Fox et al. 1977). Subgroup C Ads are considered to be endemic and are thought to establish persistent infections in lymphoid tissues. A characteristic intermittent shedding into the feces and the urinary tract has been observed that may be due to reactivation or might result from a decrease of neutralizing antibody titers to undetectable levels (Fox et al. 1977; HORWITZ 1996). How the persistent state is maintained is unclear. Evidence has been provided for a low level of continuous Ad production in lymphoid cells (HORVATH et al. 1986; MAHR and GOODING 1999) as well as for true latency that has been implicated in chronic obstructive pulmonary disease (NEUMANN et al. 1987; MATSUSE et al. 1992). In patients undergoing allogeneic bone marrow transplantations, persistent adenovirus infections cause serious problems, often associated with fatal disease.

3 The Adenovirus Replication Cycle

The first steps of the Ad infection cycle, the attachment of Ad to the host cell and its internalization, already induce host cell responses important for both the viral entry process and initiation of the innate immune defense. The fiber proteins of subgenus A, C, D, E, and F Ads bind with high affinity to a 46-kDa cell surface molecule called coxsackie/adenovirus receptor (CAR) (BERGELSON et al. 1997; TOMKO et al. 1997; ROELVINK et al. 1998). CAR belongs to the immunoglobulin superfamily and is expressed in variable amounts in most tissues (TOMKO et al. 1997). Cocrystallization and extensive mutagenesis of the fiber revealed the binding site for CAR within the fiber (BEWLEY et al. 1999; ROELVINK et al. 1999). Extended loops on the lateral surface of the C-terminal fiber knob are critical for CAR binding. Simultaneous with or subsequent to attachment, the penton base that anchors the N-terminal portion of the fiber in the capsid becomes exposed and interacts with cellular $\alpha_V\beta3$ or $\alpha_V\beta5$ integrins (WICKHAM et al. 1993; NEMEROW and STEWART 1999). During the entry process a number of signaling molecules are activated (RUSSELL 2000). Integrin ligation seems to trigger the phosphatidylinositol-3-OH kinase, inducing rapid endocytosis of the bound virions through clathrin-coated pits. After internalization of the Ad particle, the fibers are degraded and Ads escape from the endosomal/lysosomal compartment by penetration of the lipid membrane. The virions are transported along microtubules to the nuclear pore complex (GREBER et al. 1997), where the viral DNA is imported into the nucleus. Ad infection also induces interferon (IFN) production, an effect that does not require viral protein synthesis. Hence, it seems to be mediated by the early interaction of the virions with host cell components (REICH et al. 1988). Early events during the infection process also trigger Raf/mitogen-activated protein kinase (MAPK) signaling, which induces production of chemokines such as interleukin-8 (BRUDER and KOVESDI 1997). In addition, replication-defective Ad vectors, at least at high multiplicity, can activate the transcription factor NF-κB in human vascular smooth muscle cells in a capsid-dependent manner, resulting in production of a wide range of proinflammatory mediators (CLESHAM et al. 1998; BORGLAND et al. 2000). NF-κB activation by E1-deleted Ad vectors has also been observed during the first days after infection of mice, resulting in elevation of serum cytokines, lymphocyte infiltration, and hepatocellular apoptosis (LIEBER et al. 1998). At present, it is unclear whether NF-κB activation is relevant during wild-type adenovirus infections of humans.

The viral transcriptional program is initiated by the transcription of the E1A gene (Fig. 1). E1A proteins fulfill several tasks. First, they prepare the cell for efficient Ad replication in that they establish a permissive host cell environment for viral DNA synthesis. As most cells infected by Ads are quiescent, Ads induce progression into S-phase by E1A-mediated transcriptional activation of a set of growth-promoting cellular genes (CRESS and NEVINS 1996). At the same time, these

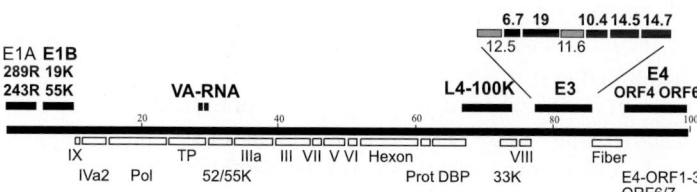

Fig. 1. Adenovirus proteins of subgenus C contributing to immune evasion. The Ad genome is drawn as a line divided in 100 map units. Gene products and transcription units are indicated along the genome as boxes and are drawn to scale. Ad products that subvert host defense mechanisms are depicted *above* and are highlighted with *bold writing* and *black bars*; those without a known immunomodulatory function are indicated with *open bars below the genome*. Virion components are *directly below the bar*, all other proteins not present in the virion are at the *bottom*. The E3 region encoding several immunomodulatory proteins is enlarged. The 12.5 and 11.6 genes are shown in *gray*, because their immunomodulatory function is not known yet or not fully established. Hexon and fiber are also named virion protein II and IV, respectively. *Prot*, protease; *DBP*, ssDNA binding protein; *Pol*, polymerase; *TP*, terminal protein. The direction of transcription and production of a precursor polypeptide is ignored

changes provoke the intrinsic cellular death program and sensitize the cells for death receptor-mediated apoptosis. Second, E1A proteins transactivate the promoters of the early transcription units E1B, E2, E3, and E4 (JONES 1995; SHENK 1996). In addition, E1A has a significant immune evasion function, namely to decapitate IFN-induced signal transduction in infected cells (see Sect. 4.2).

Each of the transcription units appears to be dedicated to a particular process. E1B proteins (E1B/19K and E1B/55K) protect cells from intrinsically triggered apoptosis (see the chapter by Reits et al., this volume). During the late phase, E1B/55K, in concert with the E4/ORF6 product, also modulates nuclear transport of viral and cellular mRNA and seems to be necessary for virus replication (IMPERIALE et al. 1995). All E2 proteins (Fig. 1) are involved in the replication of viral DNA: The viral polymerase, together with the preterminal protein and host factors, is necessary for initiation of viral DNA replication whereas the polymerase, together with the DNA-binding protein, is required for its elongation.

The early transcription unit E3 of the virus (Fig. 1) is not required for replication of the virus in tissue culture and in vivo in cotton rat lungs (GINSBERG et al. 1989) but is present in all human Ads and thus is thought to play a key role for regulating the interaction of Ads with the human host (WOLD et al. 1995; BURGERT 1996; MAHR and GOODING 1999; BURGERT and BLUSCH 2000). Several E3 proteins have been shown to counteract host defense mechanisms, such as antigen presentation, apoptosis, and the inflammatory response (see below). Interestingly, most Ad E3 proteins are transmembrane proteins, which localize to the ER, the Golgi/TGN, the plasma and nuclear membrane (WOLD et al. 1995; BURGERT and BLUSCH 2000). The E3 region is unique in this respect because no other adenoviral transcription unit encodes known transmembrane proteins. Thus, membrane integration of proteins appears to be most critical for the immunomodulatory functions of Ads. The majority of E3 proteins seem to be dedicated to prevent the lytic attack of the cellular defense system (see below).

E4 proteins regulate diverse functions, such as RNA processing, mRNA transport, and cell death. Concomitant with the onset of viral DNA replication the late transcription program is initiated. The time point at which this occurs depends on the viral strain, the multiplicity of infection, and the cell type infected. In cell culture systems the late phase usually begins at ~9–12h after infection (using 10 plaque-forming units of Ad2), whereas in primary cells this time point is substantially delayed and completion of the replication cycle might take as long as 24–40h. The major late promoter controls the synthesis of a large polycistronic primary transcript that is processed by differential splicing and differential usage of poly A sites into 18 mRNAs that encode almost exclusively structural components. All virion proteins are transported into the nucleus, where they first form empty capsids that are subsequently packaged with viral DNA (PHILIPSON 1984). Interestingly, despite the massive production of Ad particles and the major alterations of the cell physiology, cells do not lyse but remain intact for several days. Although a fraction of virus particles is released into the supernatant of producer cells, the majority remains attached to the nucleus. How the virus is set free from the nucleus has not been elucidated yet, but may involve disruption of the intermediate filament network (CHEN et al. 1993) or an active disintegration of the nuclear envelope (TOLLEFSON et al. 1996).

4 Inhibition of the IFN Response

4.1 The IFN Response

IFNs are pleiotropic cytokines playing an important role in the innate immune response. IFNs (IFN-α, IFN-β, IFN-γ) induce transcription of more than 30 IFN-stimulated genes (ISGs) whose products exhibit potent antiviral, immunomodulatory, and antiproliferative activities (STARK et al. 1998). Very early after infection, viruses often induce IFN-α/β production in the infected cell, in part by generating dsRNA-like replication intermediates. IFN-α/β stimulate the lytic potential of natural killer (NK) cells, which also belong to the first line of defense. Together with activated T cells, NK cells are an important source of IFN-γ. Upon binding of IFN-α/β and IFN-γ to specific cell surface receptors the Janus kinase/signal transducers and activators of transcription (JAK/STAT) are activated. JAKs (Tyk2 and JAK1) phosphorylate the receptors, thereby creating docking sites for STAT proteins, which themselves become phosphorylated and then transmit the signal as a dimer from the cytosol to the nucleus (STARK et al. 1998). Together with p48, STAT1 and STAT2 form a transcription-activating complex (ISGF-3) on IFN-stimulated response elements (ISRE) present in IFN-α-responsive promoters, whereas triggering of the IFN-γ receptor induces STAT1 homodimers that trans-activate genes containing γ-responsive sequences (Gas) (reviewed in STARK et al. 1998). Many IFN-induced genes encode proteins with documented antiviral effects: $2'$-$5'$ oligoadenylate synthetase upregulates RNase L to degrade RNA, protein kinase R (PKR) arrests protein translation by phosphorylating eIF2α, Mx interferes with the replication of influenza and other negative-strand RNA viruses, and major histocompatibility complex (MHC) class I/II antigens present viral antigens on the cell surface (GOODBOURN et al. 2000).

4.2 Inhibition of IFN-Induced Gene Expression by E1A

As mentioned above, Ad particles induce IFN-stimulated genes and these stimuli are suppressed by the immediate-early protein E1A (ANDERSON and FENNIE 1987; REICH et al. 1988). E1A antagonizes the IFN system by interfering with the activation of IFN-responsive genes in several ways. First, both E1A proteins, 249R and 289R, impair IFN signaling by lowering the levels of signal transducers. In E1A-expressing HeLa cells the STAT1 α protein level was decreased (LEONARD and SEN 1996), whereas in E1A[+] HT1080 cell lines the STAT1 level is normal, but neither

IFN-α nor IFN-γ was able to activate transcription of ISRE-containing genes. In these cells, p48, a protein required for IFN-α signaling, was lowered and signaling was restored by overexpression of p48 (LEONARD and SEN 1997). Second, E1A inhibits the transcriptional coactivator function of CREB binding proteins CBP/p300 that bind to STAT1 and STAT2 and thereby regulate transcription of both types of IFN-regulated genes (ZHANG et al. 1996). Therefore, E1A might directly compete with nuclear STATs for CBP/p300 binding (BHATTACHARYA et al. 1996). In addition, E1A may compete with IFN response factor 3 (IRF-3) for binding to CBP/p300 and thereby interfere with induction of IFN-α and IFN-β genes (JUANG et al. 1998). However, not all effects of E1A on STAT1-mediated transcription can be explained by competition with CBP/p300 binding, because an E1A mutant that does not bind CBP/p300 still blocks IFN-γ signaling by directly binding to the nuclear STAT1 homodimer during infection. This indicates that E1A also impairs the communication between STAT1 and adaptor molecules other than CBP/p300, which may link STAT1 to the basal transcription-activating complex on IFN-γ-responsive genes (LOOK et al. 1998).

4.3 Inhibition of IFN-Induced PKR Activity by VA-RNA

Ads also block the IFN response at the execution phase. The ubiquitously expressed double-stranded RNA activated serine/threonine protein kinase PKR is a key regulator of IFN-mediated translation inhibition (WILLIAMS 1999). Activation of PKR by dsRNA or replication intermediates results in the phosphorylation of eukaryotic translation factor eIF-2α. Phosphorylated eIF-2α traps the guanine nucleotide exchange factor eIF-2β, a rate-limiting component of the translation machinery, and thereby shuts down translation. Human Ads express a set of GC-rich, virus-associated (VA) RNAs about 160bp in length (MA and MATHEWS 1996). The VA-RNAs are abundantly transcribed by RNA polymerase III in the late phase of the infection cycle (SHENK 1996). Their stable secondary structure allows them to bind to PKR and to inhibit its activity (KITAJEWSKI et al. 1986). In recent years, it became clear that PKR can be activated independently from dsRNA, e.g., by cytokines, growth factors, and stress signals (WILLIAMS 1999). Moreover, accumulating data suggest that PKR has additional substrates that may regulate gene transcription and signal transduction. For example, PKR was shown to activate NF-κB, probably via the inhibitor of κB (IκB) kinase complex. In addition, regulated expression of PKR with either a vaccinia system or a tetracyclin system can trigger apoptosis, and both translation inhibition and NF-κB activation have been implicated in apoptosis induction (DONZE et al. 1999; WILLIAMS 1999; GIL and ESTEBAN 2000). Taken together, Ads target the IFN response at two levels: prevention of PKR activation by VA-RNA preserves viral translation and may affect the inflammatory response and apoptosis, whereas inhibition of the JAK/STAT signaling by E1A prevents upregulation of a number of interferon-responsive genes with importance for the immune response.

5 Inhibition of the E1A-Induced Intrinsic Apoptosis Program by E1B Proteins

5.1 Apoptosis Induction by E1A

The two most predominant species of E1A-derived mRNAs, the 13S and 12S mRNAs, encode E1A proteins with 289 (289R) and 243 (243R) amino acid residues, respectively. The different functional activities of E1A products can be linked to distinct regions of the protein that are conserved among different Ads and referred to as conserved regions 1–3 (CR1–3). E1A-mediated transcriptional activation requires CR3, which is only present in 289R (reviewed in JONES 1995), whereas stimulation of S-phase entry depends on the N-terminal sequences as well as CR1 and CR2, which are found in both predominant E1A forms. Entry into S-phase correlates with interactions of E1A with two families of host proteins. First, E1A binds to the members of the retinoblastoma tumor suppressor (Rb) family, inducing the release of the cellular transcription factor E2F from Rb. Active E2F plays an important role in DNA synthesis, cell cycle progression, and transcriptional activation of Ad E2 genes (for review see CRESS and NEVINS 1996). Second, E1A also binds to the p300/CBP family of histone acetyl transferases via the N-terminal sequences and CR1, thereby inhibiting their transcriptional coactivator function. This leads to transcriptional repression of genes that prevent S-phase entry, and thus E1A binding to p300 independent of Rb binding leads also to cell cycle progression (ARANY et al. 1995).

Presumably, because of these activities, E1A gene products are strong inducers of apoptosis in mammalian cells (CHINNADURAI 1998). Ad-mediated apoptosis can occur via both p53-dependent and p53-independent mechanisms. The p53 network is influenced by E1A at different checkpoints: The E1A-induced release of active E2F family members and the transcription factor ETF bind to and activate the murine p53 promoter (HALE and BRAITHWAITE 1999). E2F also induces production of p14ARF in human cells, which inhibits the ubiquitin ligase activity of murine double minute 2 (MDM2) protein that normally mediates degradation of p53 (ASHCROFT and VOUSDEN 1999). p53 is further stabilized by repression of p53/p300-dependent MDM2 transcription, disconnecting the negative feedback loop by which p53 initiates its own destruction (THOMAS and WHITE 1998). E1A may also increase p53 levels via interaction with and inhibition of SUG1, a regulatory subunit of proteasomes that are responsible for p53 degradation (TURNELL et al. 2000).

E1A can also induce apoptosis in p53-null cells (Saos-2) and, thus, in a p53-independent manner. With virus mutants that lack E1B expression, it was recently shown that the 243R form of E1A is able to induce apoptosis (CHIOU and WHITE 1997); however, conflicting data suggest that 243R can only induce apoptosis in a p53-dependent manner (TEODORO et al. 1995). The mechanism of apoptosis induction by E1A independent of p53 is not well understood. In the absence of other viral products, 243R is capable to activate a cascade of cysteine proteases, collectively called caspases, and to induce procaspase-8 processing by a mechanism independent of the adaptor protein Fas-associating protein with death domain (FADD) (NGUYEN et al. 1998), possibly via an alternative caspase-activating complex in the endoplasmic reticulum (ER), Bap31 (NG et al. 1997).

During viral infection E1A induces p53-independent apoptosis via transactivation of the E4 transcription unit (Fig. 1) and synthesis of the pro-apoptotic E4/ORF4 protein (MARCELLUS et al. 1998). Apoptosis by E4/ORF4 can be inhibited by Bcl-2 but not by zVAD-fmk, a universal inhibitor of caspases (LAVOIE et al. 1998), suggesting that E4/ORF4 may primarily perturb mitochondrial integrity. It has been proposed that E4/ORF4 may be involved in killing of infected cells at the end of the infection cycle. A similar role was suggested for the E3/11.6K protein, whose synthesis is greatly amplified late in the infection cycle. 11.6K seems to promote cell death and the release of Ads from infected cells at late time points during infection and, therefore, might facilitate rapid spread of the virus (TOLLEFSON et al. 1996). E3/11.6K was recently named the "adenovirus death protein" (ADP). The 11.6K-mediated death of infected cells did not show the typical signs of apoptosis, and its mechanism of action is still unknown. Remarkably, Ads from other subgenera lack an 11.6K homolog.

The p53-independent apoptosis induction by E1A may also be mediated by inducing sensitivity of Ad-infected cells to external proapoptotic stimuli, such as tumor necrosis factor-α (TNF) (DUERKSEN-HUGHES et al. 1989), Fas ligand (FasL) (COOK et al. 1996), or TNF-related apoptosis-inducing ligand (TRAIL) (ROUTES et al. 2000). The properties of these proapoptotic cytokines are discussed in Sects. 6.1–6.5.

As described above, the E1A-induced entry into the cell cycle, accompanied by the many changes of cellular transcription and the beginning of viral DNA replication, triggers the cellular machinery for programmed cell death. Therefore, Ads have evolved multiple mechanisms to interrupt this intrinsic cellular death program, primarily involving two proteins encoded in the E1B transcription unit, E1B/19K and E1B/55K. Both proteins are expressed early, shortly after E1A.

5.2 Inhibition of p53-Mediated Apoptosis by E1B/55K

Apart from the late functions of E1B/55K, e.g., the modulation of nuclear export of viral and cellular mRNA, the prominent activity of E1B/55K in the early phase is to counteract p53 function by direct binding to DNA-bound p53 (SHENK 1996). In vitro transcription assays demonstrate that E1B/55K contains a strong repression domain. As binding of E1B/55K to p53 profoundly increases the affinity of p53 to p53-dependent promoters, transcriptional activation of p53 target genes is efficiently blocked, presumably by affecting basal transcription (MARTIN and BERK 1999). E1B/55K also binds E4/ORF6, and together these proteins induce the rapid degradation of p53, thereby inhibiting p53-dependent apoptosis (KRATZER et al. 2000; WIENZEK et al. 2000).

5.3 Interference of E1B/19K with E1A-Induced Apoptotic Pathways

Ads with mutations in the E1B/19K gene exhibit the *cyt/deg* phenotype, which is characterized by accelerated cell death, large plaques, and degradation of viral and cellular DNA. This suggested that E1B/19K is involved in apoptosis prevention (CHINNADURAI 1998). Accumulating evidence suggests that E1B/19K interferes with both p53-dependent and p53-independent apoptosis pathways. Multiple mechanisms have been proposed by which E1B/19K counteracts apoptosis; however, the exact mechanism remains unknown.

E1B/19K shares limited homology to Bcl-2, but unlike Bcl-2, is not located to the mitochondrial membrane but rather to the nuclear and smooth ER membranes. In addition, E1B/19K associates with intermediate filaments of the cytoplasm and the nuclear lamina. The proper localization of E1B/19K to the nuclear envelope is essential for its antiapoptotic activity and appears to require interaction with lamins (RAO et al. 1997). E1B/19K binds to at least six cellular proteins, including the proapoptotic proteins Bak, Bax, and Nbk/Bik and can inhibit apoptosis induction by these proteins (BOYD et al. 1994; CHINNADURAI 1998; WHITE 1998). The inhibition of Bax-mediated apoptosis by E1B/19K has been further explored (WHITE 1998; DESAGHER and MARTINOU 2000). Triggered by an apoptotic stimulus, which can be death receptor dependent or -independent, the proapoptotic protein Bax, whose transcription is activated by p53, translocates from the cytosol to the outer mitochondrial membrane. Insertion of Bax into the membrane triggers a conformational change that is followed by rapid cytochrome *c* release. The conformational change of Bax may be triggered, or facilitated, by Bid or tBid, a truncated Bid generated by caspase-8. Cytochrome *c* together with dATP act as cofactors for the self-oligomerization of Apaf-1. Caspase-9 binding to Apaf-1 activates the caspase and induces the subsequent degradation of DNA by caspase-activated DNase (CAD) (DESAGHER and MARTINOU 2000). E1B/19K directly interacts with the Bcl-2 homology region 3 (BH3) of the conformationally changed Bax, thereby inhibiting the loss of mitochondrial membrane potential and the release of cytochrome c (PEREZ and WHITE 2000). The predominant localization of E1B/19K in the nuclear membrane and smooth ER suggests that E1B/19K might not act at the mitochondrial membrane but rather sequesters proapoptotic proteins like Bax and diverts them away from the mitochondria to the nuclear envelope and ER. It is presumed that the balance between pro- and antiapoptotic Bcl-2 family members determines the fate of the cell, survival or cell death (DESAGHER and MARTINOU 2000).

E1B/19K may also antagonize apoptosis by influencing transcription. With a yeast two-hybrid screen for E1B/19K interacting proteins, a Bcl-2-associated transcription factor (Btf) was identified that also binds apart from E1B/19K the anti-apoptotic proteins Bcl-2 and Bcl-x_L. Overexpression of Btf in HeLa cells induces apoptosis, which is abrogated by E1B/19K. In vitro transcription assays suggest that Btf has a transcription repressor activity that can be blocked by E1B/19K, Bcl-2, and Bcl-x_L, perhaps by sequestering Btf in the cytoplasm (KASOF et al. 1999). E1B/19K also seems to influence MDM2 levels by restoring transactivation of mdm2 by p53/p300, which is inhibited by E1A. Consequently, E1B/19K might reduce the amount of p53 and thereby might inhibit apoptosis induction (THOMAS and WHITE 1998). The mechanism by which E1B/19K might regulate transcription remains elusive. In transient transfection assays, E1B/19K prevented Ad5 13S E1A-induced activation of NF-κB, another transcription factor implicated in apoptosis regulation (SCHMITZ et al. 1996). Whether this mechanism is relevant during Ad infection remains to be shown.

E1B/19K is also capable of affecting caspase activity. These cysteine proteases are typically activated during apoptosis and are also activated during adenovirus infection in the absence of E1B/19K, because the *deg/cyt* phenotype is not observed upon infection with an E1B/19K mutant virus when caspase activity is blocked by a general caspase inhibitor. Thus, caspases are involved in apoptosis induction by E1A. It was reported that E1B/19K inhibits E1A-induced caspase-3 activation and poly(ADP-ribose) polymerase (PARP) cleavage (BOULAKIA et al. 1996). This indicates that E1B/19K functions at or upstream of procaspase-3 processing (PEREZ and WHITE 2000). Interestingly, nontransformed WI-38 cells do not undergo apoptosis when infected with E1B/19K mutant viruses (WHITE et al. 1986).

6 Blocking Death Receptor-Mediated Apoptosis

6.1 Apoptosis Induced by the Immune System

Many effector cells of the immune system, like activated macrophages, cytotoxic T cells (CTL), and NK cells, produce proapoptotic cytokines, such as TNF, FasL or TRAIL. These factors are also synthesized by primary target cells, e.g., epithelial and endothelial cells, upon infection. They belong to the TNF family of cytokines and are typically synthesized as type II membrane-bound precursor molecules that act by cell-to-cell contact but can also be released (BAZZONI and BEUTLER 1996). The soluble or membrane-bound ligands produced by the immune cell interact with an appropriate receptor of the TNF receptor/nerve growth factor receptor (TNFR/NGFR) family. Members of this family share a variable number of cysteine-rich extracellular domains but differ considerably in their intracellular portions. Some of them contain so-called death domains that, upon oligomerization of the receptors, recruit adaptor molecules such as FADD. This in turn triggers the association and activation of caspases, like caspase-8 (FLICE), which activates other caspases, ultimately leading to apoptosis (NAGATA 1997). Caspases are the crucial executors of apoptosis, which cleave key structural elements of the cell, resulting in the typical signs of apoptosis (EARNSHAW et al. 1999). Although the signaling pathways originating from different members of the TNFR/NGFR family share key components (ASHKENAZI and DIXIT 1998), the cellular response is distinct; therefore, we will discuss each system separately.

6.2 Modulation of TNF Activities

6.2.1 Cytolytic and Proinflammatory Activities of TNF

TNF is a pleiotropic inflammatory cytokine that is typically produced at the site of inflammation, mainly by activated macrophages and monocytes but also by many other cell types, such as T cells, B cells, and fibroblasts (BAZZONI and BEUTLER 1996). It is expressed as a 26-kDa cell surface protein from which a 17-kDa fragment is cleaved. TNF acts as a trimer by binding to two distinct TNF receptors, TNFR1 (CD120a or p55) and TNFR2 (CD120b or p75) (VANDENABEELE et al. 1995). Another ligand of these TNFRs is lymphotoxin-α (TNF-β). TNFRs are expressed on virtually all cells of the body and belong to the still-growing TNFR/NGFR family, which includes a number of death receptors, like CD95 (Fas, APO-1), death receptor 3 (DR3), DR4 (TRAIL-R1), and DR5 (TRAIL-R2) (NAGATA 1997; ASHKENAZI and DIXIT 1998; DEGLI-ESPOSTI 1999). Binding of TNF results in clustering of the TNFR and recruitment to the cytoplasmic tail of several adaptor molecules, such as FADD, TNFR-associating protein with death domain (TRADD), receptor interacting protein (RIP), and TNFR-associated factors (TRAFs). These proteins initiate several signaling pathways culminating in the activation of (a) transcription factors, such as NF-κB, AP-1, and c-jun, (b) caspases, and (c) cytosolic phospholipase A_2 ($cPLA_2$), an enzyme responsible for production of inflammatory mediators. Whereas activation of executer caspases leads to apoptosis, activation of NF-κB may serve an antiapoptotic function (WALLACH 1997). How these two opposing activities of TNF are orchestrated to generate a defined cellular response remains a challenging question.

The individual response of cells to TNF binding includes induction of proliferation, differentiation, cell survival, or cell death (WALLACH 1997). Several lines of evidence suggest that TNF exerts an antiviral effect. At high concentrations TNF can inhibit the replication of certain viruses (including Ads) in vitro, possibly by inducing lysis and/or apoptosis of infected cells. However, as TNF-mediated activation of NF-κB induces a variety of inflammatory cytokines, TNF may also exert its antiviral function indirectly (HATADA et al. 2000). Cytolysis by TNF appears to involve $cPLA_2$, which is activated by MAP kinase and possibly caspase-8 (WISSING et al. 1997; WALLACH et al. 1999) and is translocated to membranes where it can release arachidonic acid (AA).

6.2.2 Induction of TNF Susceptibility by E1A

Interestingly, Ad infection in mice induces TNF in the infected tissue (GINSBERG et al. 1991). Several years ago, it was discovered that infection of cells with Ad mutants lacking the E1B and/or the E3 region renders murine C3HA cells susceptible to TNF-mediated lysis, whereas cells infected with wild-type Ads are protected (GOODING et al. 1988). This suggested that (a) an Ad function exists that induces TNF sensitivity and (b) E1B and E3 products seem to protect against TNF-mediated lysis. Mapping studies with transfected cell lines and infection with Ad mutants revealed that induction of TNF susceptibility was mediated by the immediate-early protein E1A (CHEN et al. 1987; DUERKSEN-HUGHES et al. 1989). Essentially the same functional domains of E1A that are involved in the induction of DNA synthesis and p53 accumulation are required to sensitize infected cells to TNF lysis (SHISLER et al. 1996). E1A-mediated sensitization to TNF-induced lysis appears to involve inhibition of TNF-mediated NF-κB activation. E1A inhibits, presumably in an indirect way, IKK-mediated IκB phosphorylation and, consequently, NF-κB release to the nucleus and renders cells more sensitive to TNF-induced apoptosis (SHAO et al. 1999). However, the role of E1A in the regulation of the NF-κB system is complex and seems to depend on the cell type, the time point within the Ad life cycle, the influence of other viral gene products, and the experimental system used (SCHMITZ et al. 1996; SHISLER et al. 1996; SHAO et al. 1999).

Four adenovirus proteins have been proposed to contribute to protection of infected cells from TNF-mediated cytolysis: the E1B/19K protein (reviewed in WHITE 1998) and three E3 proteins, 14.7K, 10.4K, and 14.5K (WOLD et al. 1995) found in all human Ad subtypes examined (Fig. 1). The latter two act as a complex.

6.2.3 Inhibition of TNF Activities by E3/14.7K

As pointed out above, primary cells are resistant to TNF but become susceptible by expression of the E1A protein. However, infection with representative Ad serotypes of subgenera A–E protects murine cells from TNF-mediated apoptosis (HORTON et al. 1990; WOLD et al. 1995). By infecting mouse C3HA cells with mutant Ads containing various deletions within the E3 region the protective function was initially mapped to the E3/14.7K protein (GOODING et al. 1988). 14.7K protects most mouse cell lines and can prevent TNF cytolysis independent of other Ad proteins (HORTON et al. 1991). The anti-TNF effect of 14.7K was also demonstrated in vivo with heterologous systems. A recombinant vaccinia virus producing TNF and coexpressing the Ad2 14.7K protein (VV14.7TNF) exhibits an increased virulence compared with VVTNF by reversing the attenuating effect of TNF on VV replication (TUFARIELLO et al. 1994; WOLD et al. 1995).

The E3/14.7K proteins vary in size from 122 amino acids (aa) in Ad40 to 136 aa in Ad3. Sequence comparison of 14.7K proteins from Ads of the different subgenera reveals a relatively high degree of conservation (Fig. 2): 34 aa are identical, and the overall homology is in general >50%. Among subgenera B, D, and E the homology is even greater than 70%. This is significantly higher than that observed for the E3/14.5K and E3/19K proteins, which have an overall homology of ~32% (21%–51%) and ~38% (30%–61%), respectively. Thus, the 14.7K proteins exhibit a degree of conservation similar to that of the 12.5K proteins (53%–79%). Some 14.7K proteins from Ads belonging to the same subgenus, Ad2 and Ad6 (subgenus C) or Ad17, Ad19a, and Ad37 (subgenus D) are even identical (Fig. 2). A relatively large proportion of amino acids comprises charged residues. 14.7K is localized in the cytosol and in the nucleus (GOODING et al. 1988). A structure-function analysis of the 128-aa-long Ad5 14.7K protein using in-frame deletions and cysteine (Cys) replacement mutations suggested that functionally critical amino acids are distributed

Fig. 2. Multiple alignment of 14.7K amino acid sequences. Alignment of 14.7K protein sequences was carried out with the DNASTAR software Megalign using Clustal method with PAM250 residue weight table. Ads for which the sequences were determined and their association to a particular subgenus are depicted on the *right* of the single amino acid code. Amino acids that conform to the consensus are *shaded*. *Arrowheads* above the sequences indicate aa identical in all subgenera. The E3 sequences (this report) of Ad4 (strain RI-67; ATCC VR4) were obtained after propagating the virus four times in A549 cells. The Ad4 E3 region containing *Hin*dIII-*Bcl*I DNA fragment (map units 71.3–86.9) was cloned into the LITMUS 28 cloning vector (New England Biolabs, Frankfurt, Germany). The sequence of the Ad4 E3 region was established using the Genome Priming System (New England Biolabs, Frankfurt, Germany) and dye-terminator chemistry with the automatic DNA sequencer ABI 373A (Perkin Elmer/Applied Biosystems, Weiterstadt, Germany). Sequences from both strands were analyzed. The sequence was submitted to GenBank and obtained the accession number AF361223. References for other sequences and GenBank accession numbers are as follows: Ad2 (HERISSÉ et al. 1980; HERISSÉ and GALIBERT 1981); Ad1, Y11032; Ad6, Y16037; Ad5, M73260; Ad3 (SIGNÄS et al. 1986); Ad7 (HONG et al. 1988); Ad11 (MEI and WADELL 1992); Ad35 (BASLER et al. 1996); Ad19a, Ad37, Ad15, Ad9 and Ad8 (BURGERT and BLUSCH 2000), Ad17, AF108105; Ad12 (SPRENGEL et al. 1994) and Ad40 (DAVISON et al. 1993); Ad41 (YEH et al. 1996)

throughout the entire protein (RANHEIM et al. 1993). Three of the six Cys replacement mutants, with serines 53, 59, and 130 in our comparison (Fig. 2; corresponding to 44, 50, and 119 in Ad5) substituted for Cys, were no longer protective against TNF cytolysis. Interestingly, these three cysteines are strictly conserved whereas the exchange of nonconserved Cys at position 111, 116, and 123, (100, 105, and 112 in the Ad5 sequence) had no effect on 14.7K function (Fig. 2).

Little is known about the molecular mechanism whereby the E3/14.7K protein interferes with TNF-mediated lysis/apoptosis. Three different classes of proteins have been proposed to be targets of 14.7K. First, Wold and colleagues suggested that 14.7K interferes with the activation or function of cPLA$_2$, an enzyme that is thought to be required but is not sufficient for TNF cytolysis (WOLD et al. 1995; WALLACH et al. 1999). TNF treatment of uninfected cells results in the translocation of cPLA$_2$ from the cytosol to membranes and in the release of AA. This process requires submicromolar amounts of calcium. In Ad-infected cells translocation of cPLA$_2$ and release of AA is not observed. Mutant Ads that lack 14.7K (KRAJCSI et al. 1996) induce AA release in most murine cells (e.g., C3HA), indicating that 14.7K prevents this process. This has been confirmed by generating 14.7K transfectants. In other murine cells (e.g., C127) 10.4K-14.5K (in the absence of 14.7K) also block translocation of cPLA$_2$ to membranes and release of AA. The involvement of cPLA$_2$ in the release of AA and cell death upon infection with mutant viruses was confirmed by using specific cPLA$_2$ inhibitors and an antisense approach (THORNE et al. 1996).

Only recently were TNF-induced effects investigated with relevant human cells. In a series of experiments it was clearly demonstrated that in human A549 cells only the membrane proteins E3/10.4K-14.5K, and not 14.7K, are required to inhibit TNF-induced translocation of cPLA$_2$ and the subsequent AA release (DIMITROV et al. 1997). Precisely at which step this process is affected is not clear, but Ad infection blocks neither TNF-induced phosphorylation of cPLA$_2$ nor activation of NF-κB. Thus, there is no general block in TNFR signaling. One possibility is that the viral proteins may affect calcium mobilization required for the translocation step. By inhibition of AA release the E3 proteins might also block, apart from cytolysis, inflammation triggered by AA metabolites such as leukotrienes and prostaglandins.

A second described potential target of 14.7K is caspase-8, which is recruited to the cytoplasmic domains of the TNFR1 and Fas upon binding of the death-inducing ligands. By overexpressing 14.7K of Ad5 together with FasL, FADD, or caspase-8 via Ad vectors, it was shown that the 14.7K protein can bind to and inhibit the function of caspase-8 (CHEN et al. 1998). This suggested that 14.7K interferes with signal transduction from the death receptor. As caspase-8 has been reported to cleave and thereby activate cPLA$_2$, this activity of 14.7K could explain its inhibition of cPLA$_2$. However, a profound functional inhibition of caspase-8 during normal Ad2 infection was not observed (ELSING and BURGERT 1998; HORWITZ 2001).

Horwitz and coworkers have employed the yeast two-hybrid system to identify cellular 14.7K-interacting proteins (FIPs). The most attractive protein identified in this screen appears to be FIP-3, which is identical to NF-κB essential modulator (NEMO), a protein independently isolated by complementation cloning as an essential component of the IκB-α kinase (IKK) complex (YAMAOKA et al. 1998). NEMO (also named IKKγ or IKKAP-1) is absolutely required for NF-κB activation by TNF and IL-1. However, overexpression of FIP-3 has been reported to inhibit both basal and TNF-induced transcriptional activity of NF-κB and to induce delayed apoptosis (LI et al. 1999). FIP-3/NEMO/IKKγ binds to IKKβ, NF-κB-inducing kinase (NIK), and RIP, a protein recruited to the cytoplasmic domains of TNFR1 and Fas (LI et al. 1999). RIP also has been demonstrated to be crucial for TNF-induced NF-κB activation (KELLIHER et al. 1998). Why FIP-3 at low concentrations is essential for and upon overexpression acts as an inhibitor of NF-κB activation is currently unclear (HORWITZ 2001).

FIP-2, also called NEMO-related protein, contains two leucine zipper domains. Overexpression of FIP-2 does not cause cell death, but can reverse the protective effect of 14.7K on cell death induced by overexpression of the TNFR intracellular domain or RIP (LI et al. 1998). FIP-1 is identical to RagA and belongs to the family of small GTPases (LI et al. 1997). It does not cause cell death but forms ternary complexes with 14.7K and TCTEL, a component of the microtubule motor protein dynein (HORWITZ 2001). It will be interesting to see whether these interactions of 14.7K are also detectable during virus infection and whether they influence the TNF signal cascade.

6.2.4 Inhibition of TNF Activities by E3/10.4K–14.5K

With better-characterized virus mutants, it was subsequently shown that two other E3 proteins, E3/10.4K and 14.5K, are also able to protect cells from TNF-mediated lysis (GOODING et al. 1991b). Their activity was revealed by using murine C127 cells that remained resistant to TNF lysis on infection with mutant Ads lacking 14.7K. Protection from TNF lysis by 10.4K–14.5K is observed in 11 of 15 mouse cells tested (GOODING et al. 1991b). It is still unclear why E3/10.4K and 14.5K proteins exhibit a differential effect in different cell lines and, more importantly, whether these proteins are also able to protect human cells from TNF. Both proteins are integral membrane proteins that associate noncovalently with each other (WOLD et al. 1995). They do not appear to modulate the murine and human TNFR (SHISLER et al. 1997; Obermeier et al., manuscript in preparation). Recent experiments with human A549 cells demonstrate that these proteins inhibit TNF-induced translocation of cPLA$_2$ to membranes and the subsequent AA release without interfering with cPLA$_2$ phosphorylation. In contrast, 14.7K and E1B/19K are not required for this inhibitory effect (DIMITROV et al. 1997).

6.2.5 Influence of E1B/19K on TNF-Mediated Cytolysis

Conflicting data have been published as to the protection from TNF-mediated cytolysis by the E1B/19K protein. In several human cell lines (HEL-299 fibroblasts and cervical carcinoma cells ME-180 and HeLa), sensitized to TNF cytolysis by cycloheximide treatment or by infection with mutant viruses lacking E1B/19K, expression of E1B/19K by transfection or infection has been reported to significantly suppress TNF- and Fas-induced DNA-fragmentation and cell death (GOODING et al. 1991a; HASHIMOTO et al. 1991; WHITE 1998; see below). The mechanism of protection has not been elucidated, but protection was not accompanied by alteration of TNFR levels (or Fas). E1B/19K was unable to inhibit $cPLA_2$ translocation to membranes in human cells and did not block apoptosis induced by overexpression of caspase-8, whereas apoptosis caused by overexpression of FADD was inhibited (PEREZ and WHITE 1998). The altered subcellular localization of the *Caenorhabditis elegans* CED-4 adaptor upon overexpression of E1B/19K was taken as evidence to suggest that E1B/19K inhibits caspase-mediated cell death by interfering with the function of adaptor molecules such as FADD or Apaf-1, the presumed mammalian homolog of CED-4 (PEREZ and WHITE 1998). However, E1B/19K does not seem to interact directly with FADD, Apaf-1 or FLICE, nor does it inhibit their recruitment to the death-inducing complex (MORIISHI et al. 1999). Recent data suggest that E1B/19K blocks TNF-mediated death signaling by binding to and inhibiting a specific form of Bax, thereby preventing cytochrome *c* release and caspase-9 activation. E1B/19K expression interrupted caspase-3 processing, permitting cleavage and removal of the p12 subunit but not of the prodomain, consistent with caspase-8 but not caspase-9 enzymatic activity (PEREZ and WHITE 2000). Other reports show that E1B/19K does not profoundly inhibit TNF-mediated cytolysis. In a systematic comparison between Bcl-2, Bcl-x_L, and E1B/19K positive stable cell lines, all three proteins inhibited apoptosis induced by growth factor deprivation, but they were ineffective suppressors of Fas- or TNF-induced cell death in several mouse and human cell types (HUANG et al. 1997). Similar conclusions have been reached by COOK et al. using stable hamster and mouse transfectants (COOK et al. 1999). In the viral context, E1B/19K does not block TNF-mediated cytolysis of murine cells. Presumably, the discrepancies are due to the different cellular background and the difficulty of discriminating between cytolysis induced by the *cyt/deg* phenotype and the lysis-promoting TNF effects. More studies will be required to clarify this issue.

6.3 TNF-Induced Upregulation of E3 Protein Expression: Keeping Up with Host Cytokines

Apart from the cytolytic activity, TNF has also multiple immunoregulatory functions. For example, it induces the release of proinflammatory products like leukotrienes and prostaglandins that are thought to be involved in the recruitment of leukocytes. Like IFN-γ, TNF also stimulates transcription of MHC genes and that of other components of the antigen presentation pathway and thereby enhances T cell recognition. As discussed below, E3/19K interferes with antigen presentation of MHC class I antigens. By treating E3[+] cells with TNF, it was observed that TNF does not lyse these cells but rather leads to a further reduction of MHC antigens on the cell surface (KÖRNER et al. 1992). This effect is due to an increased synthesis of

E3/19K. Subsequent studies showed that all E3 proteins are upregulated in vitro and in vivo (FEJER et al. 1994; DERYCKERE et al. 1995) and that this effect is mediated by the cytosolic transcription factor NF-κB, which binds to and stimulates the E3 promoter (WILLIAMS et al. 1990; DERYCKERE and BURGERT 1996b). Remarkably, the E3 promoter is the only Ad promoter containing NF-κB sites and at least one site is conserved in Ads of all subgenera. The functional relevance of this positive regulatory loop is supported by data showing that Ad infection in mice induces TNF in the infected tissue (GINSBERG et al. 1991). Thus, instead of killing Ad-infected cells, TNF may actually potentiate the immunosubversive functions of E3 proteins by upregulating their expression. Combined with the TNF inhibitory activity of E3 proteins, this positive regulatory loop appears to ensure efficient virus reproduction despite the presence of TNF during the immune response. Alternatively, it may facilitate persistence in lymphocytes or lymphoid tissues where NF-κB is being constantly stimulated during lymphocyte activation (BURGERT 1996; HATADA et al. 2000). Taken together, Ads devote several proteins to counteract TNF activities.

6.4 Inhibition of Fas-Mediated Apoptosis

6.4.1 Downregulation of Fas Surface Expression by E3/10.4K–14.5K

Fas (APO-1 or CD95) belongs to the TNFR/NGFR family and is constitutively expressed in a wide variety of tissues (NAGATA 1997). In contrast, expression of the Fas ligand (FasL) is largely restricted to cells of the immune system, such as NK cells and activated T cells, and cells in immune privileged sites, such as the eye and the testes. Defective Fas or FasL genes result in accumulation of peripheral B- and T lymphocytes and the enlargement of lymph nodes. Thus, FasL expression on activated T cells is thought to be important for immune cell homeostasis and for downregulation of an immune response, but it also allows CTL and NK cells to induce Fas-mediated apoptosis of target cells (NAGATA 1997). The contribution of Fas-mediated cytolysis in vivo to the resolution of virus infections depends on the nature of the virus (HARTY et al. 2000; TRAPANI et al. 2000). In contrast to perforin, Fas is not critical for the elimination of the noncytolytic lymphocytic choriomeningitis virus, whereas Fas-mediated apoptosis is required for clearance of primary influenza infections from the lung. The Fas system, but also IFN-γ, may have a protective role in infections with vaccinia virus, Semliki forest virus, vesicular stomatitis virus, and cowpox virus, because perforin-dependent lysis was not required for resistance. However, Fas not only signals death, but, depending on the differentiation stage, can also stimulate lymphocyte proliferation (DESBARATS et al. 1999).

FasL binding induces the oligomerization of Fas via their death domains (DD) and the recruitment to the clustered receptors of the DD-containing adaptor molecule FADD. FADD contains an additional caspase recruitment domain, called death effector domain (DED), that binds to the DED of procaspase-8. Its activation initiates a cascade of proteolytic cleavages, involving downstream

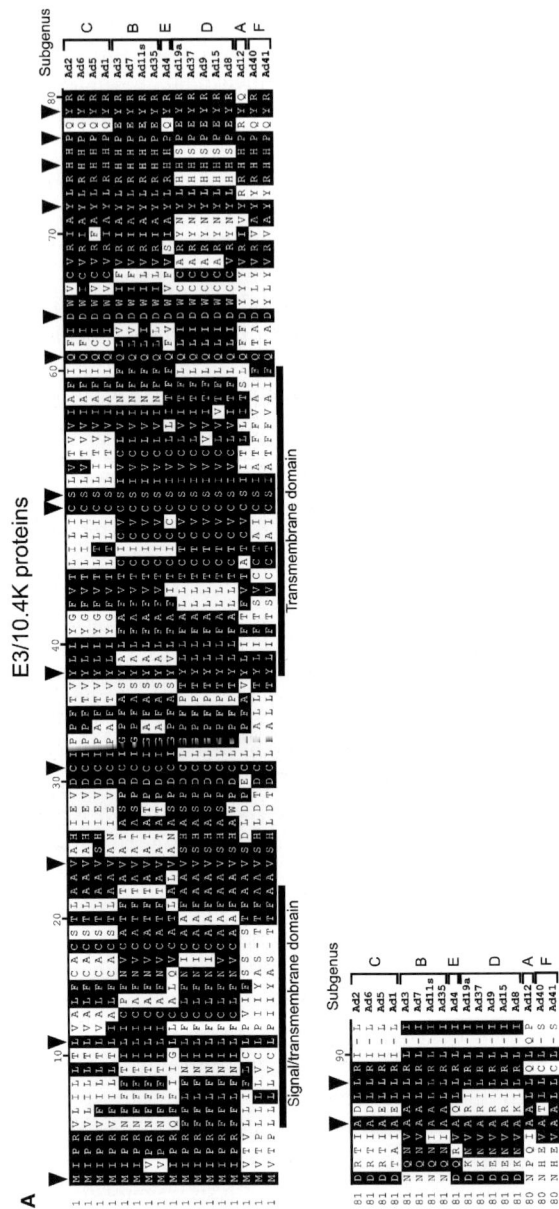

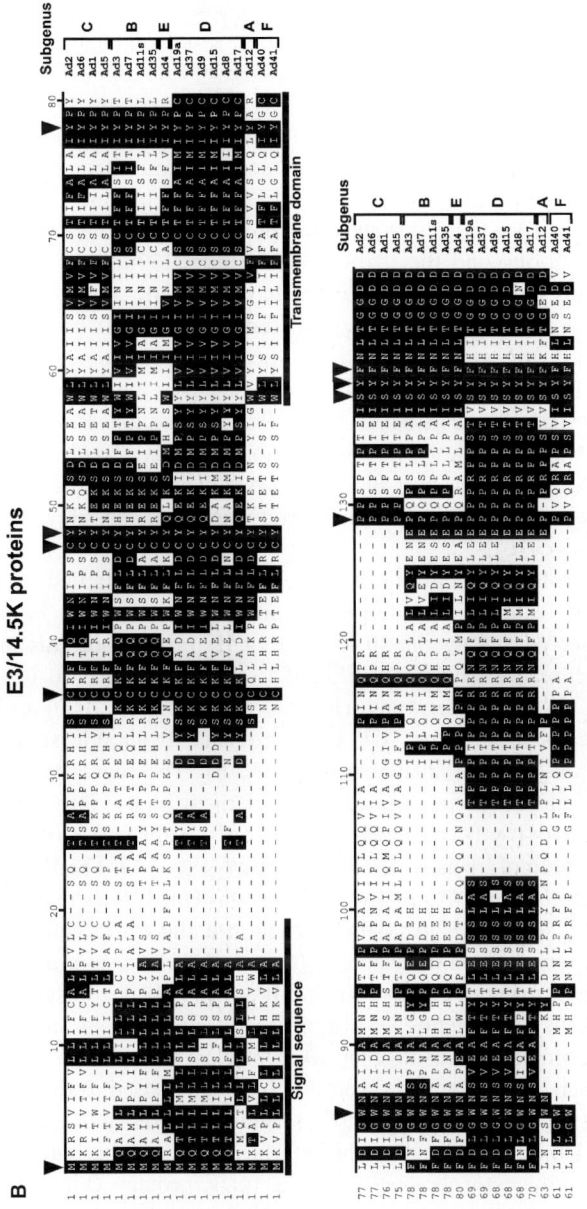

Fig. 3. Amino acid sequence comparison of E3/10.4K (**A**) and E3/14.5K (**B**) proteins. The alignments were done as described in the legend to Fig. 2, where relevant accession numbers and references are given. The 14.5K sequence alignment was further optimized manually. *Arrowheads* above the sequences indicate aa identical in all subgenera. The predicted signal sequences and transmembrane domains are *underlined*. The shading code is as in Fig. 2

effector caspases like caspase-3 and -9 whose activation causes the degradation of key structural components of the cells and ultimately leads to apoptosis (ASHKENAZI and DIXIT 1998; EARNSHAW et al. 1999). Viral interference with apoptosis of infected cells is a prerequisite for effective reproduction of viruses and may be important for establishing viral persistence (TEODORO and BRANTON 1997). Many viral antiapoptotic proteins have been described that target various steps of the apoptosis cascade. Viral FLICE-inhibitory proteins (vFLIPs), for instance, interfere with signal transduction from the cell surface receptor, whereas others block the caspases themselves (TEODORO and BRANTON 1997; O'BRIEN 1998).

Recently, several groups independently reported that the 10.4K–14.5K complex (also named RID for receptor internalization and degradation) prevents apoptosis triggered by FasL or agonist Fas antibodies. This phenomenon is accompanied by downregulation of the apoptosis receptor Fas from the cell surface of Ad-infected and E3-transfected cells (SHISLER et al. 1997; ELSING and BURGERT 1998; TOLLEFSON et al. 1998). Remarkably, downregulation of Fas is also observed after infection of primary cells (ELSING and BURGERT 1998). Thus, after the phenomenon is likely to be relevant for the efficient reproduction and, possibly, the persistence of the virus in vivo.

The 10.4K and 14.5K proteins are both transmembrane proteins. As a first step in characterizing structural features of potential functional relevance we have sought to identify aa conserved among the 10.4K and 14.5K proteins of Ads of different subgenera. Therefore, we have determined the sequences of the 10.4K–14.5K proteins of several subgenus D serotypes (BURGERT and BLUSCH 2000) and add here those for the subgenus E serotype Ad4. Thus, sequences from at least one member from each subgenus are now available (Fig. 3A, B). The sequence homology between 10.4K proteins of different subgenera is relatively high (35%–72%, average: 47.5%), and 15 aa are strictly conserved. Except for the subgenus F (Ad40 and Ad41) homologs, the length of 91 aa is strictly conserved (Fig. 3A). The 10.4K proteins contain a large number (~50%) of hydrophobic aa. Only 18 aa may be exposed on the cell surface, whereas ~30 might extrude into the cytoplasm. Interestingly, all 10.4K sequences contain dileucine motifs, motifs known to mediate transport to endosomes and lysosomes (SANDOVAL and BAKKE 1994). Two leucines (LL), or IL in 10.4K proteins of subgenus D, are present in positions –4 and –5 from the C-terminus (position 87/88). In subgenera B–E, the last two aa (IL or LI) may also constitute a dileucine-like motif. Additional leucine-based transport motifs (positions 62/63 in subgenera B and D) and tyrosine-based transport motifs (Y_{65} in subgenera A and F) are visualized at the interface between the putative transmembrane segment and the cytosolic portion of the 10.4K proteins (Fig. 3A). The latter motifs conform to the sequence YxxΦ (where Y is tyrosine, X is any amino acid, and Φ represents a bulky hydrophobic aa, e.g. L, I, F). Their close proximity to the putative transmembrane segment argues against the possibility that they represent transport determinants. It will be interesting to determine whether or not these sequences are indeed involved in targeting the proteins to endocytic compartments and, if so, whether their differential presence and position within the 10.4K proteins of different subtypes reflect distinct intracellular trafficking routes.

10.4K is expressed as two isoforms. In one of these the signal peptide sequence is cleaved, whereas in the other it remains attached and serves as a second membrane anchor (WOLD et al. 1995). Thus, the latter form is predicted to traverse the lipid bilayer twice. Both 10.4K species are linked by a disulfide bond formed between a cysteine residue at position 31, which is strictly conserved (Fig. 3A). One or both isoforms may form physical complexes with the 14.5K protein that appear to be expressed on the cell surface (STEWART et al. 1995). In support of this notion, it was recently shown that both proteins are required for Fas downregulation (SHISLER et al. 1997; ELSING and BURGERT 1998; TOLLEFSON et al. 1998). A similar requirement for both 10.4K and 14.5K proteins was shown for downregulation of the epidermal growth factor receptor (EGFR; TOLLEFSON et al. 1991; ELSING and BURGERT 1998). However, one group claims that only 10.4K is required for this activity (CARLIN et al. 1989; HOFFMAN and CARLIN 1994). The reason for these discrepancies are not entirely clear, but they may be due in part to the utilization of virus deletion/insertion mutants exhibiting altered splicing of E3 mRNAs and thus an unpredictable expression pattern for E3/10.4K and 14.5K. The implications of the EGFR modulation for virus reproduction, its potential relationship to the antiapoptotic activity of the two proteins, and the outcome of an infection in vivo are still unknown.

The sequence homology between 14.5K proteins of different subgenera is significantly lower (21%–50%, average ~30%) than that of 10.4K and 14.7K proteins. In the mature protein, only 9 of 91–127 aa are strictly conserved (Fig. 3B). The 14.5K product is a type I transmembrane protein consisting of a signal sequence, a short extracellular domain of variable length (20–40 aa), a transmembrane segment, and a cytoplasmic tail of ~46–66 aa. The length of the mature protein seems not to be critical for 14.5K function, because it varies from 91 to 127 aa (Fig. 3B). Interestingly, all 14.5K proteins contain proline-rich sequences in their cytoplasmic tail (BURGERT and BLUSCH 2000) that might serve as a protein-interacting module. The Ad5 protein is O-glycosylated and phosphorylated on serines close to the C-terminus (KRAJCSI and WOLD 1992; WOLD et al. 1995). Remarkably, 14.5K proteins also contain putative transport motifs in their cytoplasmic tail, conforming to the sequence YxxΦ. Two of these, beginning with Y in positions 78 and 139, in the comparison shown, respectively, are conserved. Y_{78} may be located within the lipid bilayer. A third YxxΦ motif (Y_{80}) is only found in 14.5K proteins of subgenus C (Fig. 3B). These "tyrosine-based signals" are involved in sorting of proteins into endosomes or lysosomes in that they are bound by adaptor protein complexes, like AP-1 or AP-2 (KIRCHHAUSEN 1999), that mediate cargo recruitment and vesicle budding at the Golgi and the plasma membrane, respectively (MARKS et al. 1997). Indeed, we recently showed that C-terminal peptides of the Ad2 14.5K and 10.4K proteins bind to AP-1 and AP-2 complexes in vitro in a Y- and LL-specific manner (Hilgendorf et al., manuscript in preparation). It will be interesting to see which of these determinants are required for 10.4K–14.5K-mediated receptor modulation.

The exact mechanism for Ad-mediated downregulation of Fas from the cell surface has not been elucidated. Our data suggest that a complex of E3/10.4K–14.5K binds to Fas on the cell surface. So far, no evidence has been presented for a direct interaction, indicating that the interaction may be short-lived or possibly mediated by other cellular proteins. The Fas-associated 14.5K protein may then be recognized by the AP-2 adaptor, and the complex is recruited into coated pits and subsequently transported into endosomes. Fas dissociates and is targeted to lysosomes where it is degraded, whereas 10.4K–14.5K might recycle to the cell surface to bind newly synthesized Fas molecules. Support for the proposed pathway comes from experiments treating infected cells with lysosomotropic agents, such as chloroquine and ammonium chloride, or bafilomycin A1, an inhibitor of the vacuolar ATPase (ELSING and BURGERT 1998; TOLLEFSON et al. 1998). Under these conditions Fas is not degraded but accumulates predominantly in vesicles staining for lysosome-associated membrane protein 2, whereas 14.5K exhibits only a limited colocalization with Fas (Hilgendorf et al., unpublished data). The kinetic of Fas disappearance from the cell surface of infected cells is significantly more rapid than after inhibition of cell surface transport of Fas by brefeldin A in mock-infected cells. This argues for an active rerouting of Fas from the cell surface rather than a direct transport from the *trans*-Golgi network to lysosomes (ELSING and BURGERT 1998). It remains to be clarified whether the mechanisms for downregulation of Fas and the EGFR are identical.

6.4.2 Influence of E3/14.7K on Fas-Mediated Signaling

Overexpression of FasL, FADD, or FLICE with Ad vectors induces apoptosis that can be inhibited by coexpression of the Ad5 14.7K protein. In this system, the 14.7K protein was shown to bind to FLICE, suggesting that 14.7K might interfere with Fas- and TNF-mediated apoptosis by inhibiting the activation of FLICE (CHEN et al. 1998). However, cells infected with a mutant virus lacking all E3 proteins except 14.7K and 12.5K are still sensitive to Fas-mediated cell death. Thus a functional inhibition of FLICE during normal Ad2 infection was not confirmed (ELSING and BURGERT 1998; HORWITZ 2001).

6.5 Downregulation of TRAIL Receptors by E3 Proteins

Interestingly, receptors belonging to the TNFR/NGFR family are differentially affected by the 10.4K–14.5K proteins. In contrast to Fas and TRAIL receptor 1 and 2 (TRAIL-R1 and -R2), murine TNFR or human CD40 are not downregulated by subgenus C Ads (SHISLER et al. 1997; ELSING and BURGERT 1998; BENEDICT et al. 2001; Obermeier et al., unpublished data). TRAIL (also named APO-2L) belongs to the TNF superfamily and displays the highest similarity to FasL. It is a type II transmembrane protein that can be cleaved from the cell surface. The functionally active form is a homotrimer. TRAIL appears to be widely expressed, because TRAIL-specific transcripts have been found in many tissues. TRAIL has also been detected on the cell surface of activated T- and B cells, dendritic cells, and monocytes (DEGLI-ESPOSTI 1999; WALCZAK and KRAMMER 2000). So far, five TRAIL-Rs, named TRAIL-R1-4 and osteoprotegerin, have been identified. TRAIL-R1 (DR4) and TRAIL-R2 (DR5) are most closely related and contain a DD in their cytoplasmic tail, whereas the other TRAIL-binding proteins lack such a domain. TRAIL-R4 has a truncated DD, TRAIL-R3 is a glycosylphosphatidylinositol-anchored protein, and osteoprotegerin is secreted. It was suggested that TRAIL-R3 and TRAIL-R4 act as decoy receptors that compete with TRAIL-R1 and TRAIL-R2 for TRAIL and protect cells from TRAIL-mediated apoptosis. However, no strict correlation was found between the expression of these receptors and TRAIL resistance of cells. Thus, sensitivity to TRAIL seems to be determined by intracellular regulators, e.g., cellular FLIPs. Upon TRAIL binding to TRAIL-R1 and TRAIL-R2, a death-inducing signaling complex is formed containing FADD and caspase-8 (KISCHKEL et al. 2000; SPRICK et al. 2000), which is followed by caspase-3 activation and the cleavage of critical cellular substrates. Binding of TRADD and RIP might be responsible for TRAIL-induced NF-κB activation (CHAUDHARY et al. 1997).

Recent reports suggest an involvement of TRAIL in the antiviral host defense. Apart from the FasL/Fas system, TRAIL is thought to be a major component of the granule-independent cytolytic machinery of CTL and NK cells (ZAMAI et al. 1998; JOHNSEN et al. 1999; KAYAGAKI et al. 1999). TRAIL was detected on $CD4^+$ T cell clones and on IFN- or T cell receptor-stimulated T cells and was utilized for target cell lysis. Moreover, it was shown that TRAIL is induced upon infection of primary fibroblasts and dendritic cells with CMV and measles virus, respectively (SEDGER et al. 1999; VIDALAIN et al. 2000). Concomitant TRAIL-R induction sensitized infected cells to TRAIL-induced killing, which was selectively augmented by IFN-γ or TNF (SEDGER et al. 1999). Likewise, reovirus infection triggers apoptosis via induced expression of TRAIL-R2 and the release of TRAIL from infected cells (CLARKE et al. 2000). Thus, TRAIL has potent antiviral activity, in particular in concert with IFN-γ.

As for TNF and FasL, transfection of Ad E1A sensitized various human tumor cell lines to TRAIL-induced killing. In contrast, no significant killing was observed when the same cells were infected with Ad5. However, apoptosis was induced by infection with viral mutants lacking the E3 region and to a lesser degree by an Ad5 mutant lacking the E1B/19K protein. Thus, Ad protects infected cells from TRAIL-induced apoptosis, and this protection is predominantly mediated by E3 proteins (ROUTES et al. 2000). Subsequent studies with Ad2 indicated that three E3 proteins are required for protection, namely, the E3/10.4K-14.5K complex and the E3/6.7K protein. Like Fas, TRAIL-Rs are downregulated and seem to be transported to endosomes/lysosomes (BENEDICT et al. 2001). Surprisingly, a tagged version of 6.7K, previously described as an ER protein (WILSON-RAWLS and WOLD 1993), was expressed on the cell surface. Therefore, it was suggested that 6.7K interacts with the 10.4K-14.5K proteins on the cell surface, conferring specificity for TRAIL-Rs. Infection of cells with 10.4K-14.5K-deficient viruses significantly increased their sensitivity to TRAIL-induced apoptosis compared with uninfected cells, suggesting that E3 proteins compensate for the increased sensitivity of infected cells caused by other viral products.

The 6.7K protein encoded in the subgenus C E3 region lacks a classic amino-terminal signal sequence but contains an internal hydrophobic region that may act as a signal-anchor domain (WILSON-RAWLS and WOLD 1993). According to this model, the N-terminus of 6.7K is located in the ER lumen and the protein can be classified as a type III transmembrane protein. On the basis of the differential extraction into the detergent phase of Triton X114 and immunofluorescence staining, it was proposed that 6.7K is an integral membrane protein of the ER. Consistent with this view, the N-terminus was shown to bear a high-mannose or hybrid-type N-glycan. Cell surface expression of a tagged 6.7K protein indicates that natural 6.7K might also be expressed at the cell surface. It will be interesting to investigate why the modulation of TRAIL-Rs (in particular TRAIL-R2) requires 6.7K in addition to 10.4K-14.5K although the modulation of Fas and the EGFR is not dependent on the presence of 6.7K. In contrast to antiapoptotic activities of other viruses (TEODORO and BRANTON 1997; O'BRIEN 1998), which target various steps of the signaling cascade originating from the death receptors, adenovirus has evolved proteins that interfere at the earliest time point possible, the interaction of the death receptor with its deadly ligand. At present, it is unclear whether the E3-mediated protection from FasL and TRAIL counteracts lysis by Ad-specific CTL or NK cells or both or impairs an as-yet unknown immune response mechanism. Nevertheless, the presence of these adenovirus functions strongly suggests that the FasL/Fas and the TRAIL/TRAIL-R systems are important host effector mechanisms targeting Ad-infected cells in vivo. It is also unclear whether the seemingly redundant viral functions that antagonize the same immune mechanism (e.g., TNF lysis) are active in the same cell or are differentially utilized in different cell types and tissues.

7 Influence of Ad on NK Cell Activity

7.1 Role of NK Cells in Antiviral Immunity

NK cells represent an important arm of the innate host defense that is thought to limit viral spread at early stages of the infection. NK cell-mediated cytotoxicity can be potently induced by viral-induced IFN-α/β. In addition to their capacity to directly lyse virus-infected cells via the granule exocytic and the death receptor pathways, NK cells secrete large amounts of antiviral cytokines like IFN-γ and TNF, which also influence the subsequent T cell response (Kos and ENGLEMAN 1996; BIRON et al. 1999). Therefore, NK cells bridge the time gap needed for induction of virus-specific CTL. Inhibitory and activating NK receptors have been described that recognize potential target cells, and both types can be expressed on the same NK cell. Although much has been learned about these NK receptors

(COLONNA et al. 2000; MORETTA et al. 2000), the exact mechanism whereby NK cells integrate opposing signals to kill virus-infected cells remains a puzzling mystery. The outcome of an NK-target cell interaction appears to depend on a delicate balance between the engagement of activating and inhibitory receptors. One important negative signal is the engagement of MHC class I molecules by MHC allele-specific inhibitory receptors. Discrimination between normal and abnormal cells with impaired MHC cell surface expression, a phenotype often triggered by virus infection (BURGERT 1996; YEWDELL and BENNINK 1999; TORTORELLA et al. 2000), is thought to be mediated predominantly by inhibitory NK receptors (MORETTA et al. 2000). Three families of inhibitory HLA (human MHC antigens) receptors on NK cells are distinguished, which share immunoreceptor tyrosine-based inhibition motifs in their cytoplasmic domains. The killer cell immunoglobulin-like receptors (KIRs) recognize groups of HLA-A, B or C alleles, whereas the leukocyte immunoglobulin-like receptors (ILT/LIRs) are more promiscuous and interact with a broad spectrum of HLA alleles. The third group belongs to the C-type lectins and is composed of CD94-NKG2 heterodimers. The latter recognize non-classic HLA-E molecules. Only in the absence of inhibitory interactions, can triggering NK receptors deliver an activating signal (MORETTA et al. 2000). These activating NK receptors can either be isoforms of the KIRs recognizing certain HLA alleles or might be specific for non-MHC ligands. The situation is further complicated by the presence of NK receptors on $CD3^+$ T cells (NKT cells). Moreover, a whole arsenal of potentially cytotoxic effector molecules are expressed and are implicated in NK cell-mediated cytotoxicity. Apart from the exocytic granule-based mechanism involving perforin and granzymes, many members of the TNF family such as TNF, TRAIL and FasL, in membrane-bound and secreted versions, seem to be utilized by NK cells for killing of different target cells (JOHNSEN et al. 1999; KASHII et al. 1999).

7.2 Inhibition of E1A-Induced NK Sensitivity During Infection

How Ads deal with the NK cell response during infection of humans in vivo is unknown. Most studies investigating the relationship between NK cells and Ads have been performed with rodent cells transformed or infected by the nontumorigenic serotypes Ad2 and Ad5 or the highly oncogenic serotype Ad12. With these systems, it was established that expression of Ad2 or Ad5 E1A sensitizes rodent cells for NK cell lysis whereas Ad12 E1A-expressing cells were resistant. It was further shown that the sensitivity of E1A-expressing rodent cells to NK cells inversely correlates with their tumorigenicity (RASKA and GALLIMORE 1982). Interestingly, although transfection of human tumor cells with Ad5 E1A renders them susceptible to NK cell lysis, infection of these cells by Ad5 does not (ROUTES and COOK 1995). Why human Ad-infected cells are resistant despite the decrease in HLA expression (see below) remains to be elucidated. Similar observations have been made for human cells infected by various strains of CMV (FLETCHER et al. 1998). Early studies using HEp-2 cells already showed no correlation between the extent of HLA modulation induced by Ads of different subgenera and the sensitivity of infected cells to NK cells (BOSSE and ADES 1991). Even though prevention of FasL and TRAIL-mediated apoptosis is expected to contribute to this resistance, infection with mutant Ads indicated that neither E1B, E2, E3, nor E4 products are responsible for blocking E1A-mediated induction of NK susceptibility (ROUTES and COOK 1995).

8 Inhibition of CTL Recognition and Lysis

8.1 Antigen Presentation by MHC Class I Molecules

Viral protein expression and turnover rapidly generate peptides that might be presented by MHC class I antigens on the cell surface of infected cells (PAMER and CRESSWELL 1998). On recognition of such a peptide-MHC class I antigen complex, $CD8^+$ CTL release their granular content including perforin and granzymes (mainly A and B) promoting lysis/apoptosis of the infected cell (TRAPANI et al. 2000). Alternatively, apoptosis can be triggered by interaction of the FasL expressed on the T cell surface with the Fas death receptor on the target cell surface (NAGATA 1997; TRAPANI et al. 2000). The relative contribution of the granule-mediated cytolysis versus death receptor-mediated apoptosis in vivo for clearing virus infections is still ill-defined and depends on the virus (HARTY et al. 2000). It seems that cytolytic viruses are predominantly controlled by noncytotoxic effector molecules like IFNs and other cytokines.

Recently, it was shown that a rather large fraction of newly synthesized proteins is directly delivered to the antigen processing machinery (REITS et al. 2000; SCHUBERT et al. 2000). This direct link between protein synthesis and antigen processing makes it of crucial importance for viruses to rapidly counteract antigen presentation and/or recognition by CTL (YEWDELL and BENNINK 1999). Antigenic peptides are predominantly generated by proteasomes in the cytosol and are translocated across the ER membrane by the transporter associated with antigen processing (TAP) (PAMER and CRESSWELL 1998). Within the ER lumen, these peptides might be further trimmed or even generated de novo by N-peptidases and the signal peptidase, respectively (LOBIGS et al. 2000). Binding of the peptides to the MHC class I heavy chain-β_2-microglobulin complex in the ER is assisted by a number of chaperones, such as calnexin, calreticulin and ERp57 (Fig. 4). The initial folding of the heavy chain, promoted by calnexin, enables binding of β_2-microglobulin. The association with ERp57 might ensure correct intramolecular disulfide bond formation, whereas calreticulin seems to aid the folding of the $\alpha 1$ and $\alpha 2$ domains and maintains them in an "open" conformation necessary for peptide binding (PAMER and CRESSWELL 1998; VAN ENDERT 1999). Depending on the HLA allele, efficient transfer of the peptide requires tapasin, a protein that appears to recruit the MHC class I antigen-chaperone complexes to TAP. Binding of the peptide completes the folding process of MHC class I molecules and triggers their

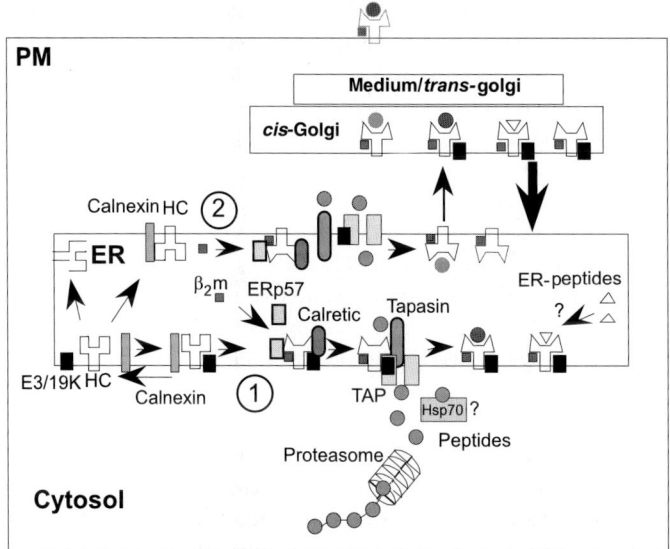

Fig. 4. Assembly of HLA molecules in the presence of E3/19K. The schematic drawing is explained in the text. *HC*, heavy chain. Two types of TAP-E3/19K complexes are shown: one complex (*1*) together with the peptide loading complex containing HC, β_2m, and the associated proteins calreticulin and tapasin and the second complex (*2*) referring to the observation that E3/19K can bind to TAP without associated HLA molecules and therefore might interfere with peptide transport or with the association of HLA molecules to TAP mediated by tapasin. The *large arrow* between Golgi and ER depicts the retrograde transport mediated by E3/19K

dissociation from TAP. Subsequently, loaded MHC molecules seem to be selectively recruited to ER exit sites and transported to the cell surface (SPILIOTIS et al. 2000).

8.2 Subversion of Antigen Presentation by the E3/19K Protein

In adenovirus infected cells, antigen presentation is impaired because of the activity of E3/19K, the most abundant E3 protein expressed by subgenus C Ads in the early phase. This protein binds to MHC class I antigens in the ER and inhibits their transport to the cell surface (ANDERSSON et al. 1985; BURGERT and KVIST 1985). As a consequence, recognition and lysis of E3/19K$^+$ target cells by allogeneic HLA- and Ad-specific CTL in vitro is drastically suppressed (BURGERT et al. 1987; WOLD et al. 1995; BURGERT 1996; FLOMENBERG et al. 1996). In vivo data obtained with animal models support an immunomodulatory role for E3/19K and the other E3 proteins during human Ad infections. Lungs of cotton rats infected with wild-type Ad show a less severe immunopathology than those infected with a mutant virus deficient for E3/19K expression (GINSBERG et al. 1989). Moreover, expression of E3 proteins in β cells of the pancreas can prevent allograft rejection of transplanted islets in certain MHC combinations and, remarkably, suppresses virus-induced diabetes in a murine model (EFRAT et al. 1995; VON HERRATH et al. 1997). Integration of E3/19K in Ad-based gene therapy vectors prolonged transgene expression in some experimental systems, but this beneficial effect depended on the mouse strain and the promoter driving E3/19K expression (LEE et al. 1995; BRUDER et al. 1997; SCHOWALTER et al. 1997).

E3/19K is a type I transmembrane glycoprotein. The mature Ad2 protein (without the signal sequence) consists of 142 aa. Several functional modules can be distinguished. HLA binding is primarily mediated by the luminal part of approximately 100 aa (PÄÄBO et al. 1986b; BURGERT 1996) and critically depends on two intramolecular disulfide bonds formed between Cys11-Cys28 and Cys22-Cys83 (SESTER and BURGERT 1994). The importance of these cysteines (Cys 18, 30, 36. and 92 in Fig. 5) for structure and function is reflected by their conservation in all known E3/19K-like proteins. Other conserved amino acids are distributed throughout the luminal domain (DERYCKERE and BURGERT 1996a; BURGERT and BLUSCH 2000). Alanine-scanning mutagenesis of these conserved amino acids revealed several amino acids essential for E3/19K function (Sester et al., manuscript in preparation). Another important feature of the protein is its ability to localize to the ER. This seems to be mediated by two structural elements: (a) an ER retention signal contained in the transmembrane segment of E3/19K (Sester, Ruzsics, and Burgert, manuscript in preparation) and (b) an ER retrieval signal in the cytoplasmic tail (GABATHULER and KVIST 1990; JACKSON et al. 1990). The motif for ER retrieval consists of two lysines, positioned either in –3 and –4 or –3 and –5 positions from the carboxy terminus (KKXX or KXKXX, where K is lysine and X represents any aa; Fig. 5). These dilysine motifs were subsequently also identified in cellular ER proteins (TEASDALE and JACKSON 1996). Proteins containing these dilysine motifs can reach the *cis*-Golgi, where they are bound by cytosolic coat proteins (COPI) that mediate their retrograde transport to the ER (JACKSON et al. 1993; COSSON and LETOURNEUR 1994; TEASDALE and JACKSON 1996). Mutation of the dilysine motif allows E3/19K to reach the cell surface, but the great majority of the protein remains in the ER. Only when the transmembrane segment of E3/19K is replaced by that of a bona fide plasma membrane protein is efficient cell surface expression observed. Thus, the transmembrane segment of E3/19K strongly contributes to ER retention and may thereby increase the efficiency of interaction with MHC molecules (Sester, Ruzsics, and Burgert, manuscript in preparation).

A schematic outline as to how E3/19K affects MHC assembly is shown in Fig. 4. Recent data show that E3/19K does not require β_2m for binding to the MHC K^d molecule. This suggests that the interaction between E3/19K and MHC molecules occurs early after translocation of both proteins into the ER, presumably before or during binding of MHC molecules to calnexin (SESTER et al. 2000). Surprisingly, this early E3/19K binding does not seem to grossly alter the further assembly process of MHC class I molecules (Fig. 4). MHC molecules are still capable of associating with calnexin and with TAP, and there is indirect evidence that MHC molecules in E3/19K$^+$ cells can bind peptides (Preckel et al., unpublished data). However, expression of E3/19K molecules in mutant cell lines lacking classic MHC molecules demonstrated that E3/19K can bind to TAP, independent

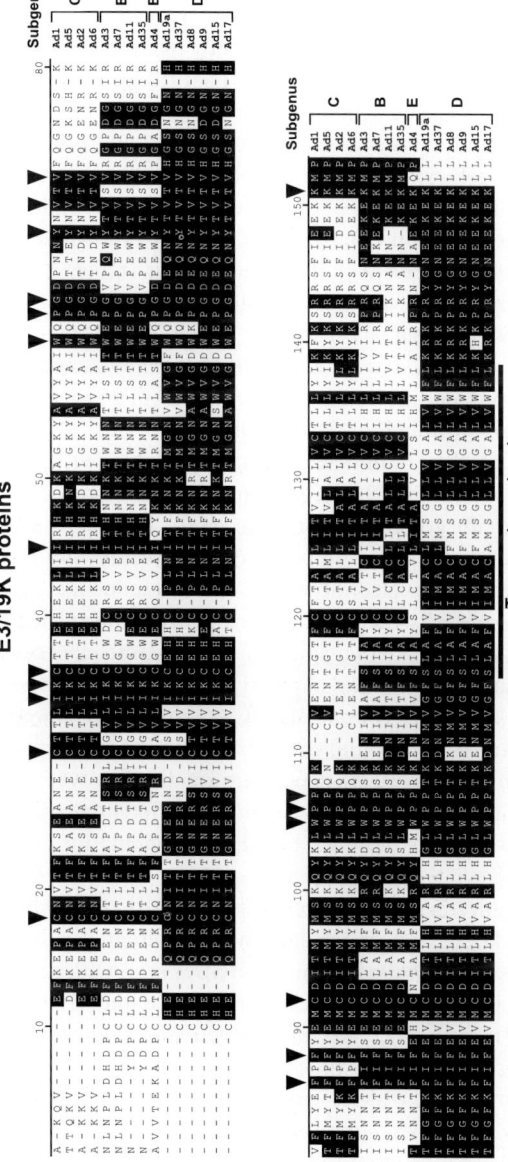

Fig. 5. Amino acid sequence comparison of E3/19K proteins. Alignment was carried out as explained in Fig. 2. The putative signal sequences as predicted by the SignalP software (NIELSEN et al. 1997) are omitted. The consensus is shaded *black*, and strictly conserved aa are marked by *arrowheads* above the sequences. References to the sequences are either given in the legend to Figs. 2 and 3 or are as follows: Ad1, Y16037; Ad6, G2828254; Ad7 (HERMISTON et al. 1993)

of these MHC molecules (Fig. 4, pathway 2). This implies that E3/19K might have a second function, namely, to interfere with peptide transfer by TAP or tapasin (BENNETT et al. 1999). Further studies will be needed to directly assess the effect of E3/19K on peptide transport and transfer. Interestingly, E3/19K also increases complex formation of the MHC K^d molecule with amyloid precursor-like protein 2, a protein previously implicated in peptide transfer to K^d (FEUERBACH and BURGERT 1993; SESTER et al. 2000). Taken together, E3/19K seems not to interfere primarily with the assembly of MHC molecules but rather abolishes egress of the completely assembled complex out of the ER/*cis*-Golgi compartment. Little is known about the fate of the retained MHC molecules except that they are rather stable.

With the exception of subgenera A and F Ads, all human Ad serotypes (subgenera B–E) express an E3/19K protein (PÄÄBO et al. 1986a; DERYCKERE and BURGERT 1996a; BURGERT and BLUSCH 2000). Depending on the number of N-linked oligosaccharides, their apparent molecular mass varies from 25 to 35kDa (DERYCKERE and BURGERT 1996a). Although all E3/19K molecules seem to share the capacity to bind HLA molecules, their amino acid sequence homology is rather low (Fig. 5). Only 20 residues of the mature E3/19K proteins are strictly conserved (DERYCKERE and BURGERT 1996a; BURGERT and BLUSCH 2000), and the overall similarity between the four subgenera is 37.7% (range: 29.6% between C and E to 61.3% between B and E). At present, it is unclear how this variation influences their ability to interact with HLA molecules. By quantitatively comparing the efficiency of transport inhibition by E3/19K of Ad2, Ad19a, and Ad3 in transfected 293 cells, it was shown that these proteins have a differential capacity to inhibit the transport of the HLA alleles present in 293 cells (DERYCKERE and BURGERT 1996a); Burgert et al., unpublished data). There are two potential interpretations: (a) the general efficacy to downregulate HLA molecules varies depending on the serotype and (b) E3/19K molecules from Ads of different subgenera exhibit a differential affinity to certain HLA alleles. To clarify this issue, more E3/19K-HLA combinations need to be examined quantitatively.

The Ad2 and Ad5 E3/19K molecules are very promiscuous in that they bind the majority if not all human HLA antigens, albeit with differential affinity (KÖRNER and BURGERT 1994; BURGERT 1996). Drastic differences exist with regard to the interaction of E3/19K with murine MHC alleles. For example, the MHC K^k and D^d alleles do not bind E3/19K and hence are not susceptible to its transport inhibition function whereas K^d and D^b bind with high affinity (BURGERT and KVIST 1987; COX et al. 1990; BURGERT 1996). Taking advantage of this differential binding capacity, hybrid MHC molecules containing domains from E3/19K-binding and nonbinding MHC alleles were used to show that the polymorphic α1 and α2 domains of MHC molecules comprising the peptide-binding pocket are essential for complex formation with E3/19K (BURGERT and KVIST 1987). Further characterization of the critical structure by using site-directed mutagenesis and antibody binding suggests that the contact site is formed or influenced by amino acids within the carboxy-terminal part of the α2 helix, the turn to α3 and the amino-terminal part of the α1 helix (FEUERBACH et al. 1994; FLOMENBERG et al. 1994). Despite the proximity to the peptide-binding pocket, there is no direct evidence as yet that E3/19K interferes with peptide binding. Given the broad reactivity of E3/19K, the structural element of HLA binding to E3/19K is expected to be rather conserved. Interestingly, the CMV protein US2, which also interferes with antigen presentation and was recently cocrystallized with HLA-A2, binds to a similar or identical region at the junction of the peptide-binding region and the α3 domain (GEWURZ et al. 2001).

Today, 15 years after the discovery of the E3/19K function (BURGERT and KVIST 1985), it has become apparent that interference with antigen presentation is a common strategy of persistent viruses (YEWDELL and BENNINK 1999). Proteins affecting this process have now been identified in human and murine CMV (HENGEL et al. 1998; see relevant chapters in this volume), Herpes simplex virus, HIV, and, most recently, in KSHV and MHV68 (COSCOY and GANEM 2000; ISHIDO et al. 2000; STEVENSON et al. 2000). Interestingly, coevolution of the individual viruses with their respective hosts seems to give rise to distinct mechanisms. Further elucidation of these mechanisms will contribute to our understanding of the antigen processing and presentation pathway.

8.3 Transcriptional Repression of Antigen Presentation Functions by Ad12 E1A

Ad12, of the highly oncogenic subgroup A, and the enteric viruses Ad40 and Ad41 (subgroup F) lack an E3/19K gene, and Ad12 is unable to retain MHC molecules in the ER during infection of 293 and HeLa cells (PÄÄBO et al. 1986a). Therefore, the E3/19K function is obviously not required for survival of these Ads in the gastrointestinal tract, where these viruses preferentially replicate. Conversely, E3/19K appears to be beneficial during infection of the respiratory tract and the other tissues favored by the majority of Ads. However, a greatly diminished cell surface expression of MHC molecules is observed in Ad12-transformed rodent and human cells. This effect is mediated by Ad12 E1A proteins and is based on their ability to interfere with the transcription of MHC molecules (see for review BLAIR and HALL 1998). Transcriptional repression affects also other components of the antigen processing machinery, such as the transporter subunits TAP1 and TAP2 and the adjacent genes LMP2 and LMP7 coding for the proteasomal subunits, and even extends to other genes within the MHC complex. The underlying molecular mechanism has not yet been fully elucidated. One report suggests a differential processing of the transcription factor NF-κB in Ad12- versus Ad5-transformed cells (SCHOUTEN et al. 1995), whereas other data point towards the binding of an active NF-κB-inhibitor to the MHC enhancer, located between 156 and 203 bp upstream of the transcriptional start site. In addition, more upstream regions of the MHC promoter (between −1200 and −1400bp) seem to participate in the transcriptional repression. In summary, pleiotropic effects seem to be responsible for the diminished cell surface expression of MHC molecules in Ad12-transformed cells (BLAIR and HALL 1998). Whether these mechanisms are operating during acute or persistent infections of Ad12 in human beings remains elusive.

8.4 Inhibition of CTL Lysis

Both mechanisms discussed above affect antigen presentation and therefore the recognition of infected cells by $CD8^+$ T cells. We do not know yet whether the anti-apoptotic viral countermeasures described in our Section 6 will protect Ad-infected cells additionally from the lytic attack by CTL, although this is likely to be the case. Recently, the late Ad protein L4-100K, which is involved in the assembly of the viral hexon capsomeres and activates late viral protein synthesis, has been shown to inhibit lymphocyte granule-mediated cell death by binding to and inhibiting granzyme B, a major serine protease of cytotoxic lymphocyte granules (ANDRADE et al. 2001). Also, E4 products were implicated in resistance of Ad vector-infected murine fibroblasts (C57BL/6) to CTL lysis. In combination with the E3 region the E4/ORF4 construct provided complete protection whereas E4/ORF6 had no effect (KAPLAN et al. 1999). This is surprising considering that E4/ORF4 was described to encode a proapototic gene (LAVOIE et al. 1998). Unfortunately, the system is rather complex, as E4 proteins may be targets for the murine CTL and no biochemical data were provided as to the expression levels of E3 and E4 proteins by these vectors.

9 Summary of Countermeasures: The Chess Game Between Virus and Host; Facts and Fiction

Immediately after the first contact between Ads and the cell, the chess game between Ads and the host is opened (Fig. 6). The entry process triggers the release of IFNs, chemokines, and presumably other cytokines like TNF. Although we do not know whether or not IL-8 is neutralized, the virus is prepared to deal with many of the antiviral effects of IFNs. Ad brings along two counteracting functions. IFNs released from infected cells, neighboring cells, or NK cells cannot signal through the IFN-R-induced JAK/STAT pathway, as transcription of IFN-responsive genes is prevented by E1A. Second, the VA-RNA prevents activation of PKR and thereby allows continuous translation of viral proteins and might also serve to inhibit PKR-mediated apoptosis. The expression of E1A drives the cell into cell cycle, concomitantly triggering the intrinsic cellular apoptosis program, which is interrupted already at the level of the controlling sensor p53 by E1B/55K in concert with E4/ORF6 and further downstream by E1B/19K. E1B/19K seems to neutralize predominantly proapoptotic Bcl-2 family members like Bax and thus might inhibit proapoptotic events originating from the mitochondria. Some of the effects ascribed to E1B/19K may not be relevant during infection of primary cells and are not shown in Fig. 6. Several hours have passed and the synthesis of early proteins is now at its peak (~12h p.i. in primary cells). E3/10.4K–14.5K (RID) are removing several proteins from the cell surface, including Fas, TRAIL-R1/2, and the EGFR. 10.4K–14.5K and 14.7K might in addition modulate the release of proinflammatory

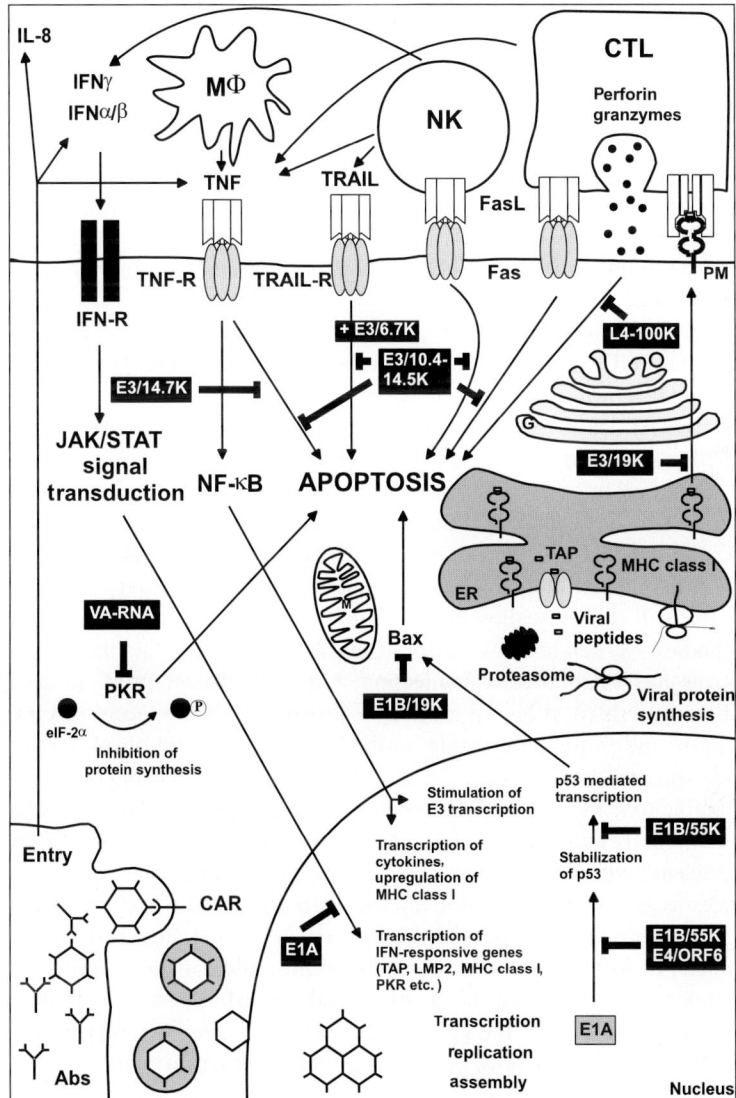

Fig. 6. Summary of antiviral host responses and Ad countermeasures. The figure is discussed in the text. Adenovirus proteins with known immunomodulatory functions are shaded *black*; the E1A activity triggering apoptosis is depicted in *gray*. For simplicity, potential effects of E1B/19K on Fas-, TRAIL-, and TNF-mediated apoptosis/cytolysis that are controversial are omitted. Also, potential stimulatory or inhibitory effects of 14.7K on NF-κB signaling is not indicated. Trimers of TNF, FasL, and TRAIL exist as membrane-bound and secreted forms. Abbreviations: *MΦ*, macrophages; *Abs*, antibodies; *PM*, plasma membrane; *G*, Golgi; *M*, mitochondrion

mediators and thus the recruitment of cells of the immune system and might impair TNF-induced cytolysis. The effect of 14.7K on NF-κB signaling and survival of infected cells is less clear. Because of the effect of E3/19K on HLA transport, no peptide-loaded HLA molecule is leaving the ER any more and HLA cell surface display slowly decreases over the next 48–72 h. Some of the HLA molecules might have reached the cell surface before E3/19K action and are presumably loaded with peptides derived from the immediate-early protein E1A or capsid components. These peptide-HLA complexes might sensitize infected cells for CTL killing after the generation of T cells (6–10 days after the initial infection) and in immune individuals; however, during a first encounter with a host who lacks preexisting Ad-specific T cells, the virus should have no problem in successfully initiating infection and spread to other cells. This is even more likely, as Ad infection does not seem to induce NK cell sensitivity.

Animal experiments suggest that Ad infection in vivo might induce TNF. In cell culture systems, TNF can induce NF-κB in the infected cells (DERYCKERE and BURGERT 1996b; DIMITROV et al. 1997), which in turn activates a number of host genes, e.g., HLA, death receptors, or cytokines that are counteracted by E3 products, but also increases expression of E3 proteins. Thereby, the NF-κB-induced host functions may be compensated by the induced expression of E3 proteins. After several days tens of thousands of infectious particles have accumulated inside the infected cell, and eventually the cell dies by apoptosis or necrosis. Cell death may result in a slow release of virus. During a primary infection this virus life style will certainly allow repeated infection of new cells. Engulfment of apoptotic bodies by neighboring cells or macrophages may facilitate virus spread. The virus has established the infection, but after the adaptive immune response has been generated, it has to reach new target cells in the face of a CTL response and more and more neutralizing antibodies. After generation of T cells, there is a short time window between 6 and 12h p.i. during which infected cells might be vulnerable to CTL attack. Thereafter, the removal of death receptors by E3/10.4–14.5K and 6.7K and the inhibition of granzyme B by the L4-100K protein may provide efficient protection from the lytic attack by CTL and NK cells. With time, the chances to escape these antiviral effector systems decrease and eventually this limits or even terminates the acute infection. However, the stealth functions may have enabled Ad to reach a tissue in which the host effector mechanisms are less effective, allowing continuing production and persistence that facilitates transmission to new hosts. Thus, the game ends *remis*. The cellular reservoir and the mechanism by which Ads achieve persistence remain a great mystery.

10 Perspectives

Ads have achieved increasing attention mostly due to their use as vectors for gene- and immunotherapy. The initial enthusiasm about Ad vectors was based on the stability of the virus, its ability to grow to high titers, and the efficiency of gene transfer in vitro. Confronted with the limited progress of Ad-mediated gene therapy over the years, it was realized that we know far too little about the interaction of the virus with its host. Therefore, it is of utmost importance to characterize these interactions in greater detail, in particular, the antiviral immune response and the countermeasures of the virus. From the administration of Ad vectors into mice, a large body of information was gained about the murine response to these crippled Ad vectors that we have not covered in this review. We refer here to a recent review that deals with some of these aspects (RUSSELL 2000). Only some findings derived from the abortive infection in mice with mostly E3-negative Ad vectors might be relevant to the interaction of wild-type Ads with the human host. In contrast to the herpes and pox viruses that apparently incorporated a large number of host sequences, we can also not delineate functional activities of the Ad stealth proteins by their sequence homology to cellular genes, because, with the exception of E1B/19K, which exhibits a very limited homology to cellular Bcl-2 family members, no significant homology has been recognized. Lacking an appropriate in vivo model and the homology to cellular genes, an alternative approach for studying host-virus interaction is to carry out experiments in tissue culture and to investigate the effects of Ad proteins on immunologically important mechanisms. From this analysis, the functional principles of many Ad proteins have been worked out that counteract host immune responses. Certainly, many gaps of knowledge remain. But, despite its limitations, the approach turned out to be very valuable. A number of viral stealth functions were discovered from which we can predict which host response mechanism the virus tries to avoid. Furthermore, by the use of primary cells and tissues instead of cell lines the relevance of these in vitro studies will be further increased.

In this review, we have focused on functions of subgenus C Ads that counteract host defense mechanisms. The authors feel that a rich unexplored reservoir of potentially interesting functions provided by other Ad subgenera awaits further investigation. The low sequence homology between Ads of different subgenera (generally <30%) let us assume that many common Ad products may not function in an identical fashion. Rather, the sequence differences are expected to influence the affinity and the specificity of the interaction with their respective host target molecules. By studying these proteins not only will we gain insight into structure-function relationships, but we also should discover new host target proteins and possibly new mechanisms of interference with host response pathways.

One of the most divergent genomic regions seems to be the early transcription unit E3. The E3 regions (~1kb) of canine and murine Ads (DRAGULEV et al. 1991; BEARD and SPINDLER 1995) have a coding potential of only ~1–2 genes, whereas the E3 region of human Ads of subgenus D, encompassing ~ 5.2kb, codes for eight genes (BURGERT and BLUSCH 2000). Therefore, it is conceivable that the E3 region is a primary site for Ad evolution and viral adaptation to the respective host. Interestingly, the size and composition of the E3 regions of human Ads classified in different subgenera also differ significantly (3–5.2kb, with 5–8 genes). Some E3 genes (10.4K, 14.5K, 14.7K) are found in all or in most (19K, 12.5K) subgenera; others seem to be unique to a specific subgenus (e.g., 6.7K, 11.6K, and 49 K). Moreover, the homologous E3 proteins of different subgenera are often poorly conserved (BURGERT and BLUSCH 2000). We hypothesize that the subgenus-specific composition of E3 genes together with the differences of the homologous E3 products contributes to subgenus-specific pathogenesis and disease. In particular, the unique E3 proteins are attractive candidates to influence the disease pattern, because it does not correlate with differential receptor usage. With the exception of subgenus B Ads, Ads from all subgenera attach to human cells in vitro via CAR, a host protein expressed in many different tissues (BERGELSON et al. 1997; NEMEROW 2000). Therefore, it is likely that the tropism/pathogenesis characteristic for the different subgenera is determined by events occurring subsequent to attachment. In light of the immunomodulatory functions of E3 products and the differential E3 coding capacity of the different subgenera, it is conceivable that E3 proteins play a prominent role for the differential disease pattern (BURGERT and BLUSCH 2000). The ongoing molecular analysis of Ad proteins from different subgenera may allow us to correlate the differential behavior of Ad products with the pathogenesis in vivo. On the basis of the knowledge gained by these "reverse functional studies" we might also be able to design better vectors that are not as immunogenic as those currently in use. This could either involve incorporation of specific immunomodulatory functions and/or elimination of others. Moreover, the studies may lead to more rational treatments in situations in which Ad infections become life threatening.

Acknowledgements. This work was supported in part by grants of the SFB 455 and the Deutsche Forschungsgemeinschaft (DFG) to H.-G. Burgert (Bu642/4-2).

References

Anderson KP, Fennie EH (1987) Adenovirus early region 1A modulation of interferon antiviral activity. J Virol 61:787–795
Andersson M, Pääbo S, Nilsson T, Peterson PA (1985) Impaired intracellular transport of class I MHC antigens as a possible means for adenoviruses to evade immune surveillance. Cell 43:215–222
Andrade F, Bull HG, Thornberry NA, Ketner GW, Casciola-Rosen LA, Rosen A (2001) Adenovirus L4-100K assembly protein is a granzyme B substrate that potently inhibits granzyme B-mediated cell death. Immunity 14:751–761
Arany Z, Newsome D, Oldread E, Livingston D, Eckner R (1995) A family of transcriptional adaptor proteins targeted by the E1A oncoprotein. Nature 374:81–84
Ashcroft M, Vousden KH (1999) Regulation of p53 stability. Oncogene 18:7637–7643
Ashkenazi A, Dixit VM (1998) Death receptors: signaling and modulation. Science 281:1305–1308
Basler CF, Droguett G, Horwitz MS (1996) Sequence of the immunoregulatory early region-3 and flanking sequences of adenovirus type-35. Gene 170:249–254
Bazzoni F, Beutler B (1996) The tumor necrosis factor ligand and receptor families. N Engl J Med 334:1717–1725
Beard CW, Spindler KR (1995) Characterization of an 11K protein produced by early region 3 of mouse adenovirus type 1. Virology 208:457–466
Benedict CA, Norris PS, Prigozy TI, Bodmer J-L, Mahr JA, Garnett CT, Martinon F, Tschopp J, Gooding LR, Ware CF (2001) Three adenovirus E3 proteins cooperate to evade apoptosis by TRAIL receptor-1 and 2. J Biol Chem 276:3270–3278
Bennett EM, Bennink JR, Yewdell JW, Brodsky FM (1999) Cutting edge: adenovirus E19 has two mechanisms for affecting class I MHC expression. J Immunol 162:5049–5052
Bergelson J, Cunningham J, Droguett G, Kurt-Jones E, Krithivas A, Hong J, Horwitz M, Crowell R, Finberg R (1997) Isolation of a common receptor for Coxsackie B viruses and adenoviruses 2 and 5. Science 275:1320–1323

Bewley MC, Springer K, Zhang YB, Freimuth P, Flanagan JM (1999) Structural analysis of the mechanism of adenovirus binding to its human cellular receptor, CAR. Science 286:1579–1583

Bhattacharya S, Eckner R, Grossman S, Oldread E, Arany Z, D'Andrea A, Livingston DM (1996) Cooperation of Stat2 and p300/CBP in signalling induced by interferon-α. Nature 383:344–347

Biron CA, Nguyen KB, Pien GC, Cousens LP, Salazar-Mather TP (1999) Natural killer cells in antiviral defense: function and regulation by innate cytokines. Annu Rev Immunol 17:189–220

Blair GE, Hall KT (1998) Human adenoviruses: Evading detection by cytotoxic T lymphocytes. Semin Virol 8:387–397

Borgland S, Bowen G, Wong N, Libermann T, Muruve D (2000) Adenovirus vector-induced expression of the C-X-C chemokine IP-10 is mediated through capsid-dependent activation of NF-κB. J Virol 74:3941–3947

Bosse D, Ades E (1991) Studies of adenovirus subtypes and down-regulation of HLA class I expression: correlations to natural-killer-mediated cytolysis. Pathobiology 59:313–315

Boulakia C, Chen G, Ng F, Teodoro J, Branton P, Nicholson D, Poirier G, Shore G (1996) Bcl-2 and adenovirus E1B 19kDa protein prevent E1A-induced processing of CPP32 and cleavage of poly (ADP-ribose) polymerase. Oncogene 12:529–535

Boyd J, Malstrom S, Subramanian T, Venkatesh L, Schaeper U, Elangovan B, D'Sa-Eipper C, Chinnadurai G (1994) Adenovirus E1B 19kDa and Bcl-2 proteins interact with a common set of cellular proteins. Cell 79:341–351

Bruder JT, Jie T, McVey DL, Kovesdi I (1997) Expression of gp19K increases the persistence of transgene expression from an adenovirus vector in the mouse lung and liver. J Virol 71:7623–7628

Bruder JT, Kovesdi I (1997) Adenovirus infection stimulates the Raf/MAPK signaling pathway and induces interleukin-8 expression. J Virol 71:398–404

Burgert H-G (1996) Subversion of the MHC class I antigen presentation pathway by adenoviruses and herpes simplex viruses. Trends Microbiol 4:107–112

Burgert H-G, Blusch J (2000) Immunomodulatory functions encoded by the E3 transcription unit of adenoviruses. Virus Genes 21:13–25

Burgert H-G, Kvist S (1985) An adenovirus type 2 glycoprotein blocks cell surface expression of human histocompatibility class I antigens. Cell 41:987–997

Burgert H-G, Kvist S (1987) The E3/19K protein of adenovirus type 2 binds to the domains of histocompatibility antigens required for CTL recognition. EMBO J 6:2019–2026

Burgert H-G, Maryanski JL, Kvist S (1987) "E3/19K" protein of adenovirus type 2 inhibits lysis of cytolytic T lymphocytes by blocking cell-surface expression of histocompatibility class I antigens. Proc Natl Acad Sci USA 84:1356–1360

Carlin CR, Tollefson AE, Brady HA, Hoffman BL, Wold WS (1989) Epidermal growth factor receptor is down-regulated by a 10,400 MW protein encoded by the E3 region of adenovirus. Cell 57:135–144

Chaudhary PM, Eby M, Jasmin A, Bookwalter A, Murray J, Hood L (1997) Death receptor 5, a new member of the TNFR family, and DR4 induce FADD-dependent apoptosis and activate the NF-κB pathway. Immunity 7:821–830

Chen MJ, Holskin B, Strickler J, Gorniak J, Clark MA, Johnson PJ, Mitcho M, Shalloway D (1987) Induction by E1A oncogene expression of cellular susceptibility to lysis by TNF. Nature 330:581–583

Chen P, Tian J, Kovesdi I, Bruder JT (1998) Interaction of the adenovirus 14.7-kDa protein with FLICE inhibits Fas ligand-induced apoptosis. J Biol Chem 273:5815–5820

Chen PH, Ornelles DA, Shenk T (1993) The adenovirus L3 23-kilodalton proteinase cleaves the amino-terminal head domain from cytokeratin 18 and disrupts the cytokeratin network of HeLa cells. J Virol 67:3507–3514

Chinnadurai G (1998) Control of apoptosis by human adenovirus genes. Sem Virol 8:399–408

Chiou SK, White E (1997) p300 binding by E1A cosegregates with p53 induction but is dispensable for apoptosis. J Virol 71:3515–3525

Chroboczek J, Bieber F, Jacrot B (1992) The sequence of the genome of adenovirus type 5 and its comparison with the genome of adenovirus type 2. Virology 186:280–285

Clarke P, Meintzer SM, Gibson S, Widmann C, Garrington TP, Johnson GL, Tyler KL (2000) Reovirus-induced apoptosis is mediated by TRAIL. J Virol 74:8135–8139

Clesham G, Adam P, Proudfoot D, Flynn P, Efstathiou S, Weissberg P (1998) High adenoviral loads stimulate NFκB-dependent gene expression in human vascular smooth muscle cells. Gene Ther 5:174–180

Colonna M, Nakajima H, Cella M (2000) A family of inhibitory and activating Ig-like receptors that modulate function of lymphoid and myeloid cells. Semin Immunol 12:121–127

Cook J, Potter T, Bellgrau D, Routes B (1996) E1 A oncogene expression in target cells induces cytolytic susceptibility at a post-recognition stage in the interaction with killer lymphocytes. Oncogene 13:833–842

Cook JL, Routes BA, Leu CY, Walker TA, Colvin KL (1999) E1A oncogene-induced cellular sensitization to immune-mediated apoptosis is independent of p53 and resistant to blockade by E1B 19kDa protein. Exp Cell Res 252:199–210

Coscoy L, Ganem D (2000) Kaposi's sarcoma-associated herpesvirus encodes two proteins that block cell surface display of MHC class I chains by enhancing their endocytosis. Proc Natl Acad Sci USA 97:8051–8056

Cosson P, Letourneur F (1994) Coatomer interaction with di-lysine endoplasmic reticulum retention motifs. Science 263:1629–1631

Cox JH, Yewdell JW, Eisenlohr LC, Johnson PR, Bennink JR (1990) Antigen presentation requires transport of MHC class I molecules from the endoplasmic reticulum. Science 247:715–718

Cress W, Nevins J (1996) Use of the E2F transcription factor by DNA tumor virus regulatory proteins. Curr Top Microbiol Immunol 208:63–78

Davison AJ, Telford EA, Watson MS, McBride K, Mautner V (1993) The DNA sequence of adenovirus type 40. J Mol Biol 234:1308–1316

Degli-Esposti M (1999) To die or not to die – the quest of the TRAIL receptors. J Leukoc Biol 65:535–542

De Jong JC, Wermenbol AG, Verweij-Uijterwaal MW, Slaterus KW, Wertheim-Van Dillen P, Van Doornum GJ, Khoo SH, Hierholzer JC (1999) Adenoviruses from human immunodeficiency virus-infected individuals, including two strains that represent new candidate serotypes Ad50 and Ad51 of species B1 and D, respectively. J Clin Microbiol 37:3940–3945

Deryckere F, Burgert H-G (1996a) Early region 3 of adenovirus type 19 (subgroup D) encodes an HLA-binding protein distinct from that of subgenera B and C. J Virol 70:2832–2841

Deryckere F, Burgert H-G (1996b) Tumor necrosis factor alpha induces the adenovirus early 3 promoter by activation of NF-κB. J Biol Chem 271:30249–30255

Deryckere F, Ebenau-Jehle C, Wold WSM, Burgert H-G (1995) Tumor necrosis factor α increases expression of adenovirus E3 proteins. Immunobiol 193:186–192

Desagher S, Martinou J (2000) Mitochondria as the central control point of apoptosis. Trends Cell Biol 10:369–377

Desbarats J, Wade T, Wade WF, Newell MK (1999) Dichotomy between naive and memory CD4(+) T cell responses to Fas engagement. Proc Natl Acad Sci USA 96:8104–8109

Dimitrov T, Krajcsi P, Hermiston TW, Tollefson AE, Hannink M, Wold WS (1997) Adenovirus E3–10.4K/14.5K protein complex inhibits tumor necrosis factor-induced translocation of cytosolic phospholipase A2 to membranes. J Virol 71:2830–2837

Donze O, Dostie J, Sonenberg N (1999) Regulatable expression of the interferon-induced double-stranded RNA dependent protein kinase PKR induces apoptosis and Fas receptor expression. Virology 256:322–329

Dragulev BP, Sira S, Abouhaidar MG, Campbell JB (1991) Sequence analysis of putative E3 and fiber genomic regions of two strains of canine adenovirus type 1. Virology 183:298–305

Duerksen-Hughes P, Wold W, Gooding L (1989) Adenovirus E1 A renders infected cells sensitive to cytolysis by tumor necrosis factor. J Immunol 143:4193–4200

Earnshaw WC, Martins LM, Kaufmann SH (1999) Mammalian caspases: structure, activation, substrates, and functions during apoptosis. Annu Rev Biochem 68:383–424

Efrat S, Fejer G, Brownlee M, Horwitz MS (1995) Prolonged survival of pancreatic islet allografts mediated by adenovirus immunoregulatory transgenes. Proc Natl Acad Sci USA 92:6947–6951

Elsing A, Burgert H-G (1998) The adenovirus E3/10.4K–14.5K proteins down-modulate the apoptosis receptor Fas/Apo-1 by inducing its internalization. Proc Natl Acad Sci USA 95:10072–10077

Fejer G, Gyory I, Tufariello J, Horwitz MS (1994) Characterization of transgenic mice containing adenovirus early region 3 genomic DNA. J Virol 68:5871–5881

Feuerbach D, Burgert H-G (1993) Novel proteins associated with MHC class I antigens in cells expressing the adenovirus protein E3/19K. EMBO J 12:3153–3161

Feuerbach D, Etteldorf S, Ebenau-Jehle C, Abastado JP, Madden D, Burgert H-G (1994) Identification of amino acids within the MHC molecule important for the interaction with the adenovirus protein E3/19K. J Immunol 153:1626–1636

Fletcher JM, Prentice HG, Grundy JE (1998) Natural killer cell lysis of cytomegalovirus (CMV)-infected cells correlates with virally induced changes in cell surface lymphocyte function-associated antigen-3 (LFA-3) expression and not with the CMV-induced down-regulation of cell surface class I HLA. J Immunol 161:2365–2374

Flomenberg P, Gutierrez E, Hogan KT (1994) Identification of class I MHC regions which bind to the adenovirus E3–19k protein. Mol Immunol 31:1277–1284

Flomenberg P, Piaskowski V, Truitt RL, Casper JT (1996) Human adenovirus-specific CD8+ T-cell responses are not inhibited by E3–19K in the presence of gamma interferon. J Virol 70:6314–6322

Fox JP, Hall CE, Cooney MK (1977) The Seattle virus watch. VII Observations of adenovirus infections. Am J Epidemiol 105:362–386

Gabathuler R, Kvist S (1990) The endoplasmic reticulum retention signal of the E3/19K protein of adenovirus type 2 consists of three separate amino acid segments at the carboxy terminus. J Cell Biol 111:1803–1810

Gewurz BE, Gaudet R, Tortorella D, Wang EW, Ploegh HL, Wiley DC (2001) Antigen presentation subverted: Structure of the human cytomegalovirus protein US2 bound to the class I molecule HLA-A2. Proc Natl Acad Sci USA 98:6794–6799

Gil J, Esteban M (2000) The interferon-induced protein kinase (PKR), triggers apoptosis through FADD-mediated activation of caspase 8 in a manner independent of Fas and TNF-α receptors. Oncogene 19:3665–3674

Ginsberg HS, Lundholm Beauchamp U, Horswood RL, Pernis B, Wold WS, Chanock RM, Prince GA (1989) Role of early region 3 (E3) in pathogenesis of adenovirus disease. Proc Natl Acad Sci USA 86:3823–3827

Ginsberg HS, Moldawer LL, Sehgal PB, Redington M, Kilian PL, Chanock RM, Prince GA (1991) A mouse model for investigating the molecular pathogenesis of adenovirus pneumonia. Proc Natl Acad Sci USA 88:1651–1655

Goodbourn S, Didcock L, Randall RE (2000) Interferons: cell signalling, immune modulation, antiviral response and virus countermeasures. J Gen Virol 81:2341–2364

Gooding LR, Aquino L, Duerksen Hughes PJ, Day D, Horton TM, Yei SP, Wold WS (1991a) The E1B 19,000-molecular-weight protein of group C adenoviruses prevents tumor necrosis factor cytolysis of human cells but not of mouse cells. J Virol 65:3083–3094

Gooding LR, Elmore LW, Tollefson AE, Brady HA, Wold WS (1988) A 14,700 MW protein from the E3 region of adenovirus inhibits cytolysis by tumor necrosis factor. Cell 53:341–346

Gooding LR, Ranheim TS, Tollefson AE, Aquino L, Duerksen-Hughes PJ, Horton TM, Wold WSM (1991b) The 10,400- and 14,500-dalton proteins encoded by region E3 of adenovirus function together to protect many but not all mouse cell lines against lysis by tumor necrosis factor. J Virol 65:4114–4123

Greber U, Suomalainen M, Stidwill R, Boucke K, Ebersold M, Helenius A (1997) The role of the nuclear pore complex in adenovirus DNA entry. EMBO J 16:5998–6007

Hale T, Braithwaite A (1999) The adenovirus oncoprotein E1a stimulates binding of transcription factor ETF to transcriptionally activate the p53 gene. J Biol Chem 274:23777–23786

Harty JT, Tvinnereim AR, White DW (2000) CD8+ T cell effector mechanisms in resistance to infection. Annu Rev Immunol 18:275–308

Hashimoto S, Ishii A, Yonehara S (1991) The E1b oncogene of adenovirus confers cellular resistance to cytotoxicity of tumor necrosis factor and monoclonal anti-Fas antibody. Int Immunol 3:343–351

Hatada EN, Krappmann D, Scheidereit C (2000) NF-κB and the innate immune response. Curr Opin Immunol 12:52–58

Hengel H, Brune W, Koszinowski UH (1998) Immune evasion by cytomegalovirus – survival strategies of a highly adapted opportunist. Trends Microbiol 6:190–197

Herissé J, Courtois G, Galibert F (1980) Nucleotide sequence of the EcoRI D fragment of adenovirus 2 genome. Nucleic Acids Res 8:2173–2192

Herisse J, Galibert F (1981) Nucleotide sequence of the EcoRI E fragment of adenovirus 2 genome. Nucleic Acids Res 9:1229–1240

Hermiston TW, Hellwig R, Hierholzer JC, Wold WS (1993) Sequence and functional analysis of the human adenovirus type 7 E3-gp19K protein from 17 clinical isolates. Virology 197:593–600

Hoffman P, Carlin C (1994) Adenovirus E3 protein causes constitutively internalized epidermal growth factor receptors to accumulate in a prelysosomal compartment, resulting in enhanced degradation. Mol Cell Biol 14:3695–3706

Hong JS, Mullis KG, Engler JA (1988) Characterization of the early region 3 and fiber genes of Ad7. Virology 167:545–553

Horton TM, Ranheim TS, Aquino L, Kusher DI, Saha SK, Ware CF, Wold WS, Gooding LR (1991) Adenovirus E3 14.7K protein functions in the absence of other adenovirus proteins to protect transfected cells from tumor necrosis factor cytolysis. J Virol 65:2629–2639

Horton TM, Tollefson AE, Wold WS, Gooding LR (1990) A protein serologically and functionally related to the group C E3 14,700-kilodalton protein is found in multiple adenovirus serotypes. J Virol 64:1250–1255

Horvath J, Palkonyay L, Weber J (1986) Group C adenovirus DNA sequences in human lymphoid cells. J Virol 59:189–192

Horwitz MS (1996) In: Fields BN, Knipe DM, Howley PM (eds) Adenoviruses. Lippincott-Raven Publishers, Philadelphia, New York, pp 2149–2171

Horwitz MS (2001) Adenovirus immunoregulatory genes and their cellular targets. Virology 279:1–8

Huang DC, Cory S, Strasser A (1997) Bcl-2, Bcl-XL and adenovirus protein E1B19kD are functionally equivalent in their ability to inhibit cell death. Oncogene 14:405–414

Imperiale MJ, Akusjärvi G, Leppard KN (1995) Post-transcriptional control of adenovirus gene expression. Curr Top Microbiol Immunol 199/II:139–171

Ishido S, Wang C, Lee BS, Cohen GB, Jung JU (2000) Downregulation of major histocompatibility complex class I molecules by Kaposi's sarcoma-associated herpesvirus K3 and K5 proteins. J Virol 74:5300–5309

Jackson MR, Nilsson T, Peterson PA (1990) Identification of a consensus motif for retention of transmembrane proteins in the endoplasmic reticulum. EMBO J 9:3153–3162

Jackson MR, Nilsson T, Peterson PA (1993) Retrieval of transmembrane proteins to the endoplasmic reticulum. J Cell Biol 121:317–333

Johnsen AC, Haux J, Steinkjer B, Nonstad U, Egeberg K, Sundan A, Ashkenazi A, Espevik T (1999) Regulation of APO-2 ligand/trail expression in NK cells-involvement in NK cell-mediated cytotoxicity. Cytokine 11:664–672

Jones N (1995) Transcriptional modulation by the adenovirus E1 A gene. Curr Top Microbiol Immunol 199 (Pt 3):59–80

Juang Y, Lowther W, Kellum M, Au WC, Lin R, Hiscott J, Pitha PM (1998) Primary activation of interferon A and interferon B gene transcription by interferon regulatory factor 3. Proc Natl Acad Sci USA 95:9837–9842

Kaplan JM, Armentano D, Scaria A, Woodworth LA, Pennington SE, Wadsworth SC, Smith AE, Gregory RJ (1999) Novel role for E4 region genes in protection of adenovirus vectors from lysis by cytotoxic T lymphocytes. J Virol 73:4489–4492

Kashii Y, Giorda R, Herberman RB, Whiteside TL, Vujanovic NL (1999) Constitutive expression and role of the TNF family ligands in apoptotic killing of tumor cells by human NK cells. J Immunol 163:5358–5366

Kasof G, Goyal L, White E (1999) Btf, a novel death-promoting transcriptional repressor that interacts with Bcl-2-related proteins. Mol Cell Biol 19:4390–4404

Kayagaki N, Yamaguchi N, Nakayama M, Kawasaki A, Akiba H, Okumura K, Yagita H (1999) Involvement of TNF-related apoptosis-inducing ligand in human CD4+ T cell-mediated cytotoxicity. J Immunol 162:2639–2647

Kelliher MA, Grimm S, Ishida Y, Kuo F, Stanger BZ, Leder P (1998) The death domain kinase RIP mediates the TNF-induced NF-κB signal. Immunity 8:297–303

Kemp MC, Hierholzer JC, Cabradilla CP, Obijeski JF (1983) The changing etiology of epidemic keratoconjunctivitis: antigenic and restriction enzyme analyses of adenovirus types 19 and 37 isolated over a 10-year period. J Infect Dis 148:24–33

Kirchhausen T (1999) Adaptors for clathrin-mediated traffic. Annu Rev Cell Dev Biol 15:705–732

Kischkel FC, Lawrence DA, Chuntharapai A, Schow P, Kim KJ, Ashkenazi A (2000) Apo2L/TRAIL-dependent recruitment of endogenous FADD and caspase-8 to death receptors 4 and 5. Immunity 12:611–620

Kitajewski J, Schneider R, Safer B, Munemitsu S, Samuel C, Thimmappaya B, Shenk T (1986) Adenovirus VAI RNA antagonizes the antiviral action of interferon by preventing activation of the interferon-induced eIF-2 alpha kinase. Cell 45:195–200

Körner H, Burgert H-G (1994) Down-regulation of HLA antigens by the adenovirus type 2 E3/19K protein in a T-lymphoma cell line. J Virol 68:1442–1448

Körner H, Fritzsche U, Burgert H-G (1992) Tumor necrosis factor alpha stimulates expression of adenovirus early region 3 proteins: implications for viral persistence. Proc Natl Acad Sci USA 89: 11857–11861

Kos FJ, Engleman EG (1996) Immune regulation: a critical link between NK cells and CTLs. Immunol Today 17:174–176

Krajcsi P, Dimitrov T, Hermiston TW, Tollefson AE, Ranheim TS, Vande Pol SB, Stephenson AH, Wold WS (1996) The adenovirus E3-14.7K protein and the E3-10.4K/14.5K complex of proteins, which independently inhibit tumor necrosis factor (TNF)-induced apoptosis, also independently inhibit TNF-induced release of arachidonic acid. J Virol 70:4904–4913

Krajcsi P, Wold WS (1992) The adenovirus E3-14.5K protein which is required for prevention of TNF cytolysis and for down-regulation of the EGF receptor contains phosphoserine. Virology 187: 492–498

Kratzer F, Rosorius O, Heger P, Hirschmann N, Dobner T, Hauber J, Stauber R (2000) The adenovirus type 5 E1B-55K oncoprotein is a highly active shuttle protein and shuttling is independent of E4orf6, p53 and Mdm2. Oncogene 17:850–857

Lavoie JN, Nguyen M, Marcellus RC, Branton PE, Shore GC (1998) E4orf4, a novel adenovirus death factor that induces p53-independent apoptosis by a pathway that is not inhibited by zVAD-fmk. J Cell Biol 140:637–645

Lee MG, Abina MA, Haddada H, Perricaudet M (1995) The constitutive expression of the immunomodulatory gp19k protein in E1(–), E3(–) adenoviral vectors strongly reduces the host cytotoxic T-cell response against the vector. Gene Ther 2:256–262

Leonard G, Sen G (1996) Effects of adenovirus E1 A protein on interferon-signaling. Virology 224:25–33

Leonard GT, Sen GC (1997) Restoration of interferon responses of adenovirus E1A-expressing HT1080 cell lines by overexpression of p48 protein. J Virol 71:5095–5101

Li Y, Kang J, Friedman J, Tarassishin L, Ye J, Kovalenko A, Wallach D, Horwitz MS (1999) Identification of a cell protein (FIP-3) as a modulator of NF-κB activity and as a target of an adenovirus inhibitor of tumor necrosis factor alpha-induced apoptosis. Proc Natl Acad Sci USA 96:1042–1047

Li Y, Kang J, Horwitz MS (1997) Interaction of an adenovirus 14.7-kilodalton protein inhibitor of tumor necrosis factor alpha cytolysis with a new member of the GTPase superfamily of signal transducers. J Virol 71:1576–1582

Li Y, Kang J, Horwitz MS (1998) Interaction of an adenovirus E3 14.7-kilodalton protein with a novel tumor necrosis factor α-inducible cellular protein containing leucine zipper domains. Mol Cell Biol 18:1601–1610

Lieber A, He C, Meuse L, Himeda C, Wilson C, Kay M (1998) Inhibition of NF-κB activation in combination with bcl-2 expression allows for persistence of first-generation adenovirus vectors in the mouse liver. J Virol 72:9267–9277

Lobigs M, Chelvanayagam G, Mullbacher A (2000) Proteolytic processing of peptides in the lumen of the endoplasmic reticulum for antigen presentation by major histocompatibility class I. Eur J Immunol 30:1496–1506

Look D, Roswit W, Frick A, Gris-Alevy Y, Dickhaus D, Walter M, Holtzman M (1998) Direct suppression of Stat1 function during adenoviral infection. Immunity 9:871–880

Ma Y, Mathews MB (1996) Structure, function, and evolution of adenovirus-associated RNA: a phylogenetic approach. J Virol 70:5083–5099

Mahr JA, Gooding LR (1999) Immune evasion by adenoviruses. Immunol Rev 168:121–130

Marcellus RC, Lavoie JN, Boivin D, Shore GC, Ketner G, Branton PE (1998) The early region 4 orf4 protein of human adenovirus type 5 induces p53-independent cell death by apoptosis. J Virol 72: 7144–7153

Marks MS, Ohno H, Kirchhausen T, Bonifacino JS (1997) Protein sorting by tyrosine-based signals: adapting to the Ys and wherefores. Trends Cell Biol 7:124–128

Martin M, Berk A (1999) Corepressor required for adenovirus E1B 55,000-molecular-weight protein repression of basal transcription. Mol Cell Biol 19:3403–3414

Matsuse T, Hayashi S, Kuwano K, Keunecke H, Jefferies WA, Hogg JC (1992) Latent adenoviral infection in the pathogenesis of chronic airways obstruction. Am Rev Respir Dis 146:177–184

Mei YF, Wadell G (1992) The nucleotide sequence of adenovirus type 11 early 3 region: comparison of genome type Ad11p and Ad11a. Virology 191:125–133

Moretta A, Biassoni R, Bottino C, Moretta L (2000) Surface receptors delivering opposite signals regulate the function of human NK cells. Semin Immunol 12:129–138

Moriishi K, Huang DC, Cory S, Adams JM (1999) Bcl-2 family members do not inhibit apoptosis by binding the caspase activator Apaf-1. Proc Natl Acad Sci USA 96:9683–9688

Nagata S (1997) Apoptosis by death factor. Cell 88: 355–365

Nemerow GR (2000) Cell receptors involved in adenovirus entry. Virology 274:1–4

Nemerow GR, Stewart PL (1999) Role of alpha(v) integrins in adenovirus cell entry and gene delivery. Microbiol Mol Biol Rev 63:725–734

Neumann R, Genersch E, Eggers HJ (1987) Detection of adenovirus nucleic acid sequences in human tonsils in the absence of infectious virus. Virus Res 7:93–97

Ng F, Nguyen M, Kwan T, Branton P, Nicholson D, Cromlish J, Shore G (1997) p28 Bap31, a Bcl-2/Bcl-XL- and procaspase-8-associated protein in the endoplasmic reticulum. J Cell Biol 139:327–338

Nguyen M, Branton P, Roy S, Nicholson D, Alnemri E, Yeh W, Mak T, Shore G (1998) E1A-induced processing of procaspase-8 can occur independently of FADD and is inhibited by Bcl-2. J Biol Chem 273:33099–33102

Nielsen H, Engelbrecht J, Brunak S, von Heijne G (1997) Identification of prokaryotic and eukaryotic signal peptides and prediction of their cleavage sites. Protein Eng 10:1–6

O'Brien V (1998) Viruses and apoptosis. J Gen Virol 79:1833–1845

Pääbo S, Nilsson T, Peterson PA (1986a) Adenoviruses of subgenera B, C, D, and E modulate cell-surface expression of major histocompatibility complex class I antigens. Proc Natl Acad Sci USA 83: 9665–9669

Pääbo S, Weber F, Nilsson T, Schaffner W, Peterson PA (1986b) Structural and functional dissection of an MHC class I antigen-binding adenovirus glycoprotein. EMBO J 5:1921–1927

Pamer E, Cresswell P (1998) Mechanisms of MHC class I-restricted antigen processing. Annu Rev Immunol 16:323–358

Perez D, White E (1998) E1B 19K inhibits Fas-mediated apoptosis through FADD-dependent sequestration of FLICE. J Biol Chem 141:1255–1266

Perez D, White E (2000) TNF-α signals apoptosis through a bid-dependent conformational change in Bax that is inhibited by E1B 19K. Mol Cell 6:53–63

Philipson L (1984) In: Ginsberg HS (ed) Adenovirus Assembly. Plenum Press, New York, pp 309–337

Ranheim TS, Shisler J, Horton TM, Wold LJ, Gooding LR, Wold WS (1993) Characterization of mutants within the gene for the adenovirus E3 14.7-kilodalton protein which prevents cytolysis by tumor necrosis factor. J Virol 67:2159–2167

Rao L, Modha D, White E (1997) The E1B 19K protein associates with lamins in vivo and its proper localization is required for inhibition of apoptosis. Oncogene 15:1587–1597

Raska KJ, Gallimore PH (1982) An inverse relation of the oncogenic potential of adenovirus-transformed cells and their sensitivity to killing by syngeneic natural killer cells. Virology 123:8–18

Reich N, Pine R, Levy D, Darnell JJ (1988) Transcription of interferon-stimulated genes is induced by adenovirus particles but is suppressed by E1A gene products. J Virol 62:114–119

Reits EA, Vos JC, Gromme M, Neefjes J (2000) The major substrates for TAP in vivo are derived from newly synthesized proteins. Nature 404:774–778

Roelvink PW, Lizonova A, Lee JGM, Li Y, Bergelson JM, Finberg RW, Brough DE, Kovesdi I, Wickham TJ (1998) The coxsackievirus-adenovirus receptor protein can function as a Cellular attachment protein for adenovirus serotypes from subgenera A, C, D, E, and F. J Virol 72: 7909–7915

Roelvink PW, Mi Lee G, Einfeld DA, Kovesdi I, Wickham TJ (1999) Identification of a conserved receptor-binding site on the fiber proteins of CAR-recognizing adenoviridae. Science 286:1568–1571

Routes J, Ryan S, Clase A, Miura T, Kuhl A, Potter T, Cook J (2000) Adenovirus E1A oncogene expression in tumor cells enhances killing by TNF-related apoptosis-inducing ligand. J Immunol 165:4522–4527

Routes JM, Cook JL (1995) E1A gene expression induces susceptibility to killing by NK cells following immortalization but not adenovirus infection of human cells. Virology 210:421–428

Rowe WP, Huebner RJ, Gillmore LK, Parrott RH, Ward TG (1953) Isolation of a cytopathogenic agent from human adenoids undergoing spontaneous degeneration in tissue culture. Proc Soc Exp Biol Med 84:570

Russell WC (2000) Update on adenovirus and its vectors. J Gen Virol 81:2573–2604

Sandoval IV, Bakke O (1994) Targeting of membrane proteins to endosomes and lysosomes. Trends Cell Biol 4:292–297

Schmitz M, Indorf A, Limbourg F, Stadtler H, Traenckner E, Baeuerle P (1996) The dual effect of adenovirus type 5 E1A 13 S protein on NF-kappaB activation is antagonized by E1B 19K. Mol Cell Biol 16:4052–4063

Schouten GJ, van der Eb AJ, Zantema A (1995) Downregulation of MHC class I expression due to interference with p105-NF kappa B1 processing by Ad12E1 A. EMBO J 14:1498–1507

Schowalter DB, Tubb JC, Liu M, Wilson CB, Kay MA (1997) Heterologous expression of adenovirus E3-gp19K in an E1a-deleted adenovirus vector inhibits MHC I expression in vitro, but does not prolong transgene expression in vivo. Gene Ther 4:351–360

Schubert U, Anton LC, Gibbs J, Norbury CC, Yewdell JW, Bennink JR (2000) Rapid degradation of a large fraction of newly synthesized proteins by proteasomes. Nature 404:770–774

Sedger LM, Shows DM, Blanton RA, Peschon JJ, Goodwin RG, Cosman D, Wiley SR (1999) IFN-gamma mediates a novel antiviral activity through dynamic modulation of TRAIL and TRAIL receptor expression. J Immunol 163:920–926

Sester M, Burgert H-G (1994) Conserved cysteine residues within the E3/19K protein of adenovirus type 2 are essential for binding to major histocompatibility complex antigens. J Virol 68:5423–5432

Sester M, Feuerbach D, Frank R, Preckel T, Gutermann A, Burgert H-G (2000) The amyloid precursor-like protein 2 associates with the major histocompatibility complex class i molecule K^d. J Biol Chem 275:3645–3654

Shao R, Hu M, Zhou B, Lin S, Chiao P, von Lindern R, Spohn B, Hung M (1999) E1A sensitizes cells to tumor necrosis factor-induced apoptosis through inhibition of IκB kinases and nuclear factor κB activities. J Biol Chem 274:21495–21498

Shenk T (1996) In: Fields BN, Knipe DM, Howley PM (eds), Adenoviridae: the Viruses and their replication. Lippincott-Raven Publishers, Philadelphia, New York, pp 2111–2148

Shisler J, Duerksen-Hughes P, Hermiston T, Wold W, Gooding L (1996) Induction of susceptibility to tumor necrosis factor by E1A is dependent on binding to either p300 or p105-Rb and induction of DNA synthesis. J Virol 70:68–77

Shisler J, Yang C, Walter B, Ware CF, Gooding LR (1997) The adenovirus E3–10.4K/14.5K complex mediates loss of cell surface Fas (CD95) and resistance to Fas-induced apoptosis. J Virol 71:8299–8306

Signäs C, Akusjärvi G, Pettersson U (1986) Region E3 of human adenoviruses; differences between the oncogenic adenovirus-3 and the non-oncogenic adenovirus-2. Gene 50:173–184

Spiliotis ET, Osorio M, Zuniga MC, Edidin M (2000) Selective export of MHC class I molecules from the ER after their dissociation from TAP. Immunity 13:841–851

Sprengel J, Schmitz B, Heuss Neitzel D, Zock C, Doerfler W (1994) Nucleotide sequence of human adenovirus type 12 DNA: comparative functional analysis. J Virol 68:379–389

Sprick MR, Weigand MA, Rieser E, Rauch CT, Juo P, Blenis J, Krammer PH, Walczak H (2000) FADD/MORT1 and caspase-8 are recruited to TRAIL receptors 1 and 2 and are essential for apoptosis mediated by TRAIL receptor 2. Immunity 12:599–609

Stark GR, Kerr IM, Williams BR, Silverman RH, Schreiber RD (1998) How cells respond to interferons. Annu Rev Biochem 67:227–264

Stevenson PG, Efstathiou S, Doherty PC, Lehner PJ (2000) Inhibition of MHC class I-restricted antigen presentation by gamma 2-herpesviruses. Proc Natl Acad Sci USA 97:8455–8460

Stewart AR, Tollefson AE, Krajcsi P, Yei SP, Wold WS (1995) The adenovirus E3 10.4K and 14.5K proteins, which function to prevent cytolysis by tumor necrosis factor and to down-regulate the epidermal growth factor receptor, are localized in the plasma membrane. J Virol 69:172–181

Straus SE (1984) In: Ginsberg HS (ed) Adenovirus infections in humans. Plenum Press, New York and London, pp 451–496

Teasdale RD, Jackson MR (1996) Signal-mediated sorting of membrane proteins between the endoplasmic reticulum and the golgi apparatus. Annu Rev Cell Dev Biol 12:27–54

Teodoro JG, Branton PE (1997) Regulation of apoptosis by viral gene products. J Virol 71:1739–1746

Teodoro JG, Shore GC, Branton PE (1995) Adenovirus E1 A proteins induce apoptosis by both p53-dependent and p53-independent mechanisms. Oncogene 11:467–474

Thomas A, White E (1998) Suppression of the p300-dependent mdm2 negative-feedback loop induces the p53 apoptotic function. Genes Dev 12:1975–1985

Thorne TE, Voelkel Johnson C, Casey WM, Parks LW, Laster SM (1996) The activity of cytosolic phospholipase A2 is required for the lysis of adenovirus-infected cells by tumor necrosis factor. J Virol 70:8502–8507

Tollefson AE, Hermiston TW, Lichtenstein DL, Colle CF, Tripp RA, Dimitrov T, Toth K, Wells CE, Doherty PC, Wold WS (1998) Forced degradation of Fas inhibits apoptosis in adenovirus-infected cells. Nature 392:726–730

Tollefson AE, Scaria A, Hermiston T, Ryerse JS, Wold LJ, Wold WSM (1996) The adenovirus death protein (E3–11.6K) is required at very late stages of infection for efficient cell-lysis and release of adenovirus from infected-cells. J Virol 70:2296–2306

Tollefson AE, Stewart AR, Yei SP, Saha SK, Wold WS (1991) The 10,400- and 14,500-dalton proteins encoded by region E3 of adenovirus form a complex and function together to down-regulate the epidermal growth factor receptor. J Virol 65:3095–3105

Tomko RP, Xu R, Philipson L (1997) HCAR and MCAR: the human and mouse cellular receptors for subgroup C adenoviruses and group B coxsackieviruses. Proc Natl Acad Sci USA 94:3352–3356

Tortorella D, Gewurz BE, Furman MH, Schust DJ, Ploegh HL (2000) Viral subversion of the immune system. Annu Rev Immunol 18:861–926

Trapani JA, Davis J, Sutton VR, Smyth MJ (2000) Proapoptotic functions of cytotoxic lymphocyte granule constituents in vitro and in vivo. Curr Opin Immunol 12:323–329

Tufariello J, Cho S, Horwitz MS (1994) The adenovirus E3 14.7-kilodalton protein which inhibits cytolysis by tumor necrosis factor increases the virulence of vaccinia virus in a murine pneumonia model. J Virol 68:453–462

Turnell A, Grand R, Gorbea C, Zhang X, Wang W, Mymryk J, Gallimore P (2000) Regulation of the 26 S proteasome by adenovirus E1A. EMBO J 19:4759–4773

van Endert PM (1999) Genes regulating MHC class I processing of antigen. Curr Opin Immunol 11:82–88

Vandenabeele P, Declercq W, Beyart R, Fiers W (1995) Two tumour necrosis factor receptors: structure and function. Trends Cell Biol 5:392–399

Vidalain P-O, Azocar O, Lamouille B, Astier A, Rabourdin-Combe C, Servet-Delprat C (2000) Measles virus induces functional TRAIL production by human dendritic cells. J Virol 74:556–559

von Herrath MG, Efrat S, Oldstone MB, Horwitz MS (1997) Expression of adenoviral E3 transgenes in beta cells prevents autoimmune diabetes. Proc Natl Acad Sci USA 94:9808–9813

Wadell G (1990) In: Zuckerman AJ, Banatvala JE, Pattison JR (eds) Adenoviruses. Wiley, New York, pp 267–297

Walczak H, Krammer PH (2000) The CD95 (APO-1/Fas) and the TRAIL (APO-2L) apoptosis systems. Exp Cell Res 256:58–66

Wallach D (1997) Cell death induction by TNF: a matter of self control. Trends Biochem Sci 22:107–109

Wallach D, Varfolomeev EE, Malinin NL, Goltsev YV, Kovalenko AV, Boldin MP (1999) Tumor necrosis factor receptor and Fas signaling mechanisms. Annu Rev Immunol 17:331–367

White E (1998) Regulation of apoptosis by adenovirus E1A and E1B oncogenes. Semin Virol 8:505–513

White E, Faha B, Stillman B (1986) Regulation of adenovirus gene expression in human WI38 cells by an E1B-encoded tumor antigen. Mol Cell Biol 6:3763–3773

Wickham T, Mathias P, Cheresh D, Nemerow G (1993) Integrins alpha v beta 3 and alpha v beta 5 promote adenovirus internalization but not virus attachment. Cell 73:309–319

Wienzek S, Roth J, Dobbelstein M (2000) E1B 55-kilodalton oncoproteins of adenovirus types 5 and 12 inactivate and relocalize p53, but not p51or p73, and cooperate with E4orf6 proteins to destabilize p53. J Virol 74:193–202

Williams BR (1999) PKR; a sentinel kinase for cellular stress. Oncogene 18:6112–6120

Williams JL, Garcia J, Harrich D, Pearson L, Wu F, Gaynor R (1990) Lymphoid specific gene expression of the adenovirus early region 3 promoter is mediated by NF-κ B binding motifs. EMBO J 9:4435–4442

Wilson-Rawls J, Wold WS (1993) The E3–6.7K protein of adenovirus is an Asn-linked integral membrane glycoprotein localized in the endoplasmic reticulum. Virology 195:6–15

Wissing D, Mouritzen H, Egeblad M, Poirier GG, Jäättelä M (1997) Involvement of caspase-dependent activation of cytosolic phospholipase A2 in tumor necrosis factor-induced apoptosis. Proc Natl Acad Sci USA 94:5073–5077

Wold WSM, Hermiston TW, Tollefson AE (1995) E3 transcription unit of adenovirus. Curr Top Microbiol Immunol 199/I:237–274

Yamaoka S, Courtois G, Bessia C, Whiteside ST, Weil R, Agou F, Kirk HE, Kay RJ, Israel A (1998) Complementation cloning of NEMO, a component of the IkappaB kinase complex essential for NF-κB activation. Cell 93:1231–1240

Yeh HY, Pieniazek N, Pieniazek D, Luftig RB (1996) Genetic organization, size, and complete sequence of early region-3 genes of human adenovirus type-41. J Virol 70:2658–2663

Yewdell JW, Bennink JR (1999) Mechanisms of viral interference with MHC class I antigen processing and presentation. Annu Rev Cell Dev Biol 15:579–606

Zamai L, Ahmad M, Bennett IM, Azzoni L, Alnemri ES, Perussia B (1998) Natural killer (NK) cell-mediated cytotoxicity: differential use of TRAIL and Fas ligand by immature and mature primary human NK cells. J Exp Med 188:2375–2380

Zhang J, Vinkemeier U, Gu W, Chakravarti D, Horvath C, Darnell JJ (1996) Two contact regions between Stat1 and CBP/p300 in interferon gamma signaling. Proc Natl Acad Sci USA 93: 15092–15096

Subject Index

A
ABC transporter 59, 60, 70, 87
acute infection 236
Ad (*see* adenovirus)
adaptive immune system 85
adenovirus (Ad) 58, 275, 296, 298, 299, 309
- Ad4 E3 proteins 275
- Ad5 295
- Ad12 60
- AD169 246
- classification 275, 276
- E1A 60
- E3/19K 68
- induced diseases 275, 276
- mediated gene therapy 308
- persistence 274
- replication cycle 276
adhesion 225
antigen
- presentation 27, 29, 106–112, 300, 305
- processing 85–89, 300, 305
α1-antitrypsin 47
apoprotein B 42
apoptosis 282, 300
- receptor 291
- p53-dependent 281
- p53-independent 281, 282
- p53-mediated by E1B/55K 282
apoptotic pathways, E1A-induced 283
β-arrestins 208
asparagine 48
aspartic acid 48
astrocytoma cells 212–219
ATP-binding cassette transporter 87

B
BAC-technology 11, 12
Bcl-2 261–268
- family 309
BiP 46, 50
blood platelets 224
BM (*see* bone marrow)
bone marrow (BM) 217, 224
- myofibroblasts 217, 227
- progenitors 224, 227
- stromal cells 224, 227
- transplantation 227
bovine herpesvirus 62, 87
Burkitt's lymphoma (BL) 61

C
Ca^{2+} 208, 217
calnexin 45, 47, 64, 65, 86
calreticulin 47, 60, 64, 86
CAR (*see* coxsackie/adenovirus receptor)
carboxypeptidase Y 42
Castleman's disease 61
CCR3 242
CD8 cytotoxic T cells 118, 119
$CD34^+$ cells 22, 227
CD94 68
CD94/NKG2 119, 122–124
cell-to-cell
- contacts 222
- fusion 218, 219, 222
- passage 228
- transmission 224, 225
cellular MHC class I proteins
- downregulation of expression by herpesviruses 132
- interactions with NK cell receptors 136–139
-- CD94/NKG2 136, 137, 142
-- gp49 138
-- ITAM 138
-- ITIM 138
-- KIR 138, 139
-- LIR 138, 139
-- Ly49 137
-- PIR 139, 146
-- Qa-1^b 136
- modulators of NK cell-mediated killing 132
- structural features 134, 135
CFTR (*see* cystic fibrosis transmembrane conductance regulator)
chaperone(s) 27, 45, 49, 50

chemoattractant 223, 224
chemokine(s) 203–235
- binding 205–207
- biology 237, 238
- C 205
- CC 205–207
- CMV-encoded 207
- CX3C 205–207, 223–225
- CXC 205–207
- ELR- 205
- ELR+ 205, 207
- evolution of 219–221
- modulation of production 217, 218
- receptors 203–234
- - autocrine stimulation 224
- - binding to 205–207
- - chemokine binding 213–217
- - CMV-encoded 207–220
- - constitutive activity 214, 224, 226
- - desensitization 208
- - endocytosis 206, 208
- - evolution of 219–221
- - expression 211–213
- - interaction 205–207
- - internalization 208
- - role in dissemination 222–226
- - role in persistence 222–226
- - sequestering 224
- - signalling 205–207, 213–217
- - transcription 209–211
- - viral 203–234
- relation to CMV dissemination 222–228
- relation to CMV persistence 222–228
- sink 205, 224
chemotaxis 207
CLTs 211
CMV-encoded MHC class I homologues
- biochemical characterisation 133, 135
- characterisation of knockout (mutant) viruses 135, 142, 144, 145
- genomic location 133, 134
- models for inhibition of NK cells 136, 146, 147
- structural features 133–135
coagulation factor VIII 42, 50
coevulotion 305
coinfection 244
co-injection (see microinjection)
cord blood progenitors 227
coreceptor (see HIV coreceptor)
coxsackie/adenovirus receptor (CAR) 276
cPLA$_2$ 214, 288–290
CPY* 43, 47, 49
CTL (see cytotoxic T lymphocyte)
CX$_3$CL 241
CXCR2 235, 249

cyclic AMP 208
cystic fibrosis transmembrane conductance regulator (CFTR) 42, 43, 50
cytomegalovirus (CMV) 155, 235
- dissemination 239
- genomes 238
- inhibition of MHC class I transcription 157
- pathogenesis 239
cytotoxic T lymphocyte (CTL, see T cell)

D
DA169 207, 209, 210
deglycosylated intermediate 41–43, 46
deglycosylation 48, 49
degradation 41, 42, 45, 50
- intermediate 48
- pathway 40, 47, 50
- proteasomal 43
- signal 25
- substrate 44, 47
dendritic cells 224, 225, 227
- HCMV infection of 140, 141
- LIR-1 expression 141
- MCMV infection of 141
- role in immune defense against viruses 141
Der3p 48
deubiquitination 25, 27, 31
dileucine motifs 294
dilysine motifs 302
disease pattern 310
dislocation 41–50
dissemination 244, 249
- mononuclear leukocytes 236
disulfide
- bonds 46
- bridges 46
DNA array technique 5–7

E
E1 47
E1A 60, 277, 279–282, 284–286, 299, 300, 305–307
E1A+ 279
E1B 278, 282, 285, 300
E1B/19K 282, 284, 286, 290, 306, 309
E2 47, 48, 300
E3 25, 26, 30, 47, 48, 275, 278, 286, 288, 291, 296, 298, 300, 301, 306, 308, 310
- products 310
- region 285, 309
E3/10.4K 296
E3/10.4K-14.5K 289, 306, 308
E3/11.6K 282–284
E3/14.7K 286, 288
E3/19K 60, 290, 291, 301–305, 308
E4 300, 306

E4/ORF4   306
E4/ORF6   306
EBNA (see Epstein-Barr virus)
EBV (see Epstein-Barr virus)
EGFP   212, 214
EGFR (see epidermal growth factor receptor)
endoplasmic reticulum   86, 88, 89, 93–96
endothelial cells   205, 222–225
- rolling and attachment to   206
epidermal growth factor receptor (EGFR)   295, 296, 306
epithelial cells   222, 224, 226
Epstein-Barr virus (EBV)   24, 29, 31, 32, 61, 259, 266, 268
- latent membrane protein 1 (LMP1)   61, 62
- latent membrane protein 2 (LMP2)   61
- EBV nuclear antigens (EBNA)   24
- - proteasomal degradation   39
equine herpesvirus type-2   62
ER retention   302
ERK2   216
Erp57   47, 60, 86
evolutionary trees   219
extraction   49

F
Fas   291, 294, 296, 300, 306
- mediated cell death   296
FasL   284, 288, 291, 297, 299, 300
Fc-receptor, MCMV   7
FIP-2   289
FIP-3   289
flavivirus   69
FLIPs   262–264, 266, 267
fractalkine   220, 223, 240

G
G protein   206–208
- coupled receptor (GPCR)   203–234
GAr (see glycine-alanine repeat)
GDP   208
gene
- therapy   33
- transfer   97
genomes   246
Gly-Ala repeat   39
glycanase   48
glycine-alanine repeat (GAr)   24, 28–31
glycosylated peptides   48
gp40   39, 40
GPCR (see G protein-coupled receptor)
granule-mediated
- cell death   306
- cytolysis   300
granulocytes, progenitors   222
granzyme   306, 308

green fluorescent protein (GFP) reporter   30, 31
GRK   208
GRP   208
GTP   208
- cycle   208

H
H. saimiri   259, 268
HCMV (see human cytomegalovirus)
hematopoietic progenitors   223
herpes simplex virus (HSV)   171–185, 259, 268
- human
- - HHV-6   209, 218
- - HHV-8   61, 213, 259, 266–268
- ICP0   179, 180
- ICP34.5   175–180
- latency   172
- reactivation   172
- type 1 (HSV1)   63
- - ICP47   62, 63
- type 2 (HSV2)   63
- US11   177, 178
- virion host shutoff   180, 181
herpes viral proteins   75–81
herpesvirus(es)   38, 39, 40, 62, 85–96
HHV (see herpes simplex virus)
HSV (see herpes simplex virus)
HIV   58
- coreceptor   218, 219
HLA (see human leukocyte antigen)
HMG-CoA (see hydroxymethylglutaryl-coenzyme A reductase)
Hodgkin/Reed-Sternberg cell (HRS)   61
Hodgkin's Disease (HD)   61
homologs of G protein-coupled receptors   4
homology search   2
HRD   48
Hrd1p   48
Hsp70   49
Hsp70s   50
Hsp90   49
HSV (see herpes simplex virus)
HSV2 (see herpes simplex virus type 2)
human cytomagalovirus (HCMV)   37, 40, 50, 62–70, 117–129
- inhibition of class I antigen presentation   102
- inhibition of class II antigen presentation   106–112
- proteins   38
- UL40   68
- US2   64, 65
- US2-US11 glycoproteins
- - effects on antigen presentation   111
- - effects on US2   107–111
- - effects on US3   112
- - mutant forms of US2   111

human cytomagalovirus (HCMV)
- US3 64, 65
- US6 62, 64–70
- US11 64, 65
human herpesvirus (*see* herpes simplex virus)
human leukocyte antigen (HLA) 301, 302, 304
- complexes 308
- HLA-A 41, 60, 68
- HLA-A2 42
- HLA-B 41, 60, 68
- HLA-C 41, 63, 68
- - effect of human cytomegalovirus US2 108, 109
- HLA-DM
- - degradation by human cytomegalovirus US2 108
- - function in antigen presentation 103, 104
- HLA-E 68, 122–124
- - binding to CD94/NKG2 142
- - HCMV gpUL40 ligand 142
- - inhibition of NK cell killing 142
- HLA-G 41, 63, 64, 68
- - modulation 300
- - molecule 304, 308
human peripheral blood neutrophils 207
hydroxymethylglutaryl-coenzyme A reductase (HMG-CoA) 42

I
IκB-α 30
ICP4 261
ICP47 39, 62, 63, 86, 89–93, 96
identification
- data bases 2
- of immune-evasive genes 2, 3, 6, 7
- by infection phenotype 4
- random genome fragments 7
IE-1 3
IE-2 3
IFN (*see* interferon)
IgM, secretory 42
IL (*see* interleukin)
ILT2 120, 122
ILT4 120
immune
- evasion 39, 89–96, 236, 275, 277, 278
- response, evasion 38
- suppressor 96
immune-evasive
- genes 2, 3, 18
- strategies
- - apoptosis 5
- - chemokines 5
- - cytokines 5
- - interferon 5
- - MHC class I downregulation 4, 8, 13–17

immunoescape 29, 38
immunomodulatory functions 235
immunoproteasome 87
infection
- acute 236
- latent 236
- persistent 38, 274, 305
interaction
- protein complexes 7, 8
- virus-host 2
interferon (IFN) 277, 280, 297, 300, 306
- counteraction of by herpes simplex virus 171–185
- IFN-α 164, 279, 280
- IFN-α/β 279, 298
- IFN-β 217, 279, 280
- IFN-γ 61, 62, 64, 87, 217, 279, 281, 290, 291, 297, 298
- IFN-mediated translation inhibition 280
- signal transduction 160
- - IFN-γ receptor 161
- - ISGF3 162
- - Janus kinases 161
- - STAT-1 161
- - STAT2 162
- - TYK2 162
interleukin
- IL-1 218
  IL-1β 217, 218
- IL-6 218
- IL-8 203–234, 249
- IL-10 62
intermediates
- deglycosylated 43
- degradation 48
internalization 277
IP3 208, 217

J
JAKs 173

K
K562 212
Kaposi's sarcoma 61
Kar2p 46
killer cell immunoglobulin-like receptors (KIRs) 68, 119, 299
KKXX 302
KXKXX 302

L
LAT (*see* latency-associated transcript)
latency 244
latency-associated transcript (LAT) 261
latent
- infection 236
- membrane protein (LMP) 24, 32

– – LMP2 62
– – LMP7 62
LIR-1
– binding to HCMV gpUL18 139, 140
– cellular distribution 139
LMP (see latent membrane protein)
loading complex 89
lymphocytes (see also T cell) 205, 228

M
m04 3, 8, 15
m06 3, 8, 13
M45 3, 12
M54 11
M55 11
m129/m130 4
m131 240
m131-m129 infusion 240
m138 3, 8
m144 3
m152 3, 8, 13, 14, 17
macrophages 239, 242
major histocompatibility complex
  (MHC; see also cellular MHC class I proteins) 24, 26, 29
– class I 37, 57–59
– – antigens 279, 290, 300, 301
– – breakdown 47
– – deglycosylated intermediate 41, 43
– – degradation 37, 38, 40, 50
– – dislocation 47
– – downregulation 40
– – heavy chain 43
– – inhibition of antigen presentation 102, 111
– – molecules 68, 86, 87, 118, 299
– – – antigen presentation 75–77
– – polyubiquitinated 47
– – retention 39, 47
– – retrograde transport 42
– – soluble cytosolic 49
– – transcription 158
– – ubiquitinated 47
– class II 62, 154
– – antigens 279
– – in antigen presentation 102, 111
– – inhibition of antigen presentation 106–112
– – mislocalization 112
– – in presentation of endogenous antigens 104–106
– – transcription 155
– evasion 39
– molecules 302, 304, 305
– peptide complex 86, 87
MCK-1 219, 220, 240, 241
– polypeptide 241

MCK-2 235, 240
– chemokine domain of 243
– glycosylation 241
– immune clearance 245
– inflammation 245
– model 246
– mutant virus 243
– permissivness 245
– proinflammatory activity 244
– recruit 245
MCMV (see murine cytomegalovirus)
MCP-1 203–234
megakaryocytes 224
MHC (see major histocompatibility complex)
MIC-A 125
$\beta 2$ microglobulin 39, 58, 60, 64, 65
microinjection 7, 8
migration
– neutrophil 223
– transendothelial 206, 224
MIP-1α 203–234
misfolded
– proteins 43, 46
– – degradation 42
model, MCMV 17
monocytes 205, 222, 224–226
– monocytic 224
– pre- 211
– progenitors 222
multiple sequence alignment 219, 220
murine cytomegalovirus (MCMV) 119, 122, 212, 219, 220, 225
myeloid cells 223–225
myelopoiesis 223

N
nasopharyngeal carcinoma 61
natural killer cells 227, 279
ND10 179
negative selection 6, 7, 9
neutrophils 205, 207, 223, 247
– chemoattraction 224
– migration 223
NF-κB 60, 62, 277, 280, 284–286, 288, 289, 291, 305, 307, 308
NK cells 63, 69, 119–129, 284, 291, 297–299, 306
– lysis 299
– mediated killing
– – in vitro
– – – inhibitory effect of gpm144 146
– – – modulation of susceptibility to lysis by HCMV infection 142, 143
– – in vivo
– – – depletion of cells bearing the NK1.1+ marker 145
– – – evasion by gpm144 144

NK cells
- - - resistance to MCMV mediated by CMV1 locus   144
- - - response to MCMV infection   144, 145
- receptors   299
- sensitivity   308
- susceptibility   300
NKG2A   68
NKG2D/DAP10   125
NMR spectroscopy   92
nuclear magnetic resonance   92
- spectroscopy   92

O
orf virus   62
oxidoreductase   46

P
p53   281
papillomavirus   60
pathogenesis, viral   247
Pdi1p   46
peptide   58–61
- transport   87–89
peptide: N-glycanase   49
peptide-MHC complex   86, 87
peripheral blood mononuclear cells   211
peritonal exudate cells   241
persistence   300
persistent
- infections   38, 274, 305
- viruses   305
phylogenetic trees   219, 220
PKR   174–178, 280, 306
PNGase   48, 49
porcine pseudorabies virus   86
positive selection   6, 7
prepro-α factor   42
primary effusion lymphoma   61
primate cytomegalovirus   62
proteasomal degradation   43, 49
proteasome   25, 26, 29, 41, 43, 49, 58, 86, 87
- inhibitors   42, 48
protein
- degradation   50
- dislocation   43, 47, 50
- disulfide isomerase (PDI)   46
- phosphatase 1α 176
γ1-34.5 protein   260, 268
proteoglycans   226
proteomics   5–7
pseudorabies virus   62

Q
quality control   45, 47

R
R33   219, 225
R78   219
Rad23p   49
random genome fragments   7
RANTES   203–234
rat cytomegalovirus (RCMV)   219, 225
receptor   242
- α-chain   47
retention signal   39
retrograde
- translocation   41, 43, 46
- transport   43, 50
retroviral infection   218, 219
ribophorin I   43, 47
ribosome   46
RNase L   173, 178, 179

S
salivary glands   222, 243
- epithelium   226
- tropism   226
SBH1   43
screening
- of deletion mutants   5, 7
- of genome fragments   7
- for loss of phenotype   9
Sec61
- complex   43
- α 43, 46
- β 43
Sec61p   42, 43, 46
Sec62p   46
Sec63p   46
secretory IgM   42
selection
- positive   6, 7
- procedures   5–7, 9
- negative   6, 7, 9
smooth muscle cells (SMC)   212, 222, 224, 225
- migration   225
species specific MHC class I downregulation   15
stabilisation signal   33
STAT   173, 279, 280, 306
STAT1   279, 280
STAT2   279, 280

T
T cell (lymphocytes)   85–87
- cytotoxic (CTL)   24, 26, 29, 38, 85–87, 284, 291, 297, 298, 300, 301, 306, 308
- - attack   308
- - lysis   306
- receptor (TCR)   42, 48
- - α chains   42

Subject Index

TAP (see transporter associated with antigen processing)
tapasin 60, 65, 69, 86
TCR (see T cell receptor)
THP-1 211, 222, 224, 242
TNF (see tumor necrosis factor)
Toledo 207, 209–212, 246
Towne 207, 209–212, 246
trafficking 249
TRAIL 282, 296, 297, 299, 307
TRAIL-induced apoptosis 297
TRAIL-mediated apoptosis 300
TRAIL-R1 285, 296
TRAIL-R1/2 306
TRAIL-R2 285, 296
TRAIL-receptors 297, 298
TRAM 43
transcriptional repression 305
translocation, retrograde 41, 43, 46
translocon 43, 44, 46
– degradation substrates 44
transport, retrograde 43
transporter associated with antigen processing (TAP) 58–70, 75–81, 87–89
– ATP binding 59, 66–69
– ATP hydrolysis 59, 66
– comformational changes 59, 66, 70
– function 77–79
– gene expression 60, 61
– inhibition 90–96
– inhibitor 90–96
– nucleotide binding domain (NBD) 59, 66
– peptide translocation 59, 60, 64
– phosphorylation 68
– selectivity 88, 89, 93
– specifity 88, 89, 93
– structure 77–79
– substrate specificity 58, 59
– transmembrane segments (TMS) 59, 70
– viral inhibitors 79–81
tumors 60
tumor necrosis factor (TNF) 284–286, 288–291, 296, 297, 299, 306–308
– α 217, 218
– lysis 286, 298, 307
– mediated lysis (apoptosis, cytolysis) 285, 288–290, 307

U

U12 213
U51 212, 218
U373 MG 212, 219
Ubc1 48
Ubc6 48
Ubc7 48
ubiquitin 47
– ligase (see E3)

ubiquitination 25, 31, 42, 47
ubiquitin-proteasome system 25–27
UL16 125
UL18 120–122
UL18 3, 4
UL33 203–234
– antibodies 211
UL33 4
UL37 3, 8
UL40 122–124
UL40 3, 4
UL78 203–234
UL78 4
UL111 3
UL146 207, 219, 220, 235, 248
UL146 3
UL147 207, 219, 220, 228, 235, 248
UL152 207
ULBP 125
US genes 118, 120, 124
US1 47
US2 38, 40–42, 44, 47
US2 3, 10
US3 39, 261
– retention 40
US3 3, 10
US5 261
US6 39, 86, 93–96
US6 3, 9, 10
US11 38, 40–42, 44
US11 3, 10
US27 203–234
US27 4
US28 203–234
US28 3, 4

V

VA-RNA 280, 306
vasuclar disease 225
vCXC-1 207, 223, 224, 228, 248
vCXC-2 248
v-FLIP 267
viral immune-evasive genes 2, 3
virus-mutant
– deletion mutant 5, 7, 10
– homologous recombination 10
– randomly 11
– site directed 11, 12
– transposon
– – insertion 2, 10, 11
– – library 7, 12
vMIA 262
vMIP-II 213

Y

YxxΦ 294, 295

Current Topics in Microbiology and Immunology

Volumes published since 1989 (and still available)

Vol. 226: **Koprowski, Hilary; Weiner, David B. (Eds.):** DNA Vaccination/Genetic Vaccination. 1998. 31 figs. XVIII, 198 pp. ISBN 3-540-63392-8

Vol. 227: **Vogt, Peter K.; Reed, Steven I. (Eds.):** Cyclin Dependent Kinase (CDK) Inhibitors. 1998. 15 figs. XII, 169 pp. ISBN 3-540-63429-0

Vol. 228: **Pawson, Anthony I. (Ed.):** Protein Modules in Signal Transduction. 1998. 42 figs. IX, 368 pp. ISBN 3-540-63396-0

Vol. 229: **Kelsoe, Garnett; Flajnik, Martin (Eds.):** Somatic Diversification of Immune Responses. 1998. 38 figs. IX, 221 pp. ISBN 3-540-63608-0

Vol. 230: **Kärre, Klas; Colonna, Marco (Eds.):** Specificity, Function, and Development of NK Cells. 1998. 22 figs. IX, 248 pp. ISBN 3-540-63941-1

Vol. 231: **Holzmann, Bernhard; Wagner, Hermann (Eds.):** Leukocyte Integrins in the Immune System and Malignant Disease. 1998. 40 figs. XIII, 189 pp. ISBN 3-540-63609-9

Vol. 232: **Whitton, J. Lindsay (Ed.):** Antigen Presentation. 1998. 11 figs. IX, 244 pp. ISBN 3-540-63813-X

Vol. 233/I: **Tyler, Kenneth L.; Oldstone, Michael B. A. (Eds.):** Reoviruses I. 1998. 29 figs. XVIII, 223 pp. ISBN 3-540-63946-2

Vol. 233/II: **Tyler, Kenneth L.; Oldstone, Michael B. A. (Eds.):** Reoviruses II. 1998. 45 figs. XVI, 187 pp. ISBN 3-540-63947-0

Vol. 234: **Frankel, Arthur E. (Ed.):** Clinical Applications of Immunotoxins. 1999. 16 figs. IX, 122 pp. ISBN 3-540-64097-5

Vol. 235: **Klenk, Hans-Dieter (Ed.):** Marburg and Ebola Viruses. 1999. 34 figs. XI, 225 pp. ISBN 3-540-64729-5

Vol. 236: **Kraehenbuhl, Jean-Pierre; Neutra, Marian R. (Eds.):** Defense of Mucosal Surfaces: Pathogenesis, Immunity and Vaccines. 1999. 30 figs. IX, 296 pp. ISBN 3-540-64730-9

Vol. 237: **Claesson-Welsh, Lena (Ed.):** Vascular Growth Factors and Angiogenesis. 1999. 36 figs. X, 189 pp. ISBN 3-540-64731-7

Vol. 238: **Coffman, Robert L.; Romagnani, Sergio (Eds.):** Redirection of Th1 and Th2 Responses. 1999. 6 figs. IX, 148 pp. ISBN 3-540-65048-2

Vol. 239: **Vogt, Peter K.; Jackson, Andrew O. (Eds.):** Satellites and Defective Viral RNAs. 1999. 39 figs. XVI, 179 pp. ISBN 3-540-65049-0

Vol. 240: **Hammond, John; McGarvey, Peter; Yusibov, Vidadi (Eds.):** Plant Biotechnology. 1999. 12 figs. XII, 196 pp. ISBN 3-540-65104-7

Vol. 241: **Westblom, Tore U.; Czinn, Steven J.; Nedrud, John G. (Eds.):** Gastroduodenal Disease and Helicobacter pylori. 1999. 35 figs. XI, 313 pp. ISBN 3-540-65084-9

Vol. 242: **Hagedorn, Curt H.; Rice, Charles M. (Eds.):** The Hepatitis C Viruses. 2000. 47 figs. IX, 379 pp. ISBN 3-540-65358-9

Vol. 243: **Famulok, Michael; Winnacker, Ernst-L.; Wong, Chi-Huey (Eds.):** Combinatorial Chemistry in Biology. 1999. 48 figs. IX, 189 pp. ISBN 3-540-65704-5

Vol. 244: **Daëron, Marc; Vivier, Eric (Eds.):** Immunoreceptor Tyrosine-Based Inhibition Motifs. 1999. 20 figs. VIII, 179 pp. ISBN 3-540-65789-4

Vol. 245/I: **Justement, Louis B.; Siminovitch, Katherine A. (Eds.):** Signal Transduction and the Coordination of B Lymphocyte Development and Function I. 2000. 22 figs. XVI, 274 pp. ISBN 3-540-66002-X

Vol. 245/II: **Justement, Louis B.; Siminovitch, Katherine A. (Eds.):** Signal Transduction on the Coordination of B Lymphocyte Development and Function II. 2000. 13 figs. XV, 172 pp. ISBN 3-540-66003-8

Vol. 246: **Melchers, Fritz; Potter, Michael (Eds.):** Mechanisms of B Cell Neoplasia 1998. 1999. 111 figs. XXIX, 415 pp. ISBN 3-540-65759-2

Vol. 247: **Wagner, Hermann (Ed.):** Immunobiology of Bacterial CpG-DNA. 2000. 34 figs. IX, 246 pp. ISBN 3-540-66400-9

Vol. 248: **du Pasquier, Louis; Litman, Gary W. (Eds.):** Origin and Evolution of the Vertebrate Immune System. 2000. 81 figs. IX, 324 pp. ISBN 3-540-66414-9

Vol. 249: **Jones, Peter A.; Vogt, Peter K. (Eds.):** DNA Methylation and Cancer. 2000. 16 figs. IX, 169 pp. ISBN 3-540-66608-7

Vol. 250: **Aktories, Klaus; Wilkins, Tracy, D. (Eds.):** Clostridium difficile. 2000. 20 figs. IX, 143 pp. ISBN 3-540-67291-5

Vol. 251: **Melchers, Fritz (Ed.):** Lymphoid Organogenesis 2000. 62 figs. XII, 215 pp. ISBN 3-540-67569-8

Vol. 252: **Potter, Michael; Melchers, Fritz (Eds.):** B1 Lymphocytes in B Cell Neoplasia. 2000. XIII, 326 pp. ISBN 3-540-67567-1

Vol. 253: **Gosztonyi, Georg (Ed.):** The Mechanisms of Neuronal Damage in Virus Infections of the Nervous System. 2001. approx. XVI, 270 pp. ISBN 3-540-67617-1

Vol. 254: **Privalsky, Martin L. (Ed.):** Transcriptional Corepressors. 2001. 25 figs. XIV, 190 pp. ISBN 3-540-67569-8

Vol. 255: **Hirai, Kanji (Ed.):** Marek's Disease. 2001. 22 figs. XII, 294 pp. ISBN 3-540-67798-4

Vol. 256: **Schmaljohn, Connie S.; Nichol, Stuart T. (Eds.):** Hantaviruses. 2001, 24 figs. XI, 196 pp. ISBN 3-540-41045-7

Vol. 257: **van der Goot, Gisou (Ed.):** Pore-Forming Toxins, 2001. 19 figs. IX, 166 pp. ISBN 3-540-41386-3

Vol. 258: **Takada, Kenzo (Ed.):** Epstein-Barr Virus and Human Cancer. 2001. 38 figs. IX, 233 pp. ISBN 3-540-41506-8

Vol. 259: **Hauber, Joachim, Vogt, Peter K. (Eds.):** Nuclear Export of Viral RNAs. 2001. 19 figs. IX, 131 pp. ISBN 3-540-41278-6

Vol. 260: **Burton, Didier R. (Ed.):** Antibodies in Viral Infection. 2001. 51 figs. IX, 309 pp. ISBN 3-540-41611-0

Vol. 261: **Trono, Didier (Ed.):** Lentiviral Vectors. 2002. 32 figs. X, 258 pp. ISBN 3-540-42190-4

Vol. 262: **Oldstone, Michael B.A. (Ed.):** Arenaviruses I. 2002, 30 figs. XVIII, 197 pp. ISBN 3-540-42244-7

Vol. 263: **Oldstone, Michael B. A. (Ed.):** Arenaviruses II. 2002, 49 figs. XVIII, 268 pp. ISBN 3-540-42705-8

Vol. 264/I: **Hacker, Jörg; Kaper, James B. (Eds.):** Pathogenicity Islands and the Evolution of Microbes. 2002. 34 figs. XVIII, 232 pp. ISBN 3-540-42081-7

Vol. 264/II: **Hacker, Jörg; Kaper, James B. (Eds.):** Pathogenicity Islands and the Evolution of Microbes. 2002. 24 figs. XVIII, 228 pp. ISBN 3-540-42682-5

Vol. 265: **Dietzschold, Bernhard; Richt, Jürgen A. (Eds.):** Protective and Pathological Immune Responses in the CNS. 2002. 21 figs. X, 278 pp. ISBN 3-540-42668-X

Vol. 266: **Cooper, Koproski (Eds.):** The Interface Between Innate and Acquired Immunity, 2002, 15 figs. XIV, 116 pp. ISBN 3-540-42894-1

Vol. 267: **Mackenzie, John S.; Barrett, Alan D. T.; Deubel, Vincent (Eds.):** Japanese Encephalitis and West Nile Viruses. 2002. 66 figs. X, 418 pp. ISBN 3-540-42783-X

Vol. 268: **Zwickl, Peter; Baumeister, Wolfgang (Eds.):** The Proteasome-Ubiquitin Protein Degradation Pathway. 2002, X, 213 pp. ISBN 3-540-43096-2

Printing (Computer to Film): Saladruck Berlin
Binding: Stürtz AG, Würzburg